NATIVE ISRAELITES

NATIVE ISRAELITES

The Search for Joseph of Egypt's Genetics
Among the Native Americans

Alexander T Paulos

Copyright © 2016 Alexander T Paulos

First printed in softback 2016

The author and publishers have made every effort to trace the owners of copyright reproduced in this book. In the event of any omission, please contact paulosrecognition@gmail.com

The right of Alexander T Paulos to be identified as author of this work has been asserted by him in accordance with the Copyright, Designs and Patents Act 1988.

All rights reserved. This book may not reproduced in whole or in part, in any form (beyond that copying permitted by Sections 107 and 108 of the U.S. Copyright Law and except by reviewers for the public press) without written permission from the publishers.

Paulos, Alexander T.
    Native Israelites: The Search for Joseph of Egypt's Genetics Among the Native Americans

    Softback ISBN: 978-1-946350-00-8

9 8 7 6 5 4 3 2 1
2018 2017 2016

For Avalon

Truth has many faces
and the truth is like
to the old road to Avalon;
it depends on your own will,
and your own thoughts,
whither the road will take you.
—**Marion Zimmer Bradley,** ***The Mists of Avalon**

Contents

Acknowledgments		xi
Author's Note		xiii
Introduction		xvii
Chapter 1	Joseph Smith and Book of Mormon Peoples	1
Chapter 2	Did Joseph Smith Change His Mind?	29
Chapter 3	The North American Story	57
Chapter 4	Yuya and Joseph of Egypt	79
Chapter 5	Locating the Seed of Joseph	101
Chapter 6	The Physical Traits of Semites and Israelites	123
Chapter 7	American Indian Depopulation	139
Chapter 8	The Lamanite Remnant	155
Chapter 9	Appearance and Genetics	193
Chapter 10	Genetic Markers of Semites, East Asians, and Siberians	203
Chapter 11	Lehi and Arabs	235
Chapter 12	North American Indians and Jews	251
Chapter 13	Ancient Artifacts Dating Back to Book of Mormon Times	269
Chapter 14	East Asian Influences in Mesoamerica	293
Chapter 15	The Destiny of the Lamanite Remnant	325
Appendix 1	Genetic Markers Expansion	337
Appendix 2	The Joseph of Egypt Connection and "Believing Blood"	355
Appendix 3	Egyptians, American Indians, and Feathers	367
Appendix 4	Old World Symbols in the New World	373
Index		399
End Notes		403

Acknowledgments

A number of amazing people have helped me to produce this book. First and foremost, my wonderful wife, Asenath (who has the same name as Joseph of Egypt's wife),[a] has been a great support to me. Without her help, this book wouldn't have materialized. I'm also grateful to Tres White, Rob Nielsen, Alexandra Decker, Rachel Rollins, David Lassetter, Brad Heitmann, and John Nielsen for providing me with helpful insights and edits to improve this book. This book is much more polished because of their efforts.

---

[a] Asenath: "follower of Neith (or Nut)." Nut is the Egyptian goddess of sky supported by Shu, the god of air.

Author's Note

This book has been a long time in the making. Looking back, my interest in the Old West began in my youth. When I was a kid I had the mountains as my playground. Like many boys do, I spent time playing Cowboys and Indians with my friends. I loved the freedom of being outside and making war cries that echoed through the canyon. And, you know, shooting at each other and all that. We were boys, after all. While our reenactments may not have been anywhere near historically accurate, we had fun.

In school I had the opportunity to further my education about Native Americans. One of the most repeated lessons was that of the First Thanksgiving. And yes, I now realize that what we were taught as first graders was not completely true. The local newspaper came to my elementary school classroom to cover our reenactment and, not to brag or anything, but I, Chief Fire Eagle (self-named), was featured in the paper—complete with my hand-made construction paper headdress.

Thinking about the past, I realize that none of these

experiences were entirely unique, but they instilled in me a sentiment and lasting curiosity about Native Americans that has never abated. Little did I know, these experiences would commence my journey of discovery to solidify connections between groups of American Indians and some important religious figures.

As a teenager, I wanted to learn as much as possible about the Old West, and not being an avid reader at the time, I turned to movies. In one of my junior high classes I watched *I Will Fight No More Forever,* which details how the US government forced the Nez Percé to leave their homeland. The film also portrays the filmmakers view of how the Nez Percé resisted the US army. I was dismayed at how the situation was handled by our government, and still remember the strong emotions the film evoked within me.

While still in my teens, I sat through the three-hour 1990 film *Dances with Wolves* starring Kevin Costner (who is genetically part Cherokee in real life). The movie was riveting. After watching it for the first time, I remember repeatedly using the Sioux word "tatanka," which means "buffalo." Since there are no tatanka in the Utah Valley, it was difficult to insert my new word into casual conversation.

My love for the romanticism of the Wild West led me to enroll in an Old West class offered at my high school. I was into cowboys and Indians at the time, and the class did not disappoint.

As I got older, my interest in history expanded to include a fascination with religious figures, particularly Joseph who was sold into Egypt. I printed up numerous pages of scriptures and Bible commentary from online sources about this Semitic man who rightfully gained his family's birthright after his older half-brother Reuben lost it. Joseph of Egypt's tale has all the elements of a good story: betrayal, mystery, a splash of romance, and the underdog ending up on top. His life was so layered with complexity and accomplishment that not even Tim Rice and Andrew Lloyd Webber's *Joseph and the Amazing Technicolor Dreamcoat* can hope to capture all its splendor.

Another Joseph I became fascinated with was Joseph Smith, Jr. In my search for knowledge about the Prophet Joseph Smith, I learned that he shared a special relationship with Joseph of Egypt. In a December 1834 blessing, Joseph Smith, Sr. relayed to his son that ancient Joseph of Egypt

> looked after his [own] posterity in the last days. ... [He] sought diligently to know ... who should bring forth the word of the Lord [to them] and his eyes beheld thee, my son [Joseph Smith, Jr.]: [and] his heart rejoiced and his soul was satisfied.[1]

According to the prophet Lehi of the Book of Mormon, Joseph of Egypt saw the Nephites in vision (See 2 Nephi 3). He wrote that Joseph of Egypt prophesied of a latter-day seer—Joseph Smith, Jr.—who would share his name, and that Joseph of Egypt was made aware of the coming forth of the Book of Mormon.

Elder Neal A. Maxwell was also cognizant of this unique relationship between the two Josephs. Maxwell wrote, "The comparisons between the two Josephs, of course, reflect varying degrees of exactitude, but they are, nevertheless, quite striking. Some similarities are situational, others are dispositional."[2] The two

Josephs shared similar dispositions, which is another way of saying they shared a number of similar temperament and character traits.

As a result of the discovery and translation of the Book of Mormon, Joseph Smith provided Latter-day Saints with additional details about Joseph of Egypt. He uncovered knowledge of previously unknown branches of Joseph of Egypt's posterity, some that belong to certain tribes of American Indians from the Lamanite remnant.

It thrills me to provide Latter-day Saints with supportive evidence to authenticate what Joseph Smith learned about Joseph of Egypt and the ancient inhabitants of the Americas. The evidence I've found and included in this book has strengthened my belief that the Prophet Joseph Smith actually translated an ancient record of Semitic people who lived on the American continent.

—*Alexander T Paulos, February 14, 2015*

Alex Paulos is Chief Fire Eagle.

Introduction

For a subject worked and reworked so often in novels, motion pictures, and television, American Indians remain probably the least understood and most misunderstood Americans of us all.... When we forget great contributions to our American history—when we neglect the heroic past of the American Indian—we thereby weaken our own heritage.... It seems a basic requirement to study the history of our American Indians. —**John F. Kennedy, 35th US President (1961)**[3]

Most American history has been written as if history were a function solely of white culture—in spite of the fact that well into the nineteenth century the Indians were one of the principal determinants of historical events. Those of us who work in frontier history are repeatedly nonplussed to discover how little has been done for us in regard to the one force bearing on our field that was active everywhere.... American historians have made shockingly little effort to understand the life, the societies, the cultures, the thinking and the feeling of the Indians, and disastrously little effort to understand how all these affected white men and their societies. —**Bernard de Voto, American historian (1957)**[4]

Book of Mormon geography is a hotly-debated topic in the LDS Church. It's so heated that it has created a schism among Latter-day Saints. If the Prophet Joseph Smith were alive today, he'd be heartbroken to observe Church members bitterly arguing with one another over *where* the Book of Mormon primarily transpired, while simultaneously contradicting *what* the Book of Mormon teaches. To the people of the New World, Jesus taught that heated arguments, especially ones about his teachings, should cease: "[T]here shall be no disputations among you, as there have hitherto been; neither shall there be disputations among you concerning the points of my doctrine, as there have hitherto been" (3 Nephi 11:28). Joseph Smith understood Christ's teachings and echoed His words with a caveat: "Avoid contention and vain disputes with men of corrupt minds, who do not desire to know the truth."[5]

I wrote *Native Israelites* for Book of Mormon enthusiasts whether in favor of Mesoamerica, North America, or South America as the primary setting of the Book of Mormon. Although I have my own opinion about Book of Mormon geography (an opinion readers will readily notice throughout this book), my focus is not on which plot of dirt the Nephites and the Lamanites actually set foot on. Instead, this book centers around the descendants of Book of Mormon peoples, especially those who belong to the seed of Joseph of Egypt. By focusing on people rather than geography, I hope to circumvent the current uproarious religious environment in the Church that uncannily resembles the contentious milieu Joseph Smith described in his day:

> In the midst of this war of words and tumult of opinions, I often said to myself: What is to be done? Who of all these parties are right; or, are they all wrong together? If any one of them be right, which is it, and how shall I know it? While I was laboring under the extreme difficulties caused by the contests of these parties of religionists, I was one day reading the Epistle of James, first chapter and fifth verse, which reads: If any of you lack wisdom, let him ask of God, that giveth to all men liberally, and upbraideth not; and it shall be given him (*Joseph Smith—History* 1:10-11).

Like Joseph Smith, I lacked wisdom, but not about which church to join; my quandary centered around the descendants of

Joseph of Egypt. I wasn't sure if the actual bloodline of Joseph of Egypt existed, and so I set out to find it. After a long search, I believe I've discovered it. I went to great lengths to learn about the many ethnicities and peoples of the world to locate possible lineages of Joseph of Egypt. These lengths include a personal journey and also research about Amerindians scattered across the American continent. I studied just about every indigenous ancient and modern group of the Americas including the Iroquois Confederacy, Algonquian-speaking tribes, Siouian-speaking tribes, Hopewellian peoples, and also the Adena, Maya, Olmec, Aztec, Guarani, and Inca peoples. On my personal journey, I traveled thousands of miles to numerous Native American reservations located in the United States and First Nations tribes in Canada to learn more about them. I can't prove beyond doubt that I've discovered the literal bloodline of Joseph of Egypt, but my claims are backed up with solid evidence. Readers can decide for themselves if I'm right.

This book helps to answer questions readers may have about Joseph of Egypt and his descendants. Questions like, "Did Joseph Smith know who the main Lamanite remnant were?" "Are all indigenous Americans the descendants of Joseph of Egypt and also related to Book of Mormon peoples?" "If not, how can one tell the difference between which natives are and which ones are not? "Does DNA evidence exist that links Joseph of Egypt to indigenous Americans?"

The prophet Mormon asserted that the Book of Mormon was primarily written to the Lamanite remnant, which means that the identification of this group is crucial to Latter-day Saint teachings.[b]

---
[b] An Account Written by the Hand of Mormon upon Plates Taken from the Plates of Nephi:

Granted, everyone needs the gospel of Jesus Christ and shouldn't be kept from receiving it just because they're not genetically related to Israelites, but that doesn't negate the fact that Mormon singled out a branch of Joseph of Egypt's lineage when he identified them as the primary group for whom the Book of Mormon was written. This actual bloodline of Joseph of Egypt matters enough that a primary goal of this book is to spark an interest in readers to discover actual Lamanite bloodlines via Joseph of Egypt's lineage, because, after all, the Book of Mormon was primarily written to them.

Covenant bloodlines and heritable spiritual traits commonly linked to genetic lineages of literal Israelites were topics often discussed by early Latter-day Saint brethren. In modern times, although still essential to a few core LDS teachings, these subjects have been left out of the general discourse of the Church. Nowadays DNA is infrequently discussed over the pulpit in favor of ultra-universalism, often due to outside pressure from a politically-correct modern world.

Heredity, although a topic lacking in vogue status, is much more important to Latter-day Saints than some may realize. This book explains why heredity should still matter to card-carrying Saints even in our politically-correct day and age. I assert that Joseph of Egypt possessed many heritable traits that he passed on to his literal descendants, and his descendants can be correctly identified when observers are able to recognize these traits. This book fleshes out this important idea.

"Wherefore, it is an abridgment of the record of the people of Nephi, and also of the Lamanites —Written to the Lamanites, who are a remnant of the house of Israel; and also to Jew and Gentile." The Title page of the Book of Mormon.

Before writing this book, I had never heard of numerous American Indian tribes such as the Lenni Lenape, Munsee,[6] and Nanticoke—all groups that once dotted the North American landscape in greater numbers than today.[7] I'm almost certain that these groups are also relatively unknown to the world at large, including many Latter-day Saints, even though historical accounts state that Joseph Smith personally told some indigenous North American tribes that they were descendants of Book of Mormon peoples.

These specific tribes of New York, and other indigenous North American groups, played integral roles in the lives of early European Americans. In the early 1600s, the Lenni Lenape were among the first tribes to make contact with European settlers. They often assisted pilgrims and other settlers who were struggling to survive in the New World, and yet few people know the tribe by name alone.[8] The Munsee also interacted with early European settlers and influenced the naming of places in New York. "Manhattan," for example, is a Munsee word for "rocky island," but who's ever heard of the Munsee?[9] And even though many Americans may recognize the name of the Iroquois of New York, it's often unknown that when European settlers first met the Iroquois, the skins of these American natives were "very white."[10] These early Iroquois would even paint their skins for fear of sunburning.[11]

After gathering large amounts of data for *Native Israelites,* I now have a greater understanding of the diversity and cultures of the various indigenous Americans living across the continent. But, of course, there is still so much more to learn.

Red Men

> European explorers saw that [North American] Indians
> wore red paint and so called them 'red.'
> —**Nancy Shoemaker, assistant Professor of History
> at the University of Wisconsin (1997)**[12]

This book features an uncommonly used term in our modern era: "red men." Although this antiquated term previously worked its way into Native American vernacular as a positive designation with which to refer to themselves, in times past and in the present it has been implemented by some ethnicities in an unseemly fashion. The ill-usage of the term is one of the main reasons why it has fallen into disuse in our more politically-correct day and age.

The term "Redman" has virtually nothing to do with actual skin coloration but merely refers to a specific color that some North American Indian tribes used to paint their skins. Thus, with their skins painted red, they became known as "red men" to European Americans.

When "red men" is used in the title, and throughout this book, it is meant to be an innocuous term. It is never implemented in a derogatory manner, and quotes with the term "red men" in them have no negative connotations. In fact, the term is used in a positive sense in every case.

The term "red men" is used in this book to connect ancient North American Indians to certain modern North American Indian tribes. Since "red men" is an expression used to describe *only* North American Indians, and ancient groups of Israelites who were painted red in the Book of Mormon, the term connects the two groups. This significant connection is the primary reason why this book often early LDS brethren who used the term. Latter-day Saint Parley P. Pratt gave an address in 1851 to certain North American Indians about their Israelitish origins: "*To the Red Men of America* . . . You are a branch of the house of Israel. You are descended from the Jews, or, rather, more generally, from the tribe of *Joseph*, which Joseph was a great prophet and ruler in Egypt."[13]

Indian

I want the white people to understand my people.
Some of you think an Indian is like a wild animal. This is a great mistake.
—**Chief Joseph, Nez Percé leader (1879)**[14]

The Prophet Joseph Smith frequently used the term "Indian" to refer to descendants of Book of Mormon peoples. On November 9, 1835, for example, he recounted his first visit with Moroni: "He told me of a sacred record which was written on plates of gold, I saw in the vision the place where they were deposited, [and] he said the indians were the literal descendants of Abraham."[15] Because of the common usage of the term by early Latter-day Saint brethren, "Indian" is used throughout this book.

The word "Indian," although a misnomer,[16] is never used in this book in a negative sense. It's implemented much more than "Native American" and "First Peoples" for an important reason.[17] Usage of the term "Indian," is always neutral and its implementation is meant to steer readers away from the assumption that all pre-Columbian inhabitants of the Americas have the same origin[18] and arrived at the same time.[19] The terms "Native American" and "First Peoples," although currently popular in North America, are used less in this book than other terms because we know that a great majority of indigenous Americans are of East Asian and Siberian origin. A more accurate made-up name for the majority of indigenous Americans could be "Americans who descended from North and East Asian peoples."[20]

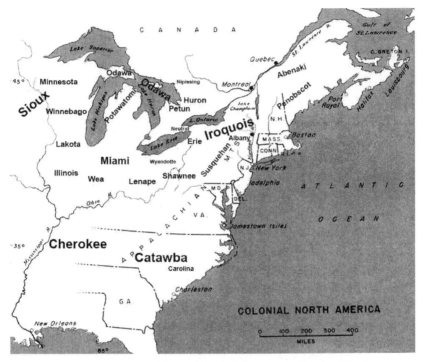

Amerindian tribes in Colonial North America (1648)

One other term used in this book to refer to indigenous Americans is "Amerindian," which is a combination of the two words "American" and "Indian." The term "Amerindian" is used chiefly in anthropological and linguistic contexts and simply refers to an indigenous American from the American continent. Since this book highlights physical anthropology, the term is used in a few sections.

The Western Hemispheric Model vs. the Limited Geography Model

> The overall picture [of the Book of Mormon] reflects before all a limited geographical and cultural point of view[.] —**Hugh W. Nibley, Latter-Day Saint scholar**[21]

This book doesn't focus much on Book of Mormon geography, and yet it's important for readers to understand two models: the Western Hemispheric and the limited geography models. Without an understanding of these two models, it's difficult to recognize the identity of the main Lamanite remnant.

Some prominent Latter-day Saints of the past have identified the Lamanites as the main indigenous people of the entire American continent and Polynesia, which supports a Western Hemispheric model for the Book of Mormon.[c] This Western Hemispheric model includes the entire American continent and often accompanies the idea that the Lamanites (and other Book of Mormon peoples) were responsible for populating the Americas.[22] This would, in effect, make Book of Mormon peoples the "first Americans."[23] However, new DNA evidence corroborates the idea that Book of Mormon peoples did not, in fact, populate the entire American continent and Polynesia.[24]

In-depth studies of the Book of Mormon have led many current Latter-day Saint scholars to support a limited geography model.[25] Said another way, after LDS scholars analyzed the distances mentioned in the Book of Mormon, a great majority of them have concluded that the major cluster of Book of Mormon settlements were not spread out across the Americas. As LDS scholar Hugh Nibley has

---

c   A Western Hemispheric view of the Book of Mormon appears to be untenable due to a limited geography model for the Book of Mormon.

expressed concerning the topic, "The overall picture [of the Book of Mormon] reflects before all a limited geographical and cultural point of view—small localized operations, with only occasional flights and expeditions into the wilderness[.]"[26]

A limited geography model for the Book of Mormon is one of several theories by LDS scholars that the book's narrative was a record of various peoples in a limited geographical region in the Americas, rather than of the entire Western Hemisphere as believed by some early Latter Day Saint brethren.[27] A good example of a limited geographical model for the Book of Mormon involves the Nephites. King Limhi, a righteous leader of the Nephites in the Book of Mormon, and his traveling party mistook the land near Cumorah for Zarahemla, which, according to the Book of Mormon, means it was not thousands of miles away in Guatemala and parts of Mexico.[d] After Limhi's caravan recognized their tactical error, they made the necessary course correction—a correction of much less than thousands of miles—to arrive at their desired destination.

Joseph Smith, Jr.

It is impossible for historians to prove that Joseph Smith was a prophet, and improbable that they will prove him a fraud. ... Similarly, historians cannot prove that the Book of Mormon was translated from golden plates and have not proven that it was simply a fiction of Joseph Smith. Instead they seek to understand its revelatory appeal, the claims it makes, and why it discloses modes of living and of believing that millions of Saints would otherwise notentertain. —**Dr. Martin E. Marty, American Lutheran religious historian (1987)**[28]

Chapter 1: Joseph Smith and Book of Mormon Peoples delves into one very important claim Joseph Smith made: the forefathers of

---

d   "I [Limhi] caused that forty and three of my people should take a journey . . . that thereby they might find the land of Zarahemla. . . . And they [Limhi and traveling companions] were lost in the wilderness . . . having traveled in a land among many waters . . . having discovered a land which had been peopled. . . . [A]nd they [Limhi and his party], having supposed it to be the land of Zarahemla, returned to the land of Nephi" (Mosiah 8:7-8; 21:26).

select tribes of American Indians wrote the Book of Mormon.[29] Without inordinately focusing on Book of Mormon geography or supporting any one specific geographical model, the first chapter also highlights connections between Nephi's vision in 1 Nephi 13 of the Book of Mormon and also United States history.

Chapter 2: Did Joseph Smith Change His Mind? discusses whether Joseph Smith shifted his attention from North American Indians to indigenous Mesoamericans as the primary Lamanite remnant. With the information provided in the second chapter, readers will most likely be able to determine for themselves if the Prophet always believed the North American Indians were predominantly the Lamanite remnant or else changed his mind about them.

Joseph of Egypt

Behold, we are a remnant of the seed of Jacob [Israel]; yea, we are a remnant of the seed of Joseph, whose coat was rent by his brethren into many pieces.
—**Moroni, Chief Captain in the Book of Mormon (Alma 46:23)**[30]

Chapter 3: The North American Story helps fill in some of the gaps between the destruction of the Nephites and the preaching of the gospel of Jesus Christ to the remnant of Joseph of Egypt in the latter days. It attempts to identify a part of Joseph of Egypt's descendants, which includes the Lamanite remnant. Joseph Smith was well-aware of the connection between the Lamanite remnant and Semitic peoples of the House of Israel through Joseph of Egypt.[31] He once wrote, "The [Nephites and Lamanites] came directly from the city of Jerusalem about six hundred years before Christ. They were principally Israelites of the descendants of Joseph [of Egypt]."[32] The third chapter discusses a few interactions between Great Lakes Indian tribes,

Joseph Smith, Jr., and other early Latter-day Saints. Also discussed is the context of the establishment of the LDS Church and the coming forth of the Book of Mormon, including missionary work performed among Great Lakes Indian tribes. Chapter 3 also explores the idea that certain North American Indians are connected to the Lamanite remnant. For example, it discusses how the Seneca of North America was the first tribe of the Lamanite remnant visited by Elders Parley P. Pratt[33] and Ziba Peterson[34] after the Prophet Joseph Smith called them to preach to the Lamanites.

Chapter 4: Yuya and Joseph of Egypt discusses a possible connection between a significant Semitic man in the Book of Mormon—Joseph of Egypt—and a man found in a tomb in the Valley of the Kings in Egypt. It covers information about how North American Indian tribes from the Great Lakes region and the midwestern United States are autosomally related to a man believed by some to be Joseph of Egypt.[e] This chapter discusses the idea that even if this Semitic man who was discovered in an Egyptian tomb isn't Joseph of Egypt, the genetic connection between North American Indians and a Semitic man buried in Egypt is significant. The connection ties North American Indians to an actual Semite via genetics and phenotype.

Chapter 5: Locating the Seed of Joseph is about the progeny of Joseph of Egypt. Chapter 5 contains photographs and quotes that tie specific Latter-day Saints and others to Joseph of Egypt. The fifth chapter talks about the significance of locating the descendants (i.e. the "seed") of Joseph of Egypt and demonstrates common physical features of his descendants.

---

e   Autosomal DNA is a term used in genetic genealogy to describe DNA which is inherited from the autosomal chromosomes.

Joseph of Egypt's Ancestry & Descendants

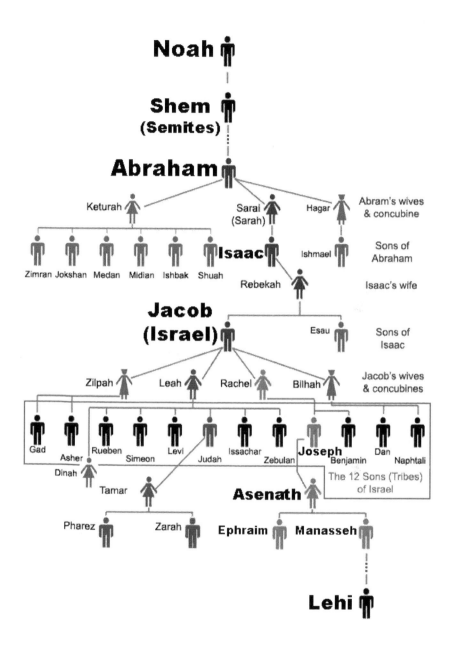

Chapter 6: The Physical Traits of Semites and Israelites uses photographs to highlight many of the common physical traits

of Israelite descendants, especially among the descendants of Joseph of Egypt. It focuses on phenotypic traits most common among the bloodlines of Israelites so readers can decide for themselves which indigenous Americans literally belong to the House of Israel.

Chapter 7: American Indian Depopulation discusses the idea that certain American Indians—including ones related to the descendants of Joseph of Egypt—have dwindled significantly in numbers due to warfare, deadly European diseases, and other causes of increased mortality rates. This depopulation of Semitic lineages can account for why few progeny of Joseph of Egypt exist today among American Indian tribes.

Lamanites

[T]herefore there began to be Lamanites again in the land.
—**Nephi, son of the disciple of Jesus Christ named Nephi (4 Nephi 1:20)**

Chapter 8: The Lamanite Remnant discusses visits made by early Latter-day Saint missionaries to the Lamanites in North America (1830-1831). Many photographs of indigenous Americans who appear to belong to the Lamanite remnant are highlighted. American Indians known to be of mixed blood were excluded from this book. To the best of my knowledge, all photographs and paintings of American Indians in this book depict full-blooded indigenous Americans. American Indians with mixed ancestry such as Quanah Parker, Sequoyah, Ohiyesa, Itawamba, and Cornplanter were passed over, even though they were notable and interesting individuals. All mixed blood Indians were left out to ensure that readers don't think indigenous Americans in this book look Semitic due to modern-day admixture with Semitic Americans.

Although the term "Lamanite" is implemented in many ways throughout the Book of Mormon,[f] and in a few ways by current Latter-day Saints,[35] only one definition for the term receives focus in this book.[36] When the term "Lamanite" is used, it refers to any group with Middle Eastern genetics once belonging to Lehi's family or any member of Lehi's caravan who traveled from the Old World to the New (unless specified otherwise).[37] The *Encyclopedia of Mormonism* definition of "Lamanite" perfectly captures the essence of the word as it is used in this book: "[A] people of Israelite origin from the Book of Mormon."[38]

Appearance and DNA

[T]he Hebrew origin of the native American races
is fundamental as testimony to the truth of the Book of Mormon.
—BH Roberts, historian, politician, and LDS leader[39]

Chapter 9: Appearance and Genetics addresses the claim made by Joseph Smith that angel Moroni told him the Indians are the literal descendants of Abraham.[40] Actual Abrahamic bloodlines receive a great deal of attention in this chapter because of this

---

f   The term "Lamanite" can refer to descendants of Laman and the people who joined his group. It can also refer to an incipient nationality based upon an ideology, with its own genealogical history and religious/political beliefs; (Mosiah 10:12-17. The name "Lamanite" referred to a religious/political faction whose distinguishing feature was its opposition to the church. See Jacob 1:13-14.) or else it can refer to one or more cultures. The Book of Mormon describes several Lamanite cultures and lifestyles, including hunting-gathering (2 Nephi 5:24), commerce (Mosiah 24:7), sedentary herding, a city-state pattern of governance (Alma 17), and nomadism (Alma 22:28). The politicized nature of early Lamanite society is reflected in the way in which dissenters from Nephite society sought refuge among Lamanites, were accepted, and came to identify themselves with them, much as some Lamanites moved in the opposite direction.

important assertion. The operative word in Joseph's claim is "literal." Because Joseph Smith claimed that Book of Mormon peoples were the literal descendants of Abraham, the statement shouldn't be glossed over or ignored, especially by Latter-day Saints. It invites further investigation. In consequence, the majority of this book addresses the idea that what you look like and your DNA are highly correlated—a correlation that is widely-known and accepted among distinguished biologists and other accredited scientists. As Richard Dawkins, British ethologist and evolutionary biologist, said in 2009 concerning the subject, "DNA links all life through the code and the more closely related the two species are physically, the more similar their codes."[41] One has to seek out this concept since it's not common knowledge, nor highly publicized. In 1865, a Jewish Rabbi named Isaac Wise seems to have understood this concept when he claimed that former US President Abraham Lincoln looked like a Jew, and therefore had Jewish genetics:

> Abraham Lincoln believed himself to be bone of our bone and flesh of our flesh. He supposed himself to be of Hebrew parentage, he said so in my presence, and indeed he possessed the common features of the Hebrew race both in countenance and features.[42]

By what seems to be no coincidence, because of his acting ability and uncanny resemblance to Abraham Lincoln, Jewish actor Daniel Day-Lewis played him in Steven Spielberg's 2012 film *Lincoln*.

The representations included in this book of the DNA-appearance correlation are meant to be user-friendly. They provide numerous visuals to help readers recognize just how essential the correlation between appearance and DNA is when attempting to understand the genetic claims of the Book of Mormon. This correlation appears to be the key to unlocking the identity of the main Lamanite remnant.

Chapter 10: Genetic Markers of Semites, East Asians, and Siberians discusses groups of genetic markers of the ancestors of Shem (son of Noah) and also East Asians and Siberians.

Chapter 11: Lehi and Arabs, Chapter 12: North American Indians and Jews, Chapter 13: Ancient Artifacts dating back to Book of Mormon Times, and *Chapter 14: East Asian Influences in Mesoamerica* all feature common physical traits among Semites and North American Indians. These chapters also demonstrate how a lack of these traits exist between Joseph of Egypt's literal descendants and other groups of indigenous Americans across the American continent. The final chapter, *Chapter 15: The Destiny of the Lamanite Remnant*, discusses the great importance of Joseph of Egypt's seed among Lehi's descendants to Latter-day Saint doctrine. Chapter 15's goal is to create a greater feeling of expectation and desire among members of the Church to fulfill Book of Mormon prophecy with the actual Lamanite remnant.

Each chapter of this book has been set up to give readers a more complete picture of what I believe actual ancestors of Book of Mormon peoples look like and the most likely genetic markers they possess. My sincere hope is that as readers explore the following chapters, they'll be able to identify the Lamanite remnant for themselves. Are they primarily the Inca, the Maya, or the Algonquin and Iroquois peoples? Let's find out.

1

Joseph Smith and Book of Mormon Peoples

When did I ever teach anything wrong from this stand? When was I ever confounded? I want to triumph in Israel before I depart hence and am no more seen. I never told you I was perfect; but there is no error in the revelations which I have taught. Must I, then, be thrown away as a thing of naught? —**Joseph Smith, Jr., Latter-day Saint Prophet (1844)**[43]

The Book of Mormon is a record of the forefathers of our western Tribes of Indians. —**Joseph Smith, Jr., Latter-day Saint Prophet (1833)**[44]

Joseph Smith . . . believed that Book of Mormon history, or at least part of it, transpired in North America. —**Donald Q. Cannon, early Mormon History professor (1995)**[45]

Two days prior to Independence Day in 1843, Latter-day Saint prophet Joseph Smith, Jr., and several of his apostles met a group of Pottawatomie chiefs who had traveled some distance to see Joseph in Illinois.[46] After patiently waiting for

Joseph to conclude some morning business, Chief Apaquachawba, one of the Pottawatomie chiefs, approached the Prophet, shook his hand, and addressed him through an interpreter:

> We as a people have long been distressed and oppressed. We have been driven from our lands many times. We have been wasted away by wars, until there are but few of us left. The white man has hated us and shed our blood, until it has appeared as though there would soon be no Indians left. We have talked with the Great Spirit, and the Great Spirit has talked with us. We have asked the Great Spirit to save us and let us live; and the Great Spirit has told us that he had raised up a great Prophet, chief, and friend, who would do us great good and tell us what to do; and the Great Spirit has told us that you are the man (pointing to the prophet [Joseph Smith]). We have now come a great way to see you, and hear your words, and to have you tell us what to do. Our horses have become poor traveling, and we are hungry. We will now wait and hear your words.[47]

According to Apostle Wilford Woodruff's account of this event, tears streamed from the eyes of the Prophet as he listened to Chief Apaquachawba.[g] It appears that Joseph's heart swelled with compassion as he contemplated the plight of these crestfallen North American Indians. With tenderness, Joseph replied to the chief via an interpreter:

> I have heard your words. They are true! The Great Spirit has told you the truth. I am your friend and brother, and I wish to do you good. Your [fore]fathers were once a great people. They worshipped the Great Spirit. The Great Spirit did them good. He was their friend, but they left the Great Spirit, and would not hear his words or keep them. The Great Spirit left them, and they began

---

g   "The Spirit of God rested upon the Lamanites, especially the orator. Joseph was much affected and shed tears." Wilford Woodruff, *Wilford Woodruff's Journal*. Ed. Scott G. Kenney. 9 vols. Midvale, UT; Signature Books, 1985.

to kill one another, and they have been poor and afflicted until now. The Great Spirit has given me a book, and told me that you will soon be blessed again. The Great Spirit will soon begin to talk with you and your children. This is the book which your [fore]fathers made (showing them the Book of Mormon). I wrote upon it. This tells what you will have to do. I now want you to begin to pray to the Great Spirit. I want you to make peace with one another, and do not kill any more Indians: it is not good. Do not kill white men; it is not good; but ask the Great Spirit for what you want, and it will not be long before the Great Spirit will bless you, and you will cultivate the earth and build good houses like white men. We will give you something to eat and to take home with you.[48]

After the Pottawatomie listened to the Prophet's sentiments, they said that what he had relayed to them brought them enjoyment. Joseph Smith kept his promise after this meeting by having an ox killed to provide them with much-needed victuals.[h] Some horses were also given to the Pottawatomie to be used as they pleased. According to one report of the event, the Pottawatomie went home mostly satisfied and contented.[49]

Some readers who pore over this story may inadvertently miss the weight of it. Unquestionably this narrative demonstrates the magnanimity of the Prophet Joseph Smith, and yet it contains much more. This story of Joseph Smith and the Pottawatomie includes a revelation from God, a prophecy made by the Prophet, and the fulfillment of important Book of Mormon prophecies. Furthermore, the location and date of this event (and other events like it in North America) have far-reaching implications for Latter-day Saints.

---

h   Later, when a Latter-day Saint emissary visited the Pottawatomies, he carried with him a hand-drawn map of the Pottawatomie lands and several fragments of Egyptian papyri left over from the Prophet Joseph Smith's translation of the Book of Abraham to be presented as gifts to the tribe. Three years later, the Pottawatomies still had Smith's letter, map, and two papyri pages. Willard Richards. Journal, 11 July 1846, LDS Church Archives.

In this story, we learn that Joseph Smith informed the Pottawatomie of the inspiration he had received from the Great Spirit (God) regarding their tribe. Joseph's revelation led him to believe that the Pottawatomie would receive blessings from the Great Spirit as soon as they changed a few questionable behaviors in their lives.[i] Joseph prophesied to the Pottawatomie that if members of their tribe would begin to pray to God, make peace with one another, and cease killing white men and Indians from other tribes, it would not be long before their lives improved. One seemingly impatient Pottawatomie chief who had listened to Joseph Smith prophesy to his people earnestly inquired, "How many moons would it be before the Great Spirit would bless our people?"[50] The Prophet replied, "Not a great many."[51]

During the meeting in Illinois with the Pottawatomie, Joseph Smith fulfilled a prophecy that he had received through revelation prior to the publication of the Book of Mormon.[j] In 1828, about two years before Joseph Smith translated the golden plates, the Lord revealed to him the main reasons the plates had been preserved for the latter days: so that the Lamanite remnant (Joseph of Egypt's descendants) could learn about their ancestors and the promises the Lord made with their forefathers. Ultimately, however, the Prophet learned from the Lord that the message produced from the golden plates would continue to live on as a tool to instruct the Lamanite

---

i   "The Great Spirit has given me a book, and told me that you will soon be blessed again." Joseph Smith, *History of the Church,* 5:480.

j   In July of 1828, the Lord told Joseph Smith that "the plates of gold had been preserved that the Lamanites might come to the knowledge of their [fore]fathers, and that they might know the promises of the Lord, and that they may believe the gospel and rely upon the merits of Jesus Christ, and be glorified through faith in his name, and that through their repentance they might be saved" (D&C 3:19-20). This particular revelation (given in July 1828) that was given to the Prophet was received after Martin Harris had lost 116 pages of the Book of Mormon manuscript, but before the Book of Mormon was first published.

remnant regarding how to exercise faith, repent of their misdeeds, and accept Jesus Christ as their Lord and Savior.[52]

Lehi, the father of Nephi and Laman, prophesied that in the latter days his descendants would "know and come to the knowledge of their forefathers" (1 Nephi 15:14).[53] When the Prophet Joseph Smith informed the Pottawatomie of their Israelite bloodline and urged them to repent, he fulfilled Book of Mormon prophecy and the prophecy that he had received in 1828.[54]

While communicating in Illinois, Chief Apaquachawba and the Prophet Joseph Smith both used the term "Great Spirit" ("Gitche Manitou" in the Algonquian language of the Pottawatomie) to refer to God.[55] Although Joseph appears to have only incorporated the term into his vernacular while speaking to native peoples, it was often used in his lifetime by select North American Indian tribes to reference a higher power. Lamoni—a Lamanite king from the Book of Mormon—also used the term "Great Spirit." This reference can be found in Alma 18 in the passages where the king converses with Ammon, a Nephite missionary. It appears that the term "Great Spirit" was used by the non-believing Lamanites at the time of the king's reign and, upon recognizing this fact, Ammon implemented the term into his missionary lesson to build on common beliefs with the king. "Believest thou that there is a Great Spirit?" Ammon asks King Lamoni in the Book of Alma. "Yea" replies the king. Ammon then states with authority, "This is God" (Alma 18: 26-28).

Like Ammon, the Prophet Joseph Smith used the indigenous North American term "Great Spirit" to teach North American Indians about the will and desires of the God of Israel. Since "Great

Spirit" is a North American Indian term that was also used by King Lamoni in the Book of Mormon, it appears to connect indigenous North Americans to Book of Mormon peoples.

A short time after the delegation of Pottawatomie Indians met with Joseph Smith in 1843, members of the tribe wrote him a letter. In their correspondence, the Pottawatomie asked the Prophet to be their "father"—an individual with whom they could rely on for counsel in temporal affairs.[56] This is another example of how a Book of Mormon prophecy was fulfilled with North American Indians. Jacob, son of Lehi, prophesied that his descendants would turn to the Gentiles like offspring who rely on a father: "Thus saith our God: . . . I will soften the hearts of the Gentiles, that they shall be like unto a father to them" (2 Nephi 10:18).[57] Joseph Smith—who was genetically part Gentile[58]—fulfilled Jacob's prophecy by assuming the role of father for these descendants of Book of Mormon peoples.[59] To accept his paternal role, Joseph Smith wrote a reply letter to the Pottawatomie on August 28, 1843:

> Dear children . . . [I]n regard to my giving you counsel and being your father . . . I shall be happy to . . . The Mormons are your friends and they are the friends of all men, and have the very best feelings to all men and especially towards you my children. . . . I feel interested in the welfare and prosperity of all my red children, and will most cheerfully do them all the good[.] . . . [—]Your Father[.]"[60]

As this heartfelt letter indicates, the Prophet Joseph Smith had a special place in his heart for his "red children" of North America, and apparently took his role seriously as the Pottawatomie's "father." In what seems a demonstration of his seriousness about this paternal role, he affectionately signed his letter to the Pottawatomie, "Your Father."[61]

This was not the first time Joseph Smith had counseled North American Indians. Less than two years before the Prophet met the Pottawatomie, he visited the Sauk & Fox Indian tribe.[62]

Early in August of 1841, Chief Keokuk and a relatively large number of his Sauk & Fox people camped for a few days along the Mississippi River near Montrose, Iowa.[63] On August 12th, about a hundred of Keokuk's people crossed the Mississippi River on a ferryboat and two flatboats to the Nauvoo landing in Illinois. Upon their arrival, they were greeted by the Nauvoo Legion Band. Disappointed to discover that the Prophet Joseph Smith had not joined the welcoming party, Chief Keokuk and the Sauk & Fox appeared upset. Insistent that Joseph Smith greet him before he and his people came ashore, Chief Keokuk had some Latter-day Saints send for the Prophet. When Joseph Smith arrived at the shore he extended a warm welcome to these indigenous North Americans. Joseph's brother Hyrum introduced him to the Sauk & Fox, and both of them escorted the tribe to a grove of trees overlooking the Mississippi River.[64]

Mary Ann Winters, an early Latter-day Saint, stood near the Prophet Joseph Smith as he preached to these Sauk & Fox Indians in a grove near the Nauvoo Temple. She recalled that as Joseph Smith spoke, his countenance lit up and his aura emanated a bright glow. His words must have penetrated the hearts of these North American Indians because, according to Winters, they looked very solemn as the Prophet spoke to them.⁶⁵

An 1841 account of Joseph Smith's address to the Sauk & Fox is recorded in the LDS *History of the Church*:

> I [Joseph Smith] accordingly went down, and met Keokuk, Kis-ku-kosh, Appenoose, and about one hundred chiefs and braves of those tribes [the Sauk & Fox],⁶⁶ with their families. I conducted them to the meeting grounds in the grove, and instructed them in many things which the Lord had revealed unto me concerning their [fore]fathers, and the promises that were made concerning them in the Book of Mormon. . . . I advised them to cease killing each other and warring with other tribes; also to keep peace with the whites; all of which was interpreted to them.⁶⁷

Chief Keokuk⁶⁸ responded to the Prophet's address with these words: "I believe you are a great and good man; I look rough, but I also am a son of the Great Spirit. I have heard your advice—we intend to quit fighting, and follow the good talk you have given us."⁶⁹ Similar to King Lamoni and Chief Apaquachawba, Chief Keokuk used the term "Great Spirit" to refer to God. Before the Prophet returned home, Chief Keokuk displayed great respect for him.

When Joseph Smith informed the Sauk & Fox of their Book of Mormon ancestry and advised them to cease killing each other and warring with other tribes, he fulfilled Book of Mormon prophecy.⁷⁰ Because of how much the Prophet counseled North American Indians such as the Sauk & Fox, it appears that he was keenly aware of who the Lamanite remnant were and the prophecies given by Book of Mormon prophets concerning them.

After translating the golden plates, Joseph Smith never ceased informing the Lamanite remnant in North America of their forefathers, nor did he stop fulfilling Book of Mormon prophecies as he spoke with them. Virtually until the day he was martyred, he went out of his way to inform North American Indian tribes of their Book of Mormon ancestry.

In July of 1831, while in Jackson County, Missouri, Joseph Smith wrote in his journal that American Indians were descendants of Shem, son of biblical Noah. He wrote that Latter-day Saint William W. Phelps preached a sermon to a group of people which included "several of the Lamanites or Indians—representative of Shem."[71] This is another way of saying that the Prophet believed American Indians were Semites (Shemites), and therefore related to Semitic groups such as the Jews.

Joseph Smith also stated that angel Moroni—the last Nephite prophet of the Book of Mormon—informed him that the American "[I]ndians were literal descendants of Abraham."[72] The Prophet did not state that this group of people were adopted into Abraham's lineage, but that they were actual relatives of the biblical Abraham. This genetic component to the covenant is significant because any individual who belongs to an actual Israelite bloodline (or else is adopted into the House of Israel) and is willing to make and keep the same covenant as Abraham of the Old Testament will receive the same blessings he received.

It appears the Prophet was certain that indigenous North Americans were genetically related to Shem, Abraham, and Israel—

important lineages of the House of Israel.ᵏ From the scriptures and revelations he had received, Joseph Smith knew of God's desire to gather scattered Israel, and thus when God informed him that groups of indigenous North Americans belonged to the House of Israel, the news must have been very exciting to him.ˡ

Joseph Smith connected the DNA of Book of Mormon peoples with Israelite genetics when he claimed: "[O]ne of the most important points in the faith of the Church of Latter-day Saints . . . is the gathering of Israel (of whom the Lamanites constitute a part)."⁷³ We know that the Prophet was referring to indigenous North Americans when he spoke of the Lamanite remnant because of the numerous times he and other early LDS brethren informed indigenous North Americans of their Book of Mormon forefathers. The Prophet informs us of the fact that this important biblical pedigree motivated his actions in North America, because shortly after learning of the heritage of North American Indians, he sent missionaries out to inform them of their Israelite bloodlines.

To solidify this idea about North American Indians, on May 23, 1844, just one month and three days shy of his martyrdom, the Prophet Joseph Smith spoke for a second time with the Sauk & Fox Indians about their genetic connections to Israel and Book of Mormon peoples.ᵐ This time the Prophet addressed approximately

---

k   The lineage of Shem, Abraham, and Israel includes the Tzadikim—a righteous branch of biblical figures such as Enoch, Melchizedek, and Joseph of Egypt. The Tzadikim are individuals who have chosen to be very righteous and belong to Shem's lineage, a genetic lineage that includes Israelites. Joseph of Egypt, an Israelite, is customarily known as the Tzadik. In Kabbalah, Joseph embodies the Sephirah of Yesod, the lower descending connection of spirituality to materiality, the social role of the Tzadik in Hasidism.

l   "We believe in the literal gathering of Israel and in the restoration of the Ten Tribes[.]" LDS Articles of Faith 10.

m   Because this visit with the Sauk & Fox Indians occurred shortly before his death, it demonstrates how he continued to believe that indigenous North Americans were blood relatives of Laman, Shem, Abraham, and Israel.

forty Sauk & Fox Indians in the back kitchen of his home. For a second time, the Prophet fulfilled Book of Mormon prophecy with the Sauk & Fox when he informed them:

> The Great Spirit has enabled me to find a book [showing them the Book of Mormon], which told me about your [fore]fathers, and Great Spirit told me, 'You must send to all the tribes that you can, and tell them to live in peace.'[74]

Joseph Smith once again linked indigenous North American DNA with Lamanite genetics when he reminded the Sauk & Fox of their Book of Mormon ancestry. He also fulfilled other Book of Mormon prophecies when he urged the tribe to live in peace and admonished them to spread the good word to other North American Indian tribes about living in peace.

In Joseph Smith's day, it appears that it was well-known to American settlers living in the Western United States that Latter-day Saints believed North American Indians were lost Israelites. A June 1836 report written by the citizens of Liberty, Missouri, criticizes Latter-day Saints who they said were

> keeping up a constant communication with our Indian tribes on our frontiers, with declaring, even from the pulpit, that the Indians are part of God's chosen people, and are destined by heaven to inherit this land, in common with themselves[.][75]

Missourians at this time apparently did not appreciate the ideas Latter-day Saints were putting into the heads of indigenous North Americans. They displayed outrage after watching Mormons inform Indians that they belonged to the House of Israel because they viewed indigenous North Americans as an inferior race rather than on par with whites. LDS religious doctrines regarding North American Indians flew in the face of the buddings of Manifest Destiny—the notion that European Americans possessed special virtues and supreme rights to American soil. Because of how North American Indians were viewed by a great majority of Americans, they were not allowed to be anything more than savages standing in the way of real estate claims given to white men by God.

Just about everything the Prophet Joseph Smith said and did with indigenous peoples of North America correlates with specific events in the vision of Nephi, son of Lehi, in the Book of Mormon (See 1 Nephi 13). Other events in the history of North America, including United States history, also fit perfectly into the vision Nephi was given by the Lord.

Nephi's Vision

Around 600 BC, after Nephi expressed a great desire to God to know more about his father Lehi's dream of the tree of life, he was given his own vision. In his vision, Nephi learns from the Lord about the symbolism of the tree of life and the other things which his father saw. Many nations and kingdoms of the Gentiles, which we can safely assume are future European countries and monarchies, are shown to Nephi (See 1 Nephi 13:1).[n] In the vision, Nephi is also shown a nefarious church in Europe that tortures, enslaves, and murders the true saints of God. Brave souls from

---

n   Latter-day Saint scholars Joseph Fielding McConkie and Robert L. Millet believe that the "nations and kingdoms of the Gentiles" are European countries. Joseph Fielding McConkie, Robert L. Millet, *Doctrinal Commentary on the Book of Mormon,* Vol. 1, p. 89.

among the tortured masses of Gentiles emerge after clamoring for freedom from this oppressive church. In desperation, these courageous yet weary Gentiles escape the iron grip of this European church by sailing the ocean to the New World.

In his vision, Nephi sees these oppressed Gentiles travel to the New World, become prosperous, and interact with the future descendants of Book of Mormon peoples. He identifies these oppressed Gentiles as groups of white people who resemble the Nephites, his future descendants.º

Nephi is also shown convoys of European ships sailing across the ocean to the New World to wage war with these white American Gentiles. According to Nephi, these white settlers who have made the New World their home are delivered by the Lord from the grasp of oppressive Europeans who have also come to the New World to claim it for their own. This portion of Nephi's vision fits perfectly with the American Revolutionary War, also known as the American War of Independence (1775-1783). Interestingly, Joseph Smith's ancestors were involved in this war.ᵖ Samuel Smith, Jr. (born in 1714), a relative of Joseph Smith on his father's side, was a distinguished community leader and a promoter of the American War of Independence.ᑫ It was said of Samuel that "[h]e was a sincere friend to the liberties of his country, and a strenuous advocate for the doctrines of Christianity."[76]

---

o   Nephi: "I beheld that they were white, and exceedingly fair and beautiful" (1 Nephi 13:15).

p   Apparently, before the United States won their freedom, Nephi witnessed in vision the American Revolutionary War (1 Nephi 13:19).

q   Great Britain oppressed European American settlers. The Declaration of Independence, which appears to be mentioned by Nephi in his vision, states: "We, therefore, the Representatives of the united States of America . . . are Absolved from all Allegiance to the British Crown, and that all political connection between them and the State of Great Britain, is and ought to be totally dissolved[.]" *The Declaration of Independence of the United States of America.*

Nephi apparently recognized that this future nation of white American Gentiles—a country that can only be interpreted as the United States of America—would be fortified by God against all other nations. Nephi prophesied that this nation was meant to be a land of liberty with no ruling kings. Concerning the subject, President Ezra Taft Benson is quoted as saying: "[George] Washington was offered a kingship, which he adamantly refused. Nephi had prophesied . . . 'there shall be no kings upon the land'" (2 Nephi 10:11). Once again, Nephi's vision fits almost seamlessly with the history of the United States of America.

Among the Gentiles in Nephi's vision who sailed to the New World in search of religious freedom were no doubt the passengers of the Mayflower. In 1620, aboard the Mayflower were 102 English Separatists, known today as Pilgrims, and about 30 crew members of the vessel. It just so happens that the Prophet Joseph Smith was a descendant of four Mayflower passengers: John Howland, Elizabeth Tilley, Edward Fuller, and Edward Fuller's wife.[r] Joseph's ancestors share the same story of escape from the oppressive church of the English monarchy in search of freedom that is found in Nephi's vision (1 Nephi 13:13).

Joseph Smith's Pilgrim ancestors settled the Plymouth Colony in present-day Plymouth, Massachusetts, with the others from their group. In 1621, the Pilgrims first met Samoset, an Abenaki lesser chief, when he introduced himself to the colony.[77] This event appears to mirror what Nephi saw in his vision (1 Nephi 13:38).[s] The

---

r   No name was provided for Edward Fuller's wife (Mrs. Fuller) on Mayflower documents.

s   "Gentiles . . . went forth upon the many waters, even unto the seed of my brethren [the Lamanite remnant], who were in the promised land" (1 Nephi 13:12). "And it came to pass that I beheld many multitudes of the Gentiles upon the land of promise" (1 Nephi 13:14).

language Samoset spoke to the Pilgrims was Algonquian, which is the same tongue spoken by the Delaware, Sauk & Fox, and Pottawatomie (all tribes personally identified by either the Prophet Joseph Smith or Elder Oliver Cowdery as members of the Lamanite remnant). This linguistic connection between Samoset's Abenaki tongue and the languages of Algonquian-speaking tribes visited by Joseph Smith and early LDS brethren suggests that the Abenaki tribe also belonged to the main Lamanite remnant.

LDS brethren in the 1930s understood that it was not by mere coincidence that the Pilgrims landed in North America—a land that is also referred to as the land of Joseph of Egypt since, Joseph of Egypt's descendants anciently ensconced in the Americas. During the LDS Church's centennial in 1930, the First Presidency declared concerning the establishment of the United States of America:

> It was not by chance that the Puritans left their native land and sailed away to the shores of New England, and that others followed later. They were . . . predestined to establish the God-given system of government under which we live, and to make of America, which is the land of Joseph [of Egypt],ᵗ the gathering place of Ephraim . . . and prepare the way for the restoration of the Gospel of Christ and the reestablishment of his Church upon earth.[78]

Nephi was made aware that groups of white American Gentiles would flourish in the New World, and that many of the Lamanite remnant would be afflicted and slain by pale-faced American Gentiles. The Lord informed Nephi of the future treatment of his descendants: "I have . . . smitten them by the hand of the

---

t   "[T]his beautiful region of country [the United States of America] is now mostly . . . the land of Joseph or the Indians[.]" WW Phelps, *Evening and Morning Star,* Vol. 1, August, 1832. No. 3, p. 22.

Gentiles" (1 Nephi 13:34). Joseph's maternal grandfather, Solomon Mack, was unwittingly involved in the fulfillment of some of this ominous declaration by the Lord through Nephi.

Solomon Mack enlisted for military service in the French and Indian War (1754-1763)—a war that provided Great Britain with a great increase in North American territory.[u] The war's costs led to colonial discontent, and eventually turned into the American Revolution. Great Britain, British America, the Iroquois Confederacy, the Catawba, and the Cherokee all fought in the French and Indian War against France, New France, the Wabanaki Confederacy, the Abenaki, a Mi'kmaq militia, Algonquins, Delaware, Ojibwe, Ottawa, Shawnee, and the Huron. Many of these specific North American Indian tribes were later visited by Latter-day Saint missionaries on missions specifically set up to convert the Lamanites to the gospel of Jesus Christ.

Mormon and Moroni's Invitation

In Mormon Chapter 7, an aged prophet Mormon invites the Lamanite remnant of the future to believe in Jesus Christ, accept His gospel, and be saved. Because the Prophet Joseph Smith and other early LDS brethren identified the Pottawatomie, Sauk & Fox, Shawnee, Seneca, Huron, and other North American Indian tribes as descendants of the Lamanites, we know that indigenous North Americans are the primary people to whom Mormon extended his invitation.

Before Mormon died, he passed on the responsibility of

---

u   The French and Indian War was the North American theater of the worldwide Seven Years' War that led to disagreements over future frontier policy.

record keeping to his son, Moroni. In Mormon Chapter 8, Moroni adds to his father's invitation to the future Lamanite remnant and includes other bits of information into his allotted portion of the golden plates. Because we know Moroni was cognizant of his father's dying wish for the Lamanite remnant to return to the fold of the House of Israel, it makes sense that he would visit Joseph Smith multiple times as an angel in Palmyra, New York. Joseph's house was located near the battlefields where countless Nephites and Lamanites perished.[v] We also cannot forget the extremely important fact that Joseph lived near the general area of the stone box containing the golden plates.[w] These facts do not appear to be coincidental and definitely tie North American Indians to the Book of Mormon.

During a visit with Joseph Smith, Moroni shared an important message with him. Moroni informed Joseph that he had been sent by the Lord "to bring the joyful tiding, that the covenant which God made with ancient Israel was at hand to be fulfilled."[79] Sure enough, Moroni was right. Shortly after he relayed this important message to the Prophet, the people to which Moroni was referring were identified and informed of their Book of Mormon ancestry.

---

v   "The great and last battle, in which several hundred thousand Nephites perished was on the hill Cumorah, the same hill from which the plates were taken by Joseph Smith, the boy about whom I spoke to you the other evening." Orson Pratt, Feb. 11, 1872, *Journal of Discourses* Vol. 14, pg. 331. "Both the Nephite and Jaredite civilizations fought their final great wars of extinction at and near the Hill Cumorah (or Ramah as the Jaredites termed it), which hill is located between Palmyra and Manchester in the western part of the State of New York." *Bruce R. McConkie, Mormon Doctrine, pp. 174-175.* "This time it will have to do with so important a matter as a war of extinction of two peoples, the Nephites and the Jaredites, on the self same battle site, with the same 'hill' marking the axis of military movements. By the Nephites this 'hill' was called the 'Hill Cumorah,' by the Jaredites the 'Hill Ramah'; it was that same 'hill,' in which the Nephite records were deposited by Mormon and Moroni, and from which Joseph Smith obtained the Book of Mormon, therefore the 'Mormon Hill,' of today—since the coming forth of the Book of Mormon—near Palmyra, New York." B.H. Roberts, *Studies of the Book of Mormon,* p. 277.

w   As Joseph Smith stated, the Hill Cumorah is in North America. "Convenient to the village of Manchester, Ontario county, New York, stands a hill of considerable size[.] . . . On the west side of this hill, not far from the top, under a stone of considerable size, lay the [golden] plates, deposited in a stone box." Joseph Smith, Jr., *Joseph Smith—History* 1:51.

Lamanite Missions

Not too long after Moroni's joyful tiding and the publication of the Book of Mormon in 1830, early Latter-day Saints expressed a great desire to fulfill the prophecies given by Book of Mormon prophets and the Prophet Joseph Smith concerning the Lamanite remnant.[80] In September 1830, just six months after the Book of Mormon was published, Elder Oliver Cowdery was the first of many Latter-day Saints to be called on missions specifically set up to find and convert the Lamanite remnant.[81] Elder Cowdery and his fellow Latter-day Saint missionaries were assigned to areas, all of which were located in North America. Interestingly, no LDS missionaries during Joseph Smith's lifetime were ever assigned to inform indigenous Mesoamericans in Guatemala or Mexico that they possessed Book of Mormon ancestry.[82] It was 44 years later that LDS missionaries were sent to Mexico, and 117 years before the Church entered Guatemala to preach the gospel of Jesus Christ.[83] In contrast, shortly after Moroni relayed his message to Joseph Smith, many North American Indian tribes were taught the gospel of Jesus Christ by Latter-day Saint missionaries and told they were the literal descendants of Lamanite peoples.

On a mission specifically set up by the Prophet Joseph Smith to convert the Lamanites, Elders Parley P. Pratt and Ziba Peterson visited the Seneca Indians of New York. In October of 1830, Joseph Smith received a revelation from the Lord stating that Elders Pratt and Peterson should continue their missions by joining Elders Oliver Cowdery and Peter Whitmer, Jr. "into the wilderness among the Lamanites" (D&C 32:1-4). In Elder Pratt's autobiography, he explains how he and his mission companions fulfilled Book of Mormon prophecy with the Seneca:

> After traveling for some days we called on an Indian nation at or near Buffalo [New York]; and spent part of a day with them, instructing them in the knowledge of the record of their forefathers. We were kindly received, and much interest was manifested by them in hearing this news.[84]

Elder Oliver Cowdery was admonished by the Lord through the Prophet Joseph Smith to "go unto the Lamanites and preach [His] gospel unto them" (D&C 8:8). After he was informed of the will of God, Elder Cowdery traveled to parts of the United States of America to preach the gospel of Jesus Christ to indigenous North Americans who belonged to the Lamanite remnant.

In late January 1831, Elders Oliver Cowdery, Parley P. Pratt, and Frederick G. Williams traveled to Kansas to preach the gospel of Jesus Christ to the Lamanites. They crossed the frozen Kansas River and walked to the Delaware (Lenni Lenape) Indian village, which was located about 12 miles west of the Missouri state line. Just as Joseph Smith did with other Algonquian-speaking tribes, Elder Cowdery and his mission companions fulfilled Book of Mormon prophecies with the Delaware Indians by informing them of their Book of Mormon ancestry. Preaching to Chief Kikthawenund, the principal chief of the Delaware Indians, and his council, Elder Cowdery said:

> [W]e . . . address you as our red brethren and friends. . . . Thousands of moons ago, when the red men's forefathers dwelt in peace and possessed this whole land, the Great Spirit talked with them and revealed His law and His will. . . . This Book [the Book of Mormon] . . . was written in the language of the forefathers of the red men. . . . Joseph Smith . . . sent us to the red men to bring some copies of [the Book of Mormon] to [you].[85]

Elder Cowdery's message to the Delaware was well-received. After the address, Chief Kikthawenund, in an appreciative and thoughtful tone, replied to him:

> We feel truly thankful to our white friends who have come so far and been at such pains to tell us good news, and especially this new news concerning the Book of our forefathers; it makes us glad[.] . . . [Y]ou shall read to us and teach us more concerning the Book of our fathers and the will of the Great Spirit.[86]

Chief Kikthawenund's positive reception of Elder Cowdery's message must have made Mormon proud. In Mormon's writings in the Book of Mormon, we learn that he believed the Lamanite remnant in the latter days would be informed of their Book of Mormon forefathers and also be numbered among the people of the first covenant.[87] This first covenant mentioned by Mormon appears to be the all-important Abrahamic Covenant.

To update the Prophet Joseph Smith concerning his missionary efforts among the Lamanites, Elder Cowdery wrote him a letter. He informed Joseph that Chief Kikthawenund and other members of the Delaware tribe had expressed belief in "every word of the Book [of Mormon]".[88] This news must have thrilled the Prophet because he knew the Lord desired for the

Lamanite remnant to learn about their forefathers and to accept the teachings of Jesus Christ contained in the Book of Mormon.[89]

While addressing the Delaware Indians, Elder Cowdery referred to them as "red men" and "red brethren," in the same way that the Prophet Joseph Smith referred to the Pottawatomie Indians as his "red children" and was known to have referred to indigenous North Americans in general as "red men."[90] The implications of Elder Cowdery's usage here of the North American slang term "red men" are notable. Elder Cowdery informed the Delaware tribe that he and his mission companions had been sent by the Prophet Joseph Smith specifically to "the red men" to provide them with their own copies of the Book of Mormon.

Other early LDS brethren also referred to North American Indians by the North American slang term "red men." In 1855, eleven years after the martyrdom of Joseph Smith, President Brigham Young wrote to Ute Indian Chief Walkara:

> [Mormons] are [the] very best friends they [the Utes] have on earth . . . It [is] because the Red Men have descended from the same fathers and are of the same family as the Mormons, and we love them, and shall continue to love them, and teach them things that may do them good.[91]

President Brigham Young connected the DNA of the Utes of North America to the genetics of members of the LDS Church—a logical connection since both groups apparently share the same genetic lineage of Joseph of Egypt.

The Origin of the Term "Red Men"

Indigenous North Americans have always been the only people known as "red men." They were the sole group of indigenous Americans in Joseph Smith's lifetime that could be connected to the term, and, according to known records, the only group of American Indians to refer to themselves by this reference. In 1854, for example, Chief Seath'tl of the Seattle tribe lamented concerning the white man's treatment of his people: "I am a red man. . . . The white man's God cannot love his red children."[92]

The use of the term "red man," which was originally neutral in tone, was first recorded in the early 17th century.[93] Around 1607-1609, Captain John Smith (no known relation to the Prophet), an English explorer who became Admiral of New England, described the red-painted bodies of natives from the Chesapeake Bay area of North America. Captain Smith said that the natives danced around a fire with their faces painted "red: but all their eyes were painted white, and some red stroakes [sic]... along their cheeks."[94]

In North America, the term "red man" applied generally to Algonquin peoples, but specifically referred to the Delaware Indians (the same tribe to which Elder Oliver Cowdery preached the gospel of Jesus Christ). The Delaware, Pottawatomie, and Sauk & Fox—all tribes visited by early Latter-day Saints—are Algonquin peoples that were also involved in wars with white American Gentiles spoken of by Nephi (the first Nephi mentioned in the Book of Mormon).

The Book of Mormon mentions groups of indigenous Americans who painted themselves red. One group, the Amlicites, painted themselves red after the manner of the Lamanites.ˣ This unique identification of indigenous North Americans has not necessarily lasted from the time of the Book of Mormon until modern times. Furthermore, the use of the term "red men" is not an attempt to prove that the two groups are seamlessly connected. Rather, it merely points to the fact that North Americans, unlike indigenous peoples of Central and South America, are "red men"—a group of people the Prophet Joseph Smith and early LDS brethren referred to as the Lamanite remnant.

The Pottawatomie, Sauk & Fox, and other indigenous North American groups in times past frequently used the color red to paint their faces and bodies, especially during times of war.[95] It is interesting to note that "Oklahoma," a location where many of the main Lamanite remnant likely reside(d), is a Choctaw Indian word, from "okla" and "humma," meaning "red people."[96] This reference further solidifies that indigenous North Americans have long referred to themselves as "red people."ʸ

Although Mayan warriors of Mesoamerica, and possibly other indigenous Mesoamerican groups, were known to have painted themselves with red colors, the specific North American slang term "red men" (and all derivations of the reference) ultimately connects specifically to indigenous North Americans. When early LDS brethren such as Joseph Smith and Oliver Cowdery used the specific

---

x   "And the Amlicites were distinguished from the Nephites, for they had marked themselves with red in their foreheads after the manner of the Lamanites; nevertheless they had not shorn their heads like unto the Lamanites. Now the heads of the Lamanites were shorn; and they were naked, save it were skin which was girded about their loins, and also their armor, which was girded about them, and their bows, and their arrows, and their stones, and their slings, and so forth" (Alma 3: 4-5).

y   The Lamanites of old and the Lamanite remnant both painted themselves red, which appears to connect the two groups.

American slang term "red men," we have no indication that they were ever using it to include indigenous Mesoamericans and/or indigenous South Americans. It appears they were only referring to North American Indians, because the term "red men" primarily denotes Algonquian-speaking tribes of North America.

Missionaries Among the Lamanites

US Map (1830) with "the borders of the Lamanites"ᶻ

On June 10, 1831, at Kirtland, Ohio, Joseph Smith received another revelation about the Lamanites that was directed towards Latter-day Saint Newel Knight. This revelation was more explicit concerning the location of the Lamanites than past revelations: "And thus you shall take your journey into the regions westward, unto the land of Missouri, unto the borders of the Lamanites[.]"[97] This reference speaks of the "borders of the Lamanites," a term not found in the Book of Mormon, but nonetheless still important. From this revelation Joseph Smith received from God, a state in the United States of America is connected to the Lamanites.

---

z   This 1830 map of the United States includes "the borders of the Lamanites." At the edge of the western US, which was adjacent to the Missouri Territory, were "the borders of the Lamanites." Joseph Smith knew that many North American Indians located in the east, including the Great Lakes region, had been forced off their homelands due to the United States Indian Removal Act of 1830.

Just beyond the United States border in the west—the "borders of the Lamanites"—was the Missouri Territory (featured in this section). This region of North America was where Indian tribes were forced to move after the Indian Removal Act of 1830 was passed by the United States Government.[aa] Some members of the Lamanite remnant had not settled in this territory until they were forced to do so. That is to say, North American Indian tribes (which included tribes from the Great Lakes region) were removed from their homelands and relocated to the western part of North America. This is why they were in the West rather than in the East (as one might expect) when the Prophet received the revelation.

In 1835, Joseph Smith established an eastern United States mission for the Twelve Apostles. He called Elder Brigham Young to "go immediately from this place to an adjacent tribe of the remnants of Joseph [of Egypt] and open the door of salvation to that long and dejected and afflicted people[.]"[98] In response to that revelation and commandment, Elder Young and other LDS brethren visited New York, Vermont, Massachusetts, New Hampshire, Maine, and parts of Canada near the Great Lakes region.

In 1837, Elder Jonathan Dunham, a Latter-day Saint Seventy,[99] and other members of the Church were called by the Prophet Joseph Smith on various North American missions to tribes in New York and elsewhere. Many of these missions were organized to follow up on the missionary efforts of Elders Cowdery and Pratt in the early 1830s. After receiving a priesthood blessing

---

aa   The Lamanites in this context were the predominant Lamanite remnant who had mostly descended from Joseph of Egypt's lineage.

that promised him a "great work" with the Lamanites, Elder Dunham and his traveling companions worked among many North American tribes including the Brotherton,[100] Oneida, Stockbridge Mohican, Allegheny, Buffalo, Catteraugus, Onondaga, Tuscarora,[101] and Tonawanda Seneca.[102] These missionaries had been called specifically to "fill Joseph [Smith's] original measures" by "proceeding from tribe to tribe, to unite the Lamanites and find a home for the Saints."[103] When the Prophet Joseph Smith instructed Latter-day Saint missionaries to unite the Lamanites on their missions to indigenous North Americans, he connected North American Indian tribes to the Lamanite remnant.

In 1840, Elder Jonathan Dunham served another mission among the Lamanites. He once again visited North American Indian tribes and recorded in his journal the conversion of an Oneida interpreter of the Iroquois Confederacy: "One Lamanite ordained and blessed by the Patriarch."[104] Elder Dunham was aware that he was fulfilling prophecy on his mission to the Lamanites since he noted: "A new scene of things are about to transpire in the west, in fulfillment of prophecy."[105]

In Joseph Smith's lifetime, the Latter-day Saints in Wisconsin who preached the gospel to the Lamanite remnant fulfilled Book of Mormon prophecy. Apostle Lyman Wight and his companions noted that the Wisconsin Mormons had become "spiritual and temporal" counselors (like fathers) to several bands of Nenomonie, Chippaway, and Winnebago Indians of North America.[106]

Latter-day Saint brethren who were serving missions in 1840 were informed of the promising success that missionaries like Dunham were having among indigenous North Americans. Elder Wilford Woodruff recorded the motivating news in his journal: "I am informed the Lamanites are beginning to embrace the work considerable."[107] Indigenous North Americans were the Lamanites to which Elder Woodruff was referring.

A Limited Geography for the Book of Mormon

Hugh Nibley, a prominent Latter-day Saint scholar, believed in a limited geography model for the Book of Mormon. Nibley remarked concerning the subject, "The overall picture [of the Book of Mormon] reflects before all a limited geographical and cultural point of view—small localized operations, with only occasional flights and expeditions into the wilderness[.]"[108] Other important LDS scholars agree with a limited geography model for the Book of Mormon. In 2013, LDS scholar John L. Sorenson published in *Mormon's Codex: An Ancient American Book* fairly specific parameters for Book of Mormon geography after analyzing the distances and measurements mentioned in the Book of Mormon. He concluded from his analyses that "The total extent of lands that Mormon knew about, based on his own words did not exceed about 600 miles (965 km) in length and half that in width."[109]

The majority of current Latter-day Saint scholars appear to agree with Sorenson and Nibley regarding the limited geography model for the Book of Mormon. However, most LDS scholars would not agree that, towards the end of Joseph Smith's life, Joseph believed indigenous North Americans were the main Lamanite remnant. This is because many of these scholars believe the Book of Mormon primarily transpired within a limited geographic model that was nowhere near the lands of the Pottawatamie, Sauk & Fox, Seneca, Huron, or Shawnee.[110]

Still, the true geographic model for the Book of Mormon must fit within a limited geography, include a Hill Cumorah located in New York (where Joseph Smith found the golden plates), coincide with the history given in Nephi's vision found in 1 Nephi 13, and also include the Pottawatomie and the Sauk & Fox tribes as descendants of the makers of the Book of Mormon. That is to say, the Book of Mormon transpired in a small "land of liberty" in the Americas where white Europeans settled after escaping the grasp of an oppressive European religion to find religious freedom—a region near the Hill Cumorah where Algonquian-speaking tribes such as the Pottawatomie and Sauk & Fox tribes (descendants of Book of Mormon writers according to Joseph Smith) resided.

In the next chapter, we will explore the probability that Joseph Smith always believed indigenous North Americans were the main Lamanite remnant. We will look at quotes from Joseph Smith during 1830—the year the Church was organized—to the year he died in 1844 to discover if he changed his mind about indigenous North Americans.

2

Did Joseph Smith Change His Mind?

I will not seek to compel any man to believe as I do, only by the force of reasoning, for truth will cut its own way. —**Joseph Smith, Jr., Latter-day Saint Prophet (1843)**[111]

Joseph [Smith, Jr.] . . . would describe the ancient inhabitants of this [American] continent . . . This he would do with as much ease, seemingly, as if he had spent his whole life among them. —**Lucy Mack Smith, Joseph Smith, Jr.'s mother (1853)**[112]

[O]ur western tribes of Indians, are descendants from that Joseph that was sold into Egypt[.] —**Joseph Smith, Jr., Latter-day Saint Prophet (1833)**[113]

The [Lamanite] remnant are the Indians that now inhabit this country. —**Joseph Smith, Jr., Latter-day Saint Prophet (1842)**[114]

All revelations, prophecies, and quotes of the Prophet Joseph Smith fit within a chronological timeline, which, when studied and understood in its entirety, provide us with a more complete story about his beliefs concerning the identity of the main Lamanite remnant. From the time the Prophet first saw visions of the ancient inhabitants of the Americas in the 1820s, to his last recorded interactions with Native Americans before his death, he provided us with clues to solve the puzzle surrounding the fate of the main Lamanite remnant.

This puzzle remains unsolved for many Latter-day Saints due to the fact that the Church has no official doctrine concerning Book of Mormon geography, and because not all LDS scholars agree on the location of the main settlement of the Book of Mormon. Many current Latter-day Saint scholars believe that the Prophet Joseph Smith was either unsure of the primary setting for the Book of Mormon—which implies that he was also unsure of who the main Lamanite remnant were—or else he originally believed indigenous North Americans were the main Lamanite remnant, but changed his mind in favor of native Mesoamericans.[115]

It is possible that Joseph Smith believed major Book of Mormon settlements spanned the entire American continent, but it seems implausible when considering all the revelations the Prophet received about North America and the natives living there. As described in Chapter 1, on numerous occasions he explicitly told indigenous North Americans that they were the literal descendants of the makers of the Book of Mormon.

Granted, at times Joseph Smith could have been ambiguous about the scope of Book of Mormon geography, and some historical records indicate that the Prophet was both specific and general about where Book of Mormon events took place. Joseph Smith, for example, definitely located the Hill Cumorah in New York, but he also used the general word "continent" when referring to the Book of Mormon—a term that could mean the Western Hemisphere (the American continent) or just North America.

It is also possible Joseph Smith became convinced in the last years of his life that the Maya belonged to the main Lamanite remnant, but there is a lack of solid evidence to support this claim. If Joseph changed his mind about indigenous North Americans in the last years, it seems strange that he would inform northern natives in 1844 that God sent him to reveal to them knowledge of their forefathers who made the Book of Mormon.

It does not appear that the Prophet Joseph Smith was unsure of the scope of Book of Mormon geography, or that he became convinced the Maya of Mesoamerica were the main Lamanite remnant. Joseph seems to have been more acquainted with Book of Mormon lands and the ancient inhabitants of the Americas than some scholars give him credit. In 1823, when Joseph Smith was only 18 years old, he was already receiving visions and instructions from God regarding ancient American natives. One example is how Joseph would describe to his family various details of the ancient inhabitants of the Americas. According to Joseph Smith's mother, Lucy Mack Smith:

> Joseph continued to receive instructions from the Lord, and we continued to get the children together every evening for the purpose of listening[.] . . . He would describe the ancient inhabitants of this continent, their dress, mode of traveling, and the animals upon

which they rode; their cities, their buildings, with every particular; their mode of warfare; and also their religious worship. This he would do with as much ease, seemingly, as if he had spent his whole life among them.[116]

Lucy's account of her son's descriptions of ancient American natives give us insights into just how well he knew Book of Mormon peoples. If Joseph Smith was able to describe the appearance and clothing styles of these ancient Americans, and could describe all the particulars of these people's cities and buildings, it does not appear he was unsure of who they were and where they lived.

To the contrary, solid evidence points to the idea that the Prophet Joseph Smith was confident regarding the scope of the Book of Mormon and never became convinced in the latter part of his life that the Book of Mormon primarily transpired in Central America. The following chronology[ab] demonstrates the lack of shift in focus of the Prophet Joseph Smith from one group of natives to another. As the timeline will indicate, from the year the Prophet published the Book of Mormon and restored the Church of Jesus Christ (1830) to the year of his death (1844), he consistently focused on one location as the primary setting for the Book of Mormon, and one group of indigenous Americans as the main Lamanite remnant.

1830-1841: North American Indian Focus

During the first eleven years of the Church, although most early Latter-day Saint brethren believed in North America as the

---

ab Granted, the consistent chronology of Joseph Smith's statements presented in this book does not include everything he purportedly ever said about Book of Mormon peoples and geography. Yet no important contradictory quotes were purposefully left out that would negate the idea that North American Indians were always viewed by Joseph Smith as the main Lamanite remnant.

primary location for the Book of Mormon, some asserted that Mesoamerica, South America, and the entire Western Hemisphere were included in the Book of Mormon narrative.[ac] However, from 1830-1841, the Prophet Joseph Smith appeared fixed concerning the primary location for the Book of Mormon and never displayed confusion about the identity of the main Lamanite remnant.

The year 1830 was monumental for the Church. The Book of Mormon was published, Jesus Christ's church was restored, and the first mission to the Lamanites was established. Joseph Smith was apparently quite certain of the identity of the Lamanite remnant because he only sent missionaries to convert Indians of North America.

In September 1830, in Seneca County, New York, the Lord admonished three Whitmers (David, Peter Jr., and John) through the Prophet Joseph Smith to "[B]uild up [His] church among the Lamanites[.]"[ad] These Whitmer men followed the Lord's counsel and assisted the building up of the Church among the natives of North America. In contrast, the first LDS mission in Mexico was established in 1879, which was approximately 49 years after the revelation was given to the Whitmers.

[ac] In April 1832, Elder Orson Pratt taught that Lehi and his caravan landed in South America. Orson Pratt, "The Orators of Mormonism," *Catholic Telegraph*, April 14, 1832. In October 1832, WW Phelps claimed that the Book of Mormon setting took place in North America and Mesoamerica. WW Phelps, *Evening and Morning Star*, October 1832, Vol. 1, No. 5,"The Far West." In September 1840, Elder Parley P. Pratt taught a Hemispheric setting for the Book of Mormon, Parley P. Pratt, *Millennial Star*, September 1840, Vol. 1, No. 4, "Book of Mormon."

[ad] Revelation given through Joseph Smith the Prophet to David Whitmer, Peter Whitmer, Jr., and John Whitmer, at Fayette, New York, September 1830, following the three-day conference at Fayette, but before the elders of the Church had separated. CR D&C 30:6.

Important early Church locations and "**the borders of the Lamanites**"

Also in September of 1830, in Fayette, Seneca County, New York, a revelation from the Lord was given through the Prophet Joseph Smith to Oliver Cowdery: "[Y]ou shall go unto the Lamanites and preach my gospel unto them[.] . . . [T]he city Zion shall be built . . . on the borders by the Lamanites[.]"[117] Elder Cowdery did as the Lord directed and preached the gospel of Jesus Christ to indigenous North Americans living near the future location of Zion in Jackson County, Missouri—a location near what the Prophet called "the borders of the Lamanites" (featured in this section).[ae] No Incas of South America or Maya of Mesoamerica were ever visited by Elder Cowdery.

In early October 1830, a revelation was given through the Prophet Joseph Smith to Elders Parley P. Pratt and Ziba Peterson, in which the Lord admonished these brethren to "[G]o with [His] servants, Oliver Cowdery and Peter Whitmer, Jun., into the

---

ae Joseph Smith explicitly referred to the line separating the United States from Indian territory as "the borders of the Lamanites" in a revelation he received in 1830 (See D&C 28:9).

wilderness among the Lamanites.[118] All of these early Latter-day Saint Elders heeded the Lord's counsel and served missions among the Lamanite remnant of North America. On the seventeenth day of October, the Ohio Star periodical recorded Elder Cowdery's response to the revelation: "I, Oliver, being commanded of the Lord God, to go forth unto the Lamanites, to proclaim glad tidings of great joy unto them, by presenting unto them the fulness of the Gospel[.]"[119]

At Buffalo, New York, Elders Pratt and Peterson found an audience of Seneca Indians of the Iroquois Confederacy to listen to their message. They taught these natives about their Book of Mormon forefathers and provided them with two copies of the Book of Mormon.

In late November 1830, Elder Frederick G. Williams, a Kirtland physician, was asked to join Elders Pratt and Peterson in the Western United States. These Latter-day Saint Elders preached the gospel of Jesus Christ to the Huron Indians at Sandusky, Ohio. Other North American tribes were also visited and informed of their Book of Mormon ancestry.

The year 1831 somewhat closely mirrored 1830 in regards to the number of mission calls to the Lamanites living in North America. The Prophet Joseph Smith received inspiration again in 1831 to send missionaries to convert the Lamanites in North America. In January 1831, Elders Oliver Cowdery, Parley P. Pratt, and Frederick G. Williams traveled to Kansas to preach the gospel of Jesus Christ to the Delaware (Lenni Lenape) Indians. While in Kansas, Elder Oliver Cowdery preached to Chief Kikthawenund—

the principal chief of the Delaware Indians—and his council: "This Book [The Book of Mormon] . . . was written in the language of the forefathers of the red men."[120] Elder Cowdery connected North American Indians to Book of Mormon peoples when he referred to them as "red men" and fulfilled Book of Mormon prophecies by informing them of their Book of Mormon ancestry.

In June of 1831, the Prophet received a revelation from the Lord directed to Newel Knight: "And thus you shall take your journey into the regions westward, unto the land of Missouri, unto the borders of the Lamanites."[121] This is yet another revelation from the Lord that locates "the borders of the Lamanites" in North America. Elder Newel Knight followed the counsel of the Lord and went among the indigenous North Americans near "the borders of the Lamanites."

In 1833, the Prophet Joseph Smith was commanded by the Lord to publish one of his hand-written letters in the *American Revivalist*. In this inspired letter, the Prophet claimed that the western tribes of Indians in North America belonged to the bloodline of Joseph of Egypt, which implies these natives had descended from Book of Mormon peoples.[af] No mention of Mesoamerican tribes was made in this letter.

Also in 1833, LDS missionaries claimed to prospective converts that the Book of Mormon came from heaven and "chiefly concerned the western [United States] Indians[.]"[ag] It appears these

---

[af] "[O]ur western tribes of Indians, are descendants from that Joseph that was sold into Egypt, and . . . the land of America is a promised land unto them." Joseph Smith, Jr. (February 1833).

[ag] Mormonism, Telegraph 2 (Feb. 15, 1831), Painsville, Ohio as quoted in H. Michael Marquardt, *The Rise of Mormonism: 1816-1814*, (Xulon Press, Longwood, Florida, 2005), p. 302. In similar verbiage, Joseph Smith wrote an *American Revivalist* article a couple of years after the previous quote was recorded: "[O]ur western tribes of Indians, are descendants from

missionaries got the idea from Joseph Smith's *American Revivalist* article, because their words almost perfectly resemble the words of Joseph Smith.

In 1834, the Prophet Joseph Smith led the Zion's camp march to help reinstate the Missouri Latter-day Saints on the Jackson County, Missouri, lands after these lands had been taken over by mobs. On this march, the Prophet Joseph Smith claimed inspiration from the "Spirit of the Almighty" to talk with his fellow travelers about an ancient white North American Lamanite who had a reputation that spanned from the Hill Cumorah to the Rocky Mountains.[ah] No mention of Mesoamerica was made by the Prophet in this story, which implies ancient indigenous Mesoamericans were unfamiliar with this widely-known Lamanite. We can assume that the ancient Maya were unaware of this Lamanite of North America because the Maya did not belong to the main group of Lamanites.

Also in 1834, while on the same march to redeem Zion, Joseph Smith wrote a letter to his wife Emma informing her that he had walked over Nephite plains in North America. Joseph's statement to his wife implies indigenous North Americans are related to Book of Mormon peoples since anything Nephite only existed during Book of Mormon times. Multiple accounts from early LDS brethren state that Joseph Smith located "the plains of the Nephites" in North America.[122] Only four "plains" are

---

that Joseph that was sold into Egypt[.]" Joseph Smith, Jr., 2 Feb. 1833: *American Revivalist*.
ah "[T]he visions of the past being opened to my understanding by the Spirit of the Almighty, I discovered that the person whose skeleton was before us was a white Lamanite, . . . who was known from the hill Cumorah or eastern sea to the Rocky Mountains." Joseph Smith, Jr., 1 March 1842: *Wentworth Letter*.

mentioned in the Book of Mormon: the plains of Heshlon, the plains of Agosh, the plains that existed near the city Mulek, and the plains of Nephihah.[123] Of these four, only the plains of Nephihah qualify as "the plains of the Nephites," because the other three plains did not belong to the people of Nephi.[124] In 1834, Joseph identified these plains as the same ones that Zion's Camp marched on. Incidentally, these were the tallgrass prairies belonging to the Great Plains of the United States.

Also in 1834, Latter-day Saint Levi Hancock recorded in his journal that Joseph Smith informed him of the location of the "land of Desolation" described in the Book of Mormon. He claimed the Prophet located it near the Illinois River—a principal tributary of the Mississippi River—in Illinois.[125] This identification of "the land of Desolation" in North America is plausible, since Joseph Smith definitely located the Hill Cumorah in New York, "the borders of the Lamanites" as the Missouri border, and "the plains of the Nephites" in the Great Plains region of North America.

In May 1835, the Prophet Joseph Smith assigned a few early Latter-day Saint brethren including Elders Brigham Young, John P. Greene, and Amos Orton to preach the gospel in North America to the Lamanites who were "the seed of Joseph [of Egypt]"[.][ai] The Prophet focused his attention on North American tribes while never mentioning indigenous Mesoamericans or indigenous South Americans.

---

ai   "[[G]o and preach the gospel to the remnants of Joseph [of Egypt—the Lamanites—in North America.]" —Joseph Smith, Jr. assigned Brigham Young, John P. Greene, and Amos Orton to serve missions to the Lamanites (May 1835), *Teachings of the Prophet Joseph Smith*, p. 75. May 2, 1835.

On January 6, 1836, in a meeting of the Latter-day Saint High Council at Kirtland, Ohio, the Prophet Joseph Smith referred to North American natives within the territorial limits of the United States of America as Israelites, and members of the Lamanite remnant.[126] His words to the brethren are recorded in his journal:

> Much has been said and done of late by the general government in relation to the Indians (Lamanites) within the territorial limits of the United States. One of the most important points in the faith of the Church of the Latter-day Saints, through the fullness of the everlasting Gospel, is the gathering of Israel (of whom the Lamanites constitute a part)[.][aj] . . . In view of its importance, . . . we feel like dropping a few ideas in connection with the official statements from the government concerning the Indians. . . . The joy that we shall feel, in common with every honest American, and the joy that will eventually fill their bosoms on account of nationalizing the Indians, will be reward enough when it is shown that the gathering them to themselves, and for themselves, to be associated with themselves, is a wise measure, and it reflects the highest honor upon our government. May they all be gathered in peace, and form a happy union among themselves, to which thousands may shout, Esto perpetua ["let it be forever"].[127]

It is definitely apparent from this quote that in 1836 Joseph Smith believed North American natives belonged to the main Lamanite remnant and the House of Israel. He tied North American Indians within the US border to Book of Mormon peoples when he said that the Indians living within the territorial limits of the United States were Lamanites. In this meeting, Joseph Smith displayed a

---

aj   The parentheses surrounding the word "(Lamanite)" and the words "(of whom the Lamanites constitute a part)" were included in Joseph Smith's journal. They were not added by the author of *Native Israelites*.

strong desire to have the Lamanite remnant living within the borders of the US to gather together in peace with other members of the House of Israel. He uttered concerning North American natives, "May they all be gathered in peace, and form a happy union among themselves[.]"[128]

In 1838, Joseph Smith apparently located an old Nephite altar (or tower) in Missouri.[129] Joseph wrote about the altar in his journal on Saturday, May 19th, of that year:

> We pursued our course up the river, mostly through timber[.] . . . [W]hen we arrived . . . at the foot of Tower Hill (a name I gave the place in consequence of the remains of an old Nephite altar or tower that stood there), where we camped for the Sabbath.[130]

The discovery of an old Nephite altar in North America makes sense because the Hill Cumorah and the borders of the Lamanites were located by Joseph Smith in North America.

In Doctrine and Covenants 28:9, we learn that Joseph Smith is informed by the Lord that he is supposed to build the city of Zion near the Lamanites. Joseph later learns this city is to be built in Independence, Missouri (See Doctrine and Covenants 57). It is important to note that the Land of Zion and the city of Zion are two different things. The Land of Zion is the American continent, but can also mean "planet earth" if Saints gather together in other parts of the world. On the other hand, the city of Zion was prophesied to be built near the Lamanites in North America.

Confusion about Central America

In 1841, a two-volume book, *Incidents of Travel in Central*

America, Chiapas, and Yucatan, was published that recounts the travels and discoveries of two men—John L. Stephens and Frederick Catherwood—in Southern Mexico, Guatemala, and Honduras. Many early Latter-day Saints read this wildly popular book that has created a rift and confusion among Latter-day Saints about Book of Mormon geography and the location of the main Lamanite remnant.

On September 16, 1841, Elder Wilford Woodruff recorded in his journal "[P]erused the 2d Vol of Stephens travels In Central America Chiapas of Yucatan & the ruins of Palenque & Copan. It is truly one of the most interesting histories I have read."[131] Early Latter-day Saint Dr. John M. Bernhisel apparently enjoyed Stephens and Catherwood's book so much that he gifted it to the Prophet Joseph Smith sometime in 1841. On November 16, 1841, Joseph Smith thanked Bernhisel for the book in a dictated letter.[132] *Incidents of Travel* is a book that discusses archaeological discoveries in North America, and the Prophet apparently commented on them: "I have read the volumes with the greatest interest & pleasure & must say that of all histories that have been written pertaining to the antiquities of this country it is the most correct [,] luminous & comprihensive [sic]."[133]

Some Latter-day Saints have read this appreciative note to Dr. Bernhisel and have assumed that Joseph Smith is referring to Mexico, Guatemala, or even Mesoamerica in general when he says "this country." However, what some may not know is that *Incidents of Travel* not only talks about antiquities in Mesoamerica, but also discusses antiquities discovered in North America. If the Prophet was so impressed by the antiquities of Mesoamerica that he became convinced the Book of Mormon primarily transpired in Mexico and Guatemala (as so many current LDS scholars suppose), then why did the Prophet not send any LDS missionaries to Mesoamerica? Since he was commanded to preach the gospel to the Lamanite remnant, this lack of action in Mesoamerica reinforces the Prophet's previous revelations and missionary efforts to share the gospel of Jesus Christ with the main Lamanite remnant in North America.

To add to this argument, distance cannot be an excuse for why indigenous Mesoamericans were never visited by Latter-day Saint missionaries in Joseph Smith's lifetime. During this time period, LDS missionaries traveled to Canada, the United Kingdom, Australia, Germany, the Netherlands, Turkey, Israel/Palestine, and French Polynesia to perform missionary work. Hence, with this additional situational evidence that missionaries were sent to locations much further away than Mesoamerica, it does not appear that Joseph Smith believed indigenous Mesoamericans comprised the main Lamanite remnant.

1842-1844

The years 1842-1844 in LDS Church history are confusing to many Latter-day Saints simply because of all the mixed messages of early LDS brethren during this three-year period. During this time, many early LDS brethren and various members of the Quorum of the 12 Apostles supported at least three different geographical models for the Book of Mormon: North American,

Mesoamerican, and Western Hemispheric. This ambiguity created by prominent members of the early Church has created a rift that continues to exist within the Church today. The main basis for this confusion comes from three unsigned *Times and Seasons* editorials which were published in 1842 by an unknown author, but are assumed by many to have been written by the Prophet Joseph Smith.

1842: The Controversial Year

The year 1842 in LDS Church history remains an unsolved mystery, especially for Latter-day Saints who wish to solidify the identity of the main Lamanite remnant and the primary setting for the Book of Mormon. Because of how important the Prophet Joseph Smith's words are on the matter, this section focuses on his actual and verifiable words. In the beginning of 1842, Joseph Smith connected the Lamanite remnant to North American Indians inhabiting the United States of America. The Prophet also claimed in the beginning of 1842 that archeological evidence found in North America supports the Book of Mormon, including the actual golden plates upon which the book was written. During the middle to latter part of 1842, however, a journal entry and a few unsigned *Times and Seasons* editorials were published that may or may not have been written by the Prophet Joseph Smith. These documents suggest that the Prophet could have changed his mind about his solid identification of indigenous North Americans as the main Lamanite remnant.

In March of 1842, Joseph Smith wrote what is known today as the *Wentworth Letter* to John Wentworth—the editor and proprietor of the *Chicago Democrat*—that claims the Lamanite remnant inhabited the United States of America. The *Wentworth Letter* reads: "The [Lamanite] remnant are the Indians that now inhabit this country."[134] The meaning of the term "this country" in this particular context has been debated over the years by a number of LDS scholars. Although it seems clear in this instance that the Prophet was referring to the United States of America, some LDS scholars who support a Mesoamerican model for the Book of Mormon tend to support the idea that Joseph Smith believed "this country" meant the entire American continent. Aside from the fact that he wrote this letter to someone who lived in the United States of America, another reason Joseph Smith never appears to have meant "the entire American continent" when he said "this country," is that—as previously discussed—North American Indians were the only natives visited and taught the gospel of Jesus Christ in Joseph's lifetime. There seems to be no viable reason to suppose the Prophet meant "the entire American continent" when he said "this country."

In May 1842, Joseph Smith appears to have written an article in the *Times and Seasons* about ancient mummies discovered in Lexington, Kentucky. It is likely that his authorship is authentic since participants in the academic Latter-day Saint community do not appear to treat it as controversial. In this article, the Prophet claims that the existence of ancient mummies in North America provided strong corroborating evidence for the Book of Mormon.[135] The article talks about embalmed mummies that were discovered in Kentucky. It reads:

On this subject Mr. [Thomas] Ash [sic]^ak has the following reflections: . . . 'I have neither read nor known of any of the North American Indians who formed catacombs for their dead, or who were acquainted with the art of preservation by embalming.' Had Mr. Ash in his researches consulted the Book of Mormon his problem would have been solved, and he would have found no difficulty in accounting for the mummies being found in the above mentioned case. . . . another strong evidence of the authenticity of the Book of Mormon.[136]

On August 24th, 1815, Dr. Samuel L. Mitchill,[137] an American naturalist, commented in a letter to Samuel M. Burnside —the Secretary of the American Antiquarian Society at the time— concerning the mummies to which the May 1842 *Times and Seasons* article was referring:

> Dear SIR: I offer you some observations on a curious piece of American antiquity now in New York. It is a human body found in one of the limestone caverns of Kentucky. It is a perfect exsiccation; all the fluids are dried up. . . . The outer envelope of the body is a deer-skin[.] . . . The innermost tegument is a mantle of cloth, like the preceding, but furnished with large brown feathers, arranged and fashioned with great art, so as to be capable of guarding the living wearer from wet and cold. . . . The scalp, with small exceptions, is covered with sorrel [reddish] or foxy hair. . . . I am equally obliged to reject the opinion that it belonged to any of the tribes of aborigines, now or lately inhabiting Kentucky. . . . This conclusion is strengthened by the consideration that su⌐ manufactures are not prepared by the actual and resident red ⌐ the present day.[138]

---

ak   Thomas Ashe (1770-1835) was an Irish writer and novelist. In Ashe's *America*, he writes of his experiences with a vast network of huge open discovered in 1783 beneath the city of Lexington, Kentucky, that stone altar for sacrifices, human skulls and bones piled high, and m⌐

The time period for which these mummies lived is uncertain; the DNA cannot be tested due to the fact that the mummies are missing. However, as the quote attests, these mummies possessed characteristics unlike the natives of the eighteenth and nineteenth centuries, which is why Mitchill probably believed they were not genetically related to modern Native Americans. (The red hair of the mummy was probably a dead giveaway.)

In June of 1842, an article was published in the *Times and Seasons* that may or may not have been written by Joseph Smith. Mesoamerican model enthusiasts believe the Prophet wrote the article, but controversy surrounds the article's authorship.[139] It reads, "[T]he Mexican records [accounts given by archaeologists who reported on Mexico] agree so well with the words of the book of Ether[.] . . . These accounts, then, precisely agree, one of which was found in Ontario county, N. Y., and the other in Mexico."[140] If Joseph Smith wrote the article, then he apparently supported a Western Hemispheric model for the Book of Mormon in June of 1842.

On June 25, 1842, speaking of John Lloyd Stephens and Frederick Catherwood, it is believed that Joseph Smith wrote the following in his personal journal:

> Messrs. Stephens and Catherwood have succeeded in collecting in the interior of America[,] a large amount of relics of the Nephites, or the ancient inhabitants of America treated of in the Book of Mormon, which relics have recently been landed in New York.[141]

Although it is unclear what is meant by the term "interior of 'rica," and whether the Prophet wrote this entry, it could support any of the three main models for Book of Mormon geography: the North American, the Mesoamerican, or the Western Hemispheric. This lack of clarity has created much confusion for groups of Latter-day Saints.

In July 1842, Elder John E. Page, an early Latter-day Saint, taught that several Book of Mormon cities (See 3 Nephi 8-9) were cities spoken of in *Incidents of Travel in Central America*.[142] It is uncertain where Elder Page learned of this idea. The idea may or may not have come from the September 15, 1842, *Times and Seasons* article that may have been written by Joseph Smith, or he could have made it up. Elder Page also wrote in July of 1842, "Let it be distinctly understood, that the Prophet Alma uttered this prophecy, not far from Guatemala or Central America, some 82 years before the birth of Christ."[143] Even among LDS scholars who support a Mesoamerican model for the Book of Mormon, there is disagreement on this topic. LDS scholar and Mesoamerican model supporter Matthew Roper commented on quotes of early LDS brethren who had made claims about Book of Mormon cities:

> Those who thought that Copan, Quirigua, Palenque, and Uxmal were the very cities named in the Book of Mormon text were mistaken. We know now, but nobody in 1842, or for a long time afterward, could date accurately the age of those ruins. [John L.] Stephens's opinion, thoughtful and informed, was still just one among many at the time. So it was not unreasonable for Joseph Smith or Latter-day Saints in 1842 to draw their own conclusions. . . . *Incidents* provided a glimpse of a civilization whose level and complexity few had witnessed, and [John L.] Stephens was keenly aware of many other cities yet to be discovered.[144]

What Roper is apparently saying here is that, although we have quotes of early LDS brethren in Church History referring to cities that they said fit within the time frame of the Book of Mormon, they were mere guesses since no one at that time knew the dates of the cities discovered in Mesoamerica.

In July 1842, an article was published in the *Times and Seasons* about the high civilization in the Americas. Much of the article appears to have been written in Joseph Smith's style, so he may have authored it. In the article, examples of ancient North American archaeological finds are given by the author as a testimony of the truthfulness of the Book of Mormon. The article talks about brass weapons that had been discovered in Canada and Florida:

> If men, in their researches into the history of this country [The US], in noticing the mounds, fortifications, statues, architecture, implements of war, of husbandry, and ornaments of silver, brass, &c.-were to examine the Book of Mormon, their conjectures would be removed, and their opinions altered; uncertainty and doubt would be changed into certainty and facts; and they would find that those things that they are anxiously prying into were matters of history, unfolded in that book [the Book of Mormon].[145]

As the author of the article mentions, the Book of Mormon unfolds events in ancient North American history. The author argues that the history of "this country" (i.e. the history of North America) is unfolded in the Book of Mormon.

After the *Times and Seasons* article talks about North ⟨...⟩, it makes a statement: "[A] great and a mighty people had ⟨...⟩is continent[.]"[146] This statement is followed by supporting evidence: "Stephens and Catherwood's researches in Central America abundantly testify of this thing (a great and a mighty people had inhabited this continent)."[147] The article adds further evidence to the initial claim:

> The stupendous ruins, the elegant sculpture, and the magnificence of the ruins of Guatamala [Guatemala], and other cities, corroborate this statement, and show that a great and mighty people-men of great minds, clear intellect, bright genius, and comprehensive designs inhabited this continent. Their ruins speak of their greatness; the Book of Mormon unfolds their history.[148]

The author (or possibly authors) of this *Times and Seasons* article dated July 15, 1842, appears to support a Hemispheric model. The article mentions that aspects of both ancient North American and ancient Mesoamerican history are unfolded in the Book of Mormon. However, most of the article is about North America, which was Joseph Smith's focus all along. It is possible that Mesoamerica is connected to the Book of Mormon, but it is unclear how much of a role it played in the book because of how little Joseph Smith referred to it, and the fact that he never sent missionaries to preach in these areas.

An unsigned editorial dated September 15, 1842, was published in the *Times and Seasons*. Although no known concrete evidence exists as to the article's authorship, many LDS scholars attribute the article to Joseph Smith. It reads: "[T]hese wonderful ruins of Palenque are among the mighty works of the Nephites."[149] If the Prophet did in fact write this article, it seems to support a Mesoamerican model for the Book of Mormon.

Another unsigned *Times and Seasons* editorial dated October 1, 1842, reads: "[T]he ruins of Quirigua are those of Zarahemla[.]"[150] It is unclear if Joseph Smith wrote this article, but a number of LDS scholars purport this as evidence that Joseph Smith placed the city of Zarahemla in Quiriguá, Guatemala.

Many current scholars support the idea that if North America was where Joseph Smith consistently believed the Book of Mormon primarily transpired, he would not have "allowed" a differing view from his own to be published in LDS Church newspapers. LDS scholar Matthew Roper has said concerning the topic:

> It is highly unlikely that the Prophet would have allowed that view [a Hemispheric model for the Book of Mormon] to receive such wide circulation for so long a time had he felt that it contradicted anything of significant doctrinal or revelatory significance to the Saints.[151]

However, as Latter-day Saint scholar Terryl Givens has noted, Joseph Smith was known to allow the publication of views contrary to his own in Church-owned newspapers:

> Joseph Smith was as likely to promote openness as to exert his authority. . . . He severely rebuked his own brother Hyrum for performing unauthorized rituals. But . . . Smith seldom exerted editorial control over the publications of church members. Associates, missionaries, editors of church newspapers, and apostles like W. W. Phelps, Orson Pratt, and Parley Pratt printed broadsides, pamphlets, and in some cases books with no supervision. When several reviewers referred to Parley Pratt's Voice of Warning as a standard work along with Mormon scriptures, neither he nor [Joseph] Smith corrected the perception. Smith allowed a virtually unconstrained flowering of Mormon theology, as his colleagues

enthusiastically joined in vigorous exploration, elaboration, and conjecture, constituting a communal project of religion making.[152]

Roper's claim that Joseph Smith would not have allowed early Latter-day Saints to publish their own views about the Book of Mormon in Church periodicals is unfounded. Elder Orson Pratt, for example, believed in a South American landing for Lehi.[153] No known records indicate that the Prophet Joseph Smith located the landing of Lehi in South America, nor is there indication that he corrected Pratt's view. On September 15, 1841, Elder Charles Wandell drew a comparison between the "glyphs of Otolum" found in Palenque, Mexico, to the Reformed Egyptian characters found on the "Anthon Manuscript"—the small piece of paper on which the Prophet Joseph Smith wrote several lines in 1828 and which Martin Harris showed to Columbia University professor Charles Anthon.[154] No known records indicate that Joseph Smith supported the idea that glyphs of indigenous Mesoamericans were similar to Book of Mormon glyphs, nor is there known evidence that the Prophet corrected this view. Orson Pratt's brother Parley believed the Book of Mormon narrative transpired from North America to South America but often emphasized findings from Mesoamerica as proof of the Book of Mormon.[155] No known records indicate Joseph Smith supported or corrected these particular ideas of Parley P. Pratt.

It is easy to be confused about Book of Mormon geography since Latter-day Saint brethren have not always agreed on the primary setting for the Book of Mormon.[156] LDS leaders have even recognized that all American Indians are not related to Book of Mormon peoples. From 1981-2007, the LDS Church officially claimed that the Lamanites were "the principal ancestors of the American Indians",[al] but later changed their original statement by saying: "[T]he Lamanites . . . are *among* the ancestors of the American Indians."[157]

Due to the varying ideas of early LDS brethren about the identity of the Lamanite remnant and Book of Mormon geography, it is difficult for Latter-day Saints to know the mind of Joseph Smith regarding these matters. If the authorship of the three controversial *Times and Seasons* articles and one controversial journal entry could be authenticated then it would be much simpler. But without these accounts in the mix, one must look to what Joseph Smith himself said and did to find answers: locating the Hill Cumorah in New York and personally identifying at least two North American Indian tribes as belonging to the Lamanite remnant.

1843-1844: North American Indian Focus

As the first chapter of this book mentioned, in 1843, the Prophet Joseph Smith informed the Pottawatomie of their Book of Mormon ancestry: "The Great Spirit has given me a book, and told me that . . . [t]his is the book which your [fore]fathers made (showing them the Book of Mormon)".[158] This address was given a year before Joseph was killed, and about a year after he read Stephens and Catherwood's travel book. The information contained in this address is one of the main reasons why it is much easier to believe Joseph Smith never changed his mind about North America.

---

al   Elder Bruce R. McConkie, Book of Mormon Introduction (1981).

In 1843, Joseph Smith gave no indication he became convinced in the last years of his life that the Maya were the main Lamanite remnant.

The year 1844 mirrored 1843 in regards to what the Prophet Joseph Smith believed and revealed about the ancestry of the Indian tribes of North America. A month before Joseph's martyrdom in 1844, he informed the Sauk & Fox tribe that he had found the Book of Mormon, and had received a revelation from the Great Spirit that told him to inform their tribe about their Book of Mormon forefathers.[159] This address made by the Prophet to indigenous North Americans shortly before his death solidifies the idea that he was not "convinced" he had been in error by identifying indigenous North Americans like the Pottawatomie and Sauk & Fox tribes as members of the Lamanite remnant.[160]

Did Joseph Know the Identity of the Lamanite Remnant?

If Joseph Smith did not know the scope of Book of Mormon geography, or he became convinced in the last years of his life that Mesoamerica was the primary setting for the Book of Mormon (like some LDS scholars have attested), then these stances negate what the Prophet said to the Whitmers in 1830 about building up Christ's church in North America. They negate what the Lord said via revelation to Oliver Cowdery in 1830 about preaching the gospel to the Lamanites of North America. Furthermore, if one of these stances is true, "the borders of the Lamanites" should have been located among the Maya, not among the natives of North America, and Elders Pratt, Peterson, Cowdery, Whitmer, and Williams should have been directed by the Prophet to travel to the Mesoamerican wilderness, not the North American wilderness. With a belief that Mesoamerica is the primary setting for the Book of Mormon, the Prophet should not have directed Newel Knight to Missouri, but to Mexico and/or Guatemala. This also would mean that the Prophet misinterpreted the revelation the Lord gave about the Lamanites and what He meant by the "regions westward." With a Mesoamerican perspective of the Book of Mormon, what the Lord must have meant by "regions westward" in the revelation to Newel Knight was not a region in the west, but a southern region from where they were at the time—a "southward" region that included Mexico and Guatemala.[161]

If the most widely accepted Mesoamerican model for the Book of Mormon among current LDS scholars is true, the Prophet should not have claimed inspiration by the "Spirit of the Almighty" in 1834 to tell his fellow travelers that an ancient white North American Lamanite had a reputation spanning from the Hill Cumorah to the Rocky Mountains. He would also have been mistaken to tell his wife Emma that he had been "wandering over the plains of the Nephites" or that the "land of Desolation" was near the Illinois river, since both would have been located in Mesoamerica.

It appears that most of the confusion and controversy between the Mesoamerican and North American models stems from a small handful of controversial unsigned *Times and Seasons* articles and one journal entry, none of which can definitively be attributed to the Prophet Joseph Smith. More convincing, however, are the previously mentioned revelations and actions taken by the Prophet that directed him specifically to the red men of North America—the main Lamanite remnant.

In the April 1929 General Conference, President Anthony W. Ivins of the First Presidency unwittingly displayed doubt that Joseph Smith placed the Book of Mormon in Mesoamerica when he spoke to Latter-day Saints about the uncertain location of Zarahemla:

> There is a great deal of talk about the geography of the Book of Mormon. Where was the land of Zarahemla? Where was the City of Zarahemla? and other geographic matters. It does not make any difference to us. *There has never been anything yet set forth that definitely settles that question. So the Church says we are just waiting until we discover the truth.* . . . *We* do not offer any definite solution. As you study the Book of Mormon keep these things in mind and do not make definite statements concerning things that have not been proven in advance to be true.[162]

If President Ivins were truly convinced Joseph Smith authored the aforementioned controversial *Times and Seasons* articles that place Zarahemla in Guatemala, he would not have said, "Where was the [c]ity of Zarahemla? . . . There has never been anything yet set forth that definitely settles that question."[163]

Similar to President Ivins, a number of other apostles and prophets have made similar statements concerning the topic of Book of Mormon geography. President Harold B. Lee demonstrated his uncertainty about Joseph Smith placing Zarahemla in Guatemala when he said, "[I]f the Lord wanted us to know where . . .

Zarahemla was, he'd have given us [the] latitude and longitude, don't you think?"[164] Elder John A. Widtsoe of the Council of the Twelve also displayed his uncertainty stating that, "As far as can be learned, the Prophet Joseph Smith, translator of the book, did not say where, on the American continent, Book of Mormon activities occurred."[165] And in 1918, President Joseph F. Smith refused to approve any Book of Mormon maps, owing to the fact that he felt "the Lord had not yet revealed it, and that if it were officially approved and afterwards found to be in error, it would affect the faith of the people."[166]

It appears that Latter-day Saint brethren do not consider the unsigned *Times and Seasons* articles as official Church doctrine, and that they do not constitute sufficient evidence to place Zarahemla in Guatemala, or prove the validity of any site specific locations upon which the Mesoamerican model is based.[167]

3

The North American Story

We found several [North American Indians] who could read, and to them we gave copies of the Book [of Mormon], explaining to them that it was the Book of their forefathers. Some began to rejoice exceedingly, and took great pains to tell the news to others, in their own language. —**Parley P. Pratt, Latter-day Saint Apostle (1831)**[168]

[B]uild up my church among the Lamanites[.] —**Revelation received through the Prophet Joseph Smith at Fayette, New York, Seneca County (1830)**[169]

[T]he City [of Zion] shall be built, but [its location] shall be given hereafter. Behold, I say unto you, that it shall be among the Lamanites. . . . [T]he land of Missouri . . . is the land of promise, and the place for the city of Zion. . . . the place which is now called Independence is the center place. —**Two separate revelations received through the Prophet Joseph Smith (1830 & 1831)**[170]

The western marches of the state of New York in the early years of the nineteenth century produced two great religious prophets . . . Joseph Smith . . . [and] an Iroquois Indian . . . Handsome Lake. —**Anthony F. C. Wallace, author (1952)**[171]

On June 15, 1799, a man prayed fervently to God and asked Him a series of questions. His prayer careened heavenward as he sought answers about life and religion. As the man supplicated earnestly, the heavens apparently opened and, through a conduit, a heavenly messenger appeared to him. In the same encounter, three more glorious and radiant beings made their appearance.[172] One of the beings addressed himself as a messenger from the heavens, called the man by name, and claimed to have a message of great import. The messenger then informed the man of a great mission he was to complete, and that they, the messengers, had been sent from above as a result of the man's ponderings about the Creator, and because he genuinely wished to better himself and mankind.[173]

Although this kind of story may sound somewhat familiar to Latter-day Saints, it actually belongs to an Onödowága Indian of the Iroquois Confederacy. Given the name Ganioda'yo at birth, he was affectionately known in English as "Handsome Lake."[174]

Handsome Lake is said to have received other visions. In each successive vision, messengers provided him with important information he was to share with his people. His tribe, the Onödowága, or Seneca, of the Great Lakes region, were quickly deteriorating, mostly due to intemperance and the ravages of European diseases sweeping the land. Drunkards and dissolute individuals were common among his tribe, and witchcraft was apparently widespread.

At the time of Handsome Lake's first vision, he was among the many unhealthy, malnourished, and addiction-enslaved Seneca. The white man's firewater had him in a death grip. Until the messengers provided him with wisdom about how the Seneca, and the Iroquois at large, could escape their awful situation, he was on the road to self-destruction.

Handsome Lake applied the teachings he had received from the heavenly messengers, and his life turned around. His health improved, and as he began teaching others the truths he received, he became known as "the Seneca Prophet." Others referred to him as "Sedwa'gowa'ne," meaning "Our Great Teacher."[175]

The message of Handsome Lake included a moral code eventually known to the world as the *Code of Handsome Lake*.[am] Once the teachings spread across the land, Handsome Lake's religion became more successful than most religious institutions of the time. It appears that because his code incorporated traditional Iroquois religion with Christian values, it extended beyond the Seneca to many tribes within the Iroquois Confederacy in Canada and the United States. His religion emphasized survival without jettisoning a cherished Iroquois identity. It also stressed the need to make healthy adjustments to thrive in an ever-changing world.

Handsome Lake cared deeply about his tribe and was involved in relations with the United States Government. On November 11, 1794, he signed a US treaty known as the Pickering Treaty (also known as the Treaty of Canandaigua) with the Six Nations of Iroquois. President George Washington represented the United States of America in the treaty, which was signed after the American Revolutionary War. The treaty established peace and friendship between the US and the Six Nations of Iroquois. It also affirmed Iroquois land rights.

---

am   In 1826, the women Faithkeepers of the Tonawanda Seneca asked Handsome Lake's grandson to recall the words of the teacher. From these recollections, they created the "Code of Handsome Lake," which has since evolved into the "Gaiwiio" (Good Word), the syncretic American religion practiced today as the longhouse religion of the Haudenosaunee.

Thaonawyuthe, "**Chainbreaker**," a **Seneca** (b. 1737 - 1760, d. 1859)[176]

Handsome Lake's religion lived on after its founder's death in 1815. The teachings of his religion, which are often referred to as "Gaiiwo," continued to be taught by Handsome Lake's disciples including Thaonawyuthe—known to many as "Chainbreaker" (featured above)—who was a nephew of Handsome Lake.

It was largely due to Handsome Lake's influence that the Seneca, and other affiliates of the Iroquois Confederacy, were receptive to the Christian teachings of the Latter-day Saint missionaries. Mission companions Parley P. Pratt[177] and Ziba Peterson[178] were the first to visit the Seneca in 1830—the same year the Prophet Joseph Smith established the church to which they belonged. Because Chainbreaker was alive when Elders Pratt and Peterson preached the gospel of Jesus Christ to the Seneca, it is likely he conversed with them.

Elders Pratt and Peterson shared the message of the Book of Mormon with the Seneca. Members of the tribe seemed amenable to their words.[179] In his journal, Elder Pratt reminisced about this encounter with members of the Lamanite remnant: "We were kindly received, and much interest was manifested by them on hearing this news [of the Book of Mormon]."[180] Two copies of the Book of Mormon were presented to the Seneca prior to the missionaries' departure. This fulfillment of Book of Mormon prophecy with the Seneca Indians is one of the reasons why the Book of Mormon narrative fits well with the North American story (i.e. the history of North America).

Elders Pratt and Peterson—some of the first Latter-day Saint missionaries called to serve among the Lamanite remnant—had both been baptized prior to their mission calls in Seneca Lake (a lake named after indigenous North Americans who they would later identify as descendants of the Lamanites), New York. Furthermore, Elders Pratt and Peterson were both ordained as Elders in the Church of Christ—an early name for The Church of Jesus Christ of Latter-day Saints—in Seneca County, New York,[181] after the Church was established in the same county (Seneca County).[182] It is interesting that the Church was organized on lands so close to the Hill Cumorah (one of the only confirmed Book of Mormon lands), which is located in a county named after a group of Lamanite remnants to which the Book of Mormon was primarily written. These connections between the Seneca, Book of Mormon peoples, and Latter-day Saints appear to include divine correlation. Concerning divine correlation, Elder Neal A. Maxwell once said, "[T]he Book of Mormon plates were not buried in Belgium, only to have Joseph Smith born centuries later in distant Bombay."[183] They were buried in New York so Joseph Smith could find them in his youth.

Long before Elders Pratt and Peterson met with the Seneca, the Prophet Joseph Smith had interacted in his youth with North American Indian tribes, maybe even before his First Vision. Young Joseph appears to have been cognizant of, and possibly even made well-versed in, Handsome Lake's religious teachings and strong spiritual influence on the Seneca.[184] The Prophet may have been conscious of some of the cultural practices and spiritual beliefs of many Seneca (and other tribes of the Iroquois Confederacy) since he interacted with them on numerous occasions in his youth.

Written accounts attest that young Joseph Smith was known to have frequented "Mud Creek"—a main tributary which feeds the Erie Canal and Clyde River in Wayne County, New York, and an area where Seneca, and other Iroquois travelers would often camp.[185] It was primarily a stopover point for Indians from the Iroquois Confederacy on their trade routes. Because of the time Joseph Smith spent among the Iroquois, and due to his curious nature, it is not a stretch to think that he engaged in conversations with these North American Indians at Mud Creek.

It appears from historical records that Handsome Lake's influence planted a seed in Joseph Smith's mind long before he sent Latter-day Saint missionaries to visit members of the Iroquois Confederacy. Joseph's interactions with Handsome Lake's people may have influenced his future decisions. For example, when Latter-day Saint missionaries were sent on missions by Joseph Smith to convert *the* Lamanite remnant to the gospel of Jesus Christ, the first group of Lamanites the missionaries visited in this dispensation were none other than Handsome Lake's beloved Seneca.

Red Jacket (2 photos),^{an} a Seneca chief (one on left, c. 1853,[186] right one, c. 1828[187])

Red Jacket (featured above), another nephew of Handsome Lake, gained some prominence within the United States. Red Jacket, also known as "Sagoyewatha" (Keeper Awake), was a conscientious man and a skilled public speaker. As one of his names attests, he apparently had the ability to captivate an audience.

Red Jacket was arguably the most widely-known Seneca of the region during Joseph Smith's youth. Known for his public speaking, he gave multiple speeches in the New York area during this time. Consequently, he arrived in Palmyra in 1822 to deliver a speech.[188] As a teenager in New York, Joseph belonged to the juvenile debating club in Palmyra. Because of Joseph's interest in public speaking, it is possible that he could have attended this event. During his speech, Red Jacket attempted to clear his name of any

---

an   Red Jacket was known as "*Otetiani*" in his youth and "*Sagoyewatha*" (Keeper Awake) because of his oratorical skills. His talk on "Religion for the White Man and the Red" (1805) has been preserved as an example of his great oratorical style.

slander spoken about him in the region.[ao] He refers to his reputation among his "red brethren" while attempting to exonerate himself:

> I have lived many years, and have always been beloved and respected by my red brethren. . . . I have round my neck a silver plate, presented to me by General [George] Washington, which he told me to preserve and wear so long as I felt friendly to him and the United States, as an evidence of his friendship for me. If I have ever violated any treaty or any agreement made by me, why has this not been taken from me[?] You see it here yet. I say I never have so done.[ap]

After the white man broke treaties with his people and attempted to destroy his reputation, Red Jacket was loathe to accept any religions of the race. He appears to have been so jaded by the actions of many white men in his lifetime, that he even distanced himself from his uncle Handsome Lake's religion. This distance was kept because the religion combined traditional spiritual teachings of the Iroquois with the white man's Christian beliefs. It is likely that Red Jacket viewed this integration as a mélange of sorts due to the fact that he cherished his own traditional religious Iroquois teachings and detested the lies, acquisitiveness, and gross hypocrisy he observed among white "Christians." Red Jacket appears to have decided that the twain should forever remain separate because of his stance about blending religion and race.

[ao] "Red Jacket" is a term of endearment given to him by Americans for his embroidered scarlet jacket presented to him by a British officer during the Revolution. Red Jacket sided with the Americans in the War of 1812.

[ap] Red Jacket's Speech (1822), Palmyra, NY. *Republican Advocate*, Batavia: November 15, 1822. Red Jacket received the medal to which he referred directly from the hands of General George Washington in 1792. He wore it often with pride. Today, the Tonawanda Senecas who follow the way of Handsome Lake say that in the afterlife Red Jacket is forever pushing a wheelbarrow of dirt from one place to another, never completing his task. This is his punishment for supporting the sale of Seneca lands during his lifetime.

In his 1822 Palmyra speech, besides attempting to exonerate himself, Red Jacket discussed various ethnicities, the fact that he viewed the religions of each group as separate, and that each religion should most likely remain apart:

> To the White Man[:] . . . [The Great Good Spirit] gave one way to worship him and certain customs; to the Red Man another, and his customs and way to live; and to the Black Man others still. . . . [S]o long as the Great Good Spirit will suffer me to live among his red children, I know it is my duty, (for a certain something within me tells me so) to watch over their interest, and as far as I am capable to protect them, from the cunning and avarice of the white men.[189]

Red Jacket's views most likely surfaced after relations had changed between the US Government and the Six Nations of the Iroquois Confederacy.

The Seed of Joseph of Egypt

Other Latter-day Saint brethren interacted with the Seneca and the various tribes of the Iroquois Confederacy. On May 2, 1835, five years after the Seneca were first visited by Latter-day Saint missionaries, an LDS Church conference was held. During the conference, apostle Brigham Young and Elders John P. Greene and Amos Orton were assigned the special role of converting the Lamanites. They were appointed to "go and preach the gospel to the remnants of Joseph"[190] and promised that their missionary efforts would "open the door to the whole house of Joseph."[191]

Latter-day Saint missionaries Brigham Young and John P. Greene (Young's brother-in-law) met with the tribe to which Red

Jacket (Handsome Lake's nephew) belonged. Elders Young and Greene traveled to the Seneca Indian reservation, which was located on the banks of the Allegheny River in Southwestern New York. Elder Young summed up the experience in his journal, "We there saw many of the seed of Joseph, among them were two chiefs, one a Presbyterian, the other a pagan."[192]

Elder Brigham Young consistently referred to North American Indians as the "seed of Joseph [of Egypt]."[aq] This assertion strengthens the connections between the Seneca and other North American Indian tribes, Book of Mormon peoples, and Latter-day Saints. Elder Young's words support Joseph Smith's claim that North American Indians belonged to the main Lamanite remnant (and possibly other remnants of Book of Mormon peoples) and possessed the literal lineage of Joseph of Egypt. When Young used the word "seed" to refer to the North American Indians, he clearly implied that they were genetically connected to Joseph of Egypt.[ar]

Brigham Young also testified that while the Latter-day Saints were settled in Nauvoo, the Prophet Joseph Smith received visitors from among the seed of Joseph of Egypt who belonged to North American Indian tribes.[193] In April of 1844, just a couple of months before his death, Joseph addressed a large crowd of about

---

aq   Moroni said, "[W]e are a remnant of the seed of Joseph, whose coat was rent by his brethren into many pieces" (Alma 46:23). Ether reads: "[T]he Lord brought a remnant of the seed of Joseph [of Egypt] out of the land of Jerusalem" (Ether 13:7).

ar   "[P]arallels [exist] between Mormon rituals and those of the Hopi [Indians of North America]. . . . Parallels appear between the language of the Mormon temple ceremony and the Hopi myths of origin . . . . Responding to someone who asked about similarities between the Mormon temple endowment and the Masonic ceremony, [Hugh] Nibley wrote that the parallels between the Mormon endowment and the rites of the Hopi 'come closest of all as far as I have been able to discover—and where did they get theirs?'" Boyd J. Peterson, *Hugh Nibley: A Consecrated Life* (Salt Lake City: Kofford Books, 2002), p. 282.

20,000 people at a general conference held shortly after the funeral service of Elder King Follett.[as] Elder Brigham Young, who was in attendance at the sermon, noted in an account the he saw eleven North American Indian chiefs (he referred to as "Lamanite chiefs and braves") on the stand with the Prophet Joseph Smith while he addressed the large audience.[194]

On numerous occasions, the Prophet Joseph Smith claimed that North American Indians and Joseph of Egypt's descendants from among the Lamanite remnant were one and the same. For example, on January 4, 1833, at Kirtland, Ohio, Joseph Smith sent a letter with an important message about the Lamanite remnant to be published in the *American Revivalist.* He sent the letter to Noah C. Saxton, editor and proprietor of both the *American Revivalist* and the *Rochester Observer*, at Rochester, New York, with the hope that all of his letter would be published. Upon hearing that his letter was not published in its entirety, Joseph Smith responded with a follow-up note to Mr. Saxton dated February 12, 1833 to express his dismay. It reads,

> The letter which I wrote . . . for publication, I wrote by the commandment of God and I am quite anxious to have it all laid before the public for it is of importance to them[.] . . . I now say unto you that if you wish to clear your garments from the blood of your readers I exhort you to publish that letter entire[.][195]

The Prophet clearly underscored the significance of the contents of his hand-written letter when he emphatically stated it

---

as   King Follett died on March 9, 1844, of accidental injuries. Literary critic Harold Bloom call the Prophet's sermon at the conference "one of the truly remarkable sermons ever preached in America." Bushman, Richard Lyman (2005), *Joseph Smith: Rough Stone Rolling,* New York: Knopf, p. 533. "The King Follett Sermon (part 1)", Ensign, April 1971: 13.

was penned "by commandment of God" and that Saxton would have the blood of his readers upon his hand for not publishing it. The original letter Joseph Smith sent to Noah C. Saxton, which focused on the United States and scattered Israel, mentioned the importance of Joseph of Egypt's genetic lineage and DNA connection with North American Indians.[at] Joseph Smith's original letter stated: "For some length of time I have been carefully viewing the state of things as now appearing throughout our Christian Land . . . the United States[.]"[196] Joseph's letter continues:

> The time has at last arrived when the God of Abraham of Isaac and of Jacob has set his hand again the second time to recover the remnants of his people[.] . . . The Book of Mormon is a record of the forefathers of our western tribes of Indians[.] . . . By it we learn that our western tribes of Indians, are descendants from that Joseph that was sold into Egypt and that the Land of America is a promised land unto them, and unto it all the tribes of Israel will come. . . . The people of the Lord . . . have already commenced gathering together to Zion[,] which is in the state of Missouri.[197]

Similarly, Elder Wilford Woodruff also identified North American Indians as the seed of Joseph of Egypt. In an 1852 reminiscent account given by Elder Wilford Woodruff,[au] he relayed the following prophecy he had heard spoken by the Prophet Joseph Smith:

[at] To learn more about Joseph of Egypt's DNA and inborn traits, refer to *Appendix 2: The Joseph of Egypt Connection and "Believing Blood."*

[au] Although records indicate that Elder Woodruff may have supported a Western Hemispheric model for the Book of Mormon, these separate identifications of the Lamanite remnant link these tribes to Book of Mormon peoples of North America (but only if the ancestors of these particular tribes had actually lived in North America from the time of the Book of Mormon to when they met Joseph Smith and other LDS brethren).

> This work will fill the Rocky Mountains with tens of thousands of Latter-day Saints, and there will be joined with them the Lamanites who dwell in those mountains, who will receive the Gospel of Christ at the mouth of Elders of Israel, and they will be united with the Church and the kingdom of God, and bring forth much good.[198]

In Woodruff's reminiscent account, Joseph Smith identified North American Indians who dwell in the Rocky Mountains as members of the Lamanite remnant. In the same 1852 address, Woodruff commented on Joseph Smith's prophecy about the Lamanites living in the Rocky Mountains:

> I little thought, when I listened to those words, that I should never live to see the fulfillment of these words of the Prophet. I little thought that I should ever visit the Rocky Mountains, or ever see the Lamanites of whom he then was speaking. These men before me today bring to my mind sayings of the Prophet. His mind expanded on that occasion, and he had a good deal to say with regard to the progress of this work, what the Elders of Israel would have to pass through, and the work that God would require at their hands in the redemption of the Lamanites and the honest and meek through the world, and in the building up of the Zion of God on earth. But I have lived to see these days. I have lived to see the Lamanites in these mountains. I have visited a great many of them—the Zunis, Lagunas, Moquis, Navajos, Apaches, and a great many of these Indian tribes. I have preached the Gospel to them, in connection with my brethren, through interpreters. I have spent many interesting days with these Lamanites in the mountains of Israel. I spent three days in the wilderness in Arizona with Petone, the great war chief of the nation. I preached the gospel to him. He called his tribe together, stood upon his feet some two hours or more, and told his tribe all that we had said to him.[199]

Evidence of Joseph of Egypt's Descendants

Multiple Native American and First Nations tribes have oral histories that recount their peoples inhabiting North America for thousands of years. According to the mythologies found in most Native American cultures, their people settled in similar locations that their ancestors first settled.[200] These histories, if historically accurate, support the notion that certain tribes have remained in North America for a long time. Granted, numerous migrations of Indian tribes have occurred over the last few millennia, and much shifting has transpired anciently and in modern times, but no known evidence can support the migration of ancient North American Indian tribes such as the Pottawatmie and Sauk & Fox from Mesoamerica to North America.[201] The following are descriptions of evidence that has been discovered in North America that supports the oral histories of these indigenous peoples who have remained in place for at least hundreds of years, and also demonstrates strong connections that tie the ancient inhabitants of North America to Joseph of Egypt.

Metal Plates in North America

Metal plates are one example of evidence that has been discovered in North America which supports the idea that the Hill Cumorah mentioned in the Book of Mormon is located in North America. In 1775, James Adair, an American Indian trader, described two brass plates and five copper plates that were discovered by the Tuccabatche Indians of North America.[202] According to Adair, an Indian informant described the following to a man named William Bolsover on July 27, 1759:

> [H]e was told by his forefathers that those plates were given to them by the man we call God; that there had been many more of other shapes, some as long as he could stretch with both his arms, and some had writing upon them which were buried with particular men; and that they had instructions given with them, viz. [that is to say], they must only be handled by particular people, and those fasting; and no unclean woman must be suffered to come near them or the place where they are deposited. He said none but this town's people had any such plates given them, and that they were different people from the Creeks [a North American Indian tribe]. He only remembered three more, which were buried with three of his family, and he was the man of the family now left. He said, there were two copper plates under the king's cabin, which had lain there from the first settling of the town.[203]

As Adair's Indian informant attested, ancient metal plates were not completely uncommon in North America. Other people besides Adair and his Native American friends were cognizant of advanced metallurgy, including ancient artifacts such as metal plates, that had been discovered in North America. New Yorker Orsamus Turner reported in 1809 that a New York farmer plowed up an "Ancient Record, or Tablet."[204] This plate, according to Turner, was made of copper and "had engraved upon one side of it . . . what would appear to have been some record, or as we may well imagine some brief code of laws."[205]

Other accounts of metal plate discoveries exist. In 1823, John Haywood, a historian known as "The father of Tennessee history," authored *The Natural and Aboriginal History of Tennessee*.[206] In this book, he remarks that "human bones of large size" and "two or three plates of brass, with characters inscribed resembling letters"[207] were found in one West Virginia mound, and a circular piece of brass with letter-like characters was discovered in North Carolina.

No known prominent archaeologist currently disputes the fact that advanced metallurgy existed during Book of Mormon time periods in North America. Since advanced metallurgy has been discovered in North America, it increases the likelihood that indigenous North Americans are the main Lamanite remnant.[208] Metal plates with ancient writings on them are more common in North America than in Mesoamerica. Advanced metallurgy in Mesoamerica during Book of Mormon time periods is scant or non-existent (especially items like metal books).[209]

The Practice of Scalping

When the Pottawatomie tribe visited Joseph Smith in 1843, they were a group of "red men" who sought for help even though they knew they were not well liked by most pale-faced European Americans. This tribe had sided with the British in the War of 1812 (also known as the "forgotten war") and were known among European Americans as a bellicose people. The fierceness and warlike tendencies of the Pottawatomie often mirrored those same traits of Lamanites of old, but for whatever reason, in their estimation, Joseph Smith was a different kind of white man than the ones to whom they were accustomed.

The Pottawatomie were vicious toward American troops during the War of 1812, including American prisoners of war (POWs). As Diana Childress, author of *The War of 1812*, has said:

> After the battle at Fort Meigs, 600 US prisoners were marched to the British camp. There[,] indigenous North Americans yanked off their hats, clothing, money, and watches. These indians forced the Americans through a double line of warriors who prodded them and struck them with clubs. When a British officer tried to protect the Americans, a Native American shot him down. As the prisoners watched in horror, Potawatomi warriors began calmly shooting and scalping one American [prisoner] after another in gruesome fashion.[210]

Although this account of the Pottawatomie focuses on their cruel acts towards American prisoners of war, it includes an important connection to Book of Mormon peoples that may be overlooked by many Latter-day Saints. This connection has to do with the practice of scalping.[av]

According to known records, of all the indigenous groups of the Americas (North, Central, and South American), the practice of scalping, which is consistent with war practices found in the Book of Mormon, was only ever used by North American Indians. No known records convey the idea that Mayans, Olmecs, Aztecs, and other indigenous Mesoamericans, ever scalped anyone. Conversely, the Pottawatomie and the Sauk & Fox—two tribes the Prophet Joseph Smith personally identified as descendants of Book of Mormon peoples—both scalped foes, just as the Lamanites of old did in the Book of Mormon.

Alma Chapter 44 recounts a battle between the Nephites and the Lamanites in which a Nephite warrior scalps a Lamanite leader. During the conflict, Moroni—a Nephite commander—speaks with Zerahemnah, the main Lamanite military leader. Moroni's troops have the upper hand in the skirmish, and yet Moroni assures Zerahemnah

---

av   Scalping is the act of removing the scalp, or a portion of the scalp, either from a dead body or living person, as a trophy of battle or portable proof of a combatant's prowess in war.

that his people have no desire to "shed [their] blood for power" (Alma 44:2). After Moroni relays more of his message, Zerahemnah angrily lunges at Moroni with a raised sword. A Nephite soldier then steps in front of his leader and knocks Zerahemnah's sword to the ground, breaking it by the hilt with the tremendous blow. With a swift hand action, the soldier then scalps Zerahemnah and the top of his head falls to the ground. The soldier then picks up the scalp by the hair from off the ground and sticks it atop the tip of his own sword and states that unless they surrender their weapons, make a covenant of peace, and return to their own land, all of them will suffer the same scalpless fate (Alma 44:1-14).[211]

Independent of this account given in the Book of Mormon that includes the practice of scalping, various researchers have found that scalping was used in ancient North America at the time the Book of Mormon was written, which is solid supporting evidence in favor of the continuity between Book of Mormon peoples and indigenous North Americans. Trophy skulls discovered in several Hopewellian burials[aw] dating back to around 500 BC-400 AD—a date that of course fits within the Book of Mormon timeline—"frequently exhibit superficial cuts and scratches, apparently made by flint knives in the process of removing the flesh. There are many examples where cut marks only appear that are consistent with those which would be caused by customary techniques of scalping."[212]

Scalping was also an Old World[ax] practice implemented

---

aw   The Hopewell tradition (also called the Hopewell Culture) describes the common aspects of the North American Indian culture that flourished along rivers in the northeastern and midwestern United States from 200 BC- 500 AD, in the Middle Woodland Period.

ax   The Old World consists of places such as the Middle East, the Mediterranean, and the Far East, and is regarded collectively as the part of the world known to Europeans before contact with the Americas. It is used in the context of, and contrast with, the New World (the Americas).

among certain Semitic groups.[ay] Ancient Scythians of Eurasia,[213] for example, would scalp their enemies in battle. First century Jewish-Roman historian Flavius Josephus confirmed in his works the Semitic nature of the Scythians when he depicted Parthian-Sacae-Scythian peoples as descendants of Semites. He described them as most likely being from a group of the tribes of Israel who were carried away into captivity by Assyrians more than 700 years earlier.[214]

Fourth-century Roman soldier and historian Ammianus Marcellinus described scalping by the Alans, a nomadic people of Iranian origin (who are often Semitic) and the ancestors of the Ossetians. Scalping is mentioned in Ossetian folklore.

Because scalping was only practiced by ancient Semitic cultures, ancient cultures at the time of the Book of Mormon, and modern era North American Indians, it is logical, based on the evidence, to connect the three groups by a single thread. It appears that the well-known North American Indian scalping practice originated in the Old World, made its way from the Old World to the New World at the time of the Book of Mormon, and continued to be practiced in North America by indigenous peoples up until the modern era.

A Land of Liberty

Another important reason why the Book of Mormon narrative fits within the North American story has to do with the idea that the Book of Mormon predominantly transpired in "a land of liberty" (See 2 Nephi 10:11). No other nation in the Americas other than the United

---

ay   To learn more about ways in which the New World is connected to the Old World, refer to *Appendix 4: Old World Symbols in the New World.*

States of America fits this description provided by Book of Mormon prophets. Religious freedom provided by the United States Constitution allowed Joseph Smith to found a religion in a land that provided religious freedom. Book of Mormon prophets foretold that America would be "a land of liberty unto [the remnant of Joseph of Egypt]" (2 Nephi 1:7). President Joseph F. Smith corroborated this idea of the United States of America as "a land of liberty" when he said, "This great American nation [is where] . . . the fullest freedom and liberty of conscience [is] possible for intelligent men to exercise in the earth."[215]

US minted Indian head $5 gold coin (1910)

Featured above is a $5 Indian head gold coin minted in 1910 by the United States of America. The coin highlights a North

American Indian with a high nasal bridge[az] and the word "LIBERTY." Because of how interconnected the United States of America, liberty, the Book of Mormon, Joseph of Egypt's remnant, and indigenous North Americans appear to be, it seems uncoincidental that this coin was set up the way it was.

The United States of America has always been considered a "land of liberty" for many of its citizens.[ba] The Statue of Liberty in New York is affectionately known as "Lady Liberty." The US Pledge of Allegiance states that the United States of America is "one Nation under God, indivisible, with liberty and justice for all." It appears to be no coincidence that liberty, the United States, and indigenous North Americans—who the Prophet Joseph Smith said made the Book of Mormon—are all connected. Mexico and Guatemala are not usually considered "lands of liberty," which is another reason why the North American story seems to dovetail the narrative of the Book of Mormon.[216] The early United States of America allowed sufficient religious freedom and liberty to pave the way for the restoration of the gospel of Jesus Christ through the Prophet Joseph Smith.

---

az High nasal bridges are common among individuals with the actual blood of Israel.
ba Regrettably, indigenous North Americans, African Americans, and other minorities have not always experienced true liberty under the flag of the United States of America.

4

Yuya and Joseph of Egypt

He [Yuya of Egypt] was a person of commanding presence, whose powerful character showed itself in his face. One must picture him now as a tall man . . . [with] a great hooked nose like that of a Syrian; . . . strong lips; and a prominent, determined jaw. He has the face of an ecclesiastic, and there is something about his mouth that reminds one of the late Pope, Leo III. One feels on looking at his well-preserved features, that there may be found the originator of the great religious movement.
—**Arthur Weigall, English Egyptologist (1910)**[217]

Joseph [of Egypt] is a fruitful bough. —**Genesis 49:22**

So Joseph died . . . and they embalmed him, and he was put in a coffin in Egypt. —**Old Testament writer (Genesis 50:26)**

Before the discovery of Tutankhamen's lavish treasures, the tomb of Yuya and Tuyu was one of the most significant burials to be located in the Valley of the Kings in Egypt. Discovered on February 5, 1905, by James Quibell and Theodore M. Davis, the tomb contained one of the most complete and beautifully assembled sets of funerary equipment then known. Many intact items from the tomb indicate the high status of Yuya and his wife, Tuya. It is thought by at least one Egyptian Egyptologist—Ahmed Osman—that Yuya is Joseph of Egypt. Other Egyptologists like Arthur Weigall, a renowned English Egyptologist and archaeologist, have claimed that Yuya had "the face of an ecclesiastic[.]" and that "One feels on looking at his well-preserved features, that there may be found the originator of the great religious movement."[218]

Joseph of Egypt's bloodline is part and parcel with the Lamanite remnant. If Yuya is Joseph of Egypt, we can trace his bloodline and authenticate the idea that a number of American Indians are his descendants.

Yuya's outer coffin

As mentioned, in 1905, the mummy of a Semitic-looking man was found in the Valley of the Kings—a region set apart for

Egyptian royalty and the highest-ranking officers in the land.[219] Known among scholars as "Yuya," this possible descendant of Shem[bb] had become vizier of ancient Egypt and was buried in an ancient tomb (his outer coffin is featured above). Based on evidence from several sources, a strong case can be made that Yuya is, in fact, Joseph of Egypt. Egyptian scholar Ahmed Osman has also made this claim.[220]

This chapter is divided into seven sections, with each section touching upon the following seven main points that support the hypothesis that Yuya is Joseph of Egypt and also possibly related to North American Indians (the main Lamanite remnant):

1) Yuya was a Semite—an extremely uncommon race buried among Egyptian royalty.

2) The god Yuya worshipped was not of Egyptian origin and was named "Ya," which is likely the contracted form of the Hebrew word "Yahweh."

3) Records indicate that Yuya served two consecutive pharaohs, which is also true of Joseph of Egypt.

4) The Bible and the Koran both state that Joseph of Egypt was handsome. Yuya was also a handsome man, and photos prove it.

5) Many titles bestowed upon Yuya in his funerary papers match those given to Joseph of Egypt.

6) Yuya's DNA matches the autosomal ancestry[221] of a number of North American Indians living in the Great Lakes region and on the Great Plains.

7) Yuya has unique facial characteristics that are strikingly similar to those belonging to certain North American Indian tribes,

---

bb Shem of the Bible was the son of Noah and the father of all Semitic peoples.

particularly in the Great Lakes region.

In this chapter, and throughout this book, readers are shown many photographs of different types of faces accompanied by descriptions of various facial characteristics. While these pictures and their descriptions may on the surface seem repetitive, the ultimate result of this process is to help readers gain the ability to automatically spot these characteristics in real-world scenarios. With very little conscious effort, readers will be provided with the tools to observe the characteristics of various genetic lineages of the House of Israel, potentially even amongst their own colleagues, friends and family members.

-1-

Yuya was a Semite

Yuya was a Semitic-looking man who was buried in the Valley of the Kings in Egypt. For Yuya to be a Hebrew and buried among Egyptian royalty was unheard of, except in the case of Joseph of Egypt.

King Tut (an **Egyptian Pharaoh**), a **Semitic** man[222] & **Yuya** of Egypt (1390 BC)

Featured above are three profiles: Egyptian Pharaoh King Tut, a Semitic man from Iran, and Yuya of Egypt. Yuya and the Semitic man both share prototypical Semitic physical features including high nasal bridges, long and aquiline noses, narrow and rectangular-shaped faces, and long-headed (dolichocephalic) skulls, whereas King Tut's face differs from the other two.[223] After examining Yuya's mummy in the early 1900s, Egyptologist Henri Naville noted that Yuya "might be Semitic[.]"[224] Naville was uncertain about Yuya's ethnicity in his report, but he knew Yuya's appearance was uncommon in Egypt. It appears he lacked the necessary fortitude to solidify the fact that Yuya's physical features clearly resemble those of a prototypical Semite.

Archaeologist Arthur Weigall, a man involved in the discovery of Yuya's tomb, came even closer than Naville to pinpointing Yuya's ethnic origin. He wrote,

> His [Yuya's] nose is prominent, aquiline and high-bridged ... [T]he nostrils have been dilated in the process of packing the nasal fossae with linen. . . . [T]he nose is 56 millimeters long (2.2 inches). . . . The jaw is moderately square. . . . The eyebrows and eyelashes are well-preserved . . . [and] are of a dark brown colour. . . . When we come to enquire into the racial character of the body of Yuaa [Yuya], there is very little we can definitely seize on as a clear indication of his origin and affinities. . . . Unlike the ears of most of the royal mummies of the New Empire, Yuaa's [Yuya's] were not pierced [which was a common practice among Egyptians]. . . . The form of the face (and especially of the nose) is such as we find more commonly in Europe than in Egypt.[225]

To the trained eye, photographs of Yuya's mummy indicate his facial features are non-Egyptian. A likely reason for why Yuya looked European to Weigall is that many Europeans possess Semitic features and ancestry because they originated from the Middle East.[bc] In the quote above, Weigall described Yuya's "nose [as] prominent, aquiline and high-bridged."[226] These are physical features commonly found among Semites and other Middle Easterners.[227] As previously mentioned, a number of Europeans have these heritable physical traits because their genetics originally came from the Near East.

-2-

The God Yuya Worshipped was not Egyptian

The god Yuya worshipped was not of Egyptian origin and was named "Ya," which is likely the contracted form of the Hebrew word "Yahweh." It was customary for Egyptians to include the names of the gods on the sarcophagi and other funerary furniture of the deceased to indicate whose protection the dead person was placed under. Ra-mos, Ptah-hotep, and Tutankh-amun were common names of gods included on funerary items. Yuya's sarcophagi[228] and other funerary furniture were unique: they did not have the names of Egyptian gods on them, which means his items did not follow the common pattern of Egyptian royalty. According to Yuya's funerary texts and the writings on his furniture, he worshipped a non-Egyptian god named "Ya"—a name that appears to be the contracted form of the God of Israel. Joseph of Egypt also worshipped "Yah" or "Yahweh." The Bible reads, "And [Yah] was with Joseph, and he was a prosperous man; and he was in the house of his master the Egyptian" (Genesis 39:2).

---

bc   As archaeologist Arthur Weigall pointed out, Yuya looks more European than Egyptian, which implies that a European man became vizier of Egypt rather than a Semite. A possible reason for why Yuya looked European to Weigall is that many Europeans possess Semitic features and ancestry.

The name of Yuya's god was placed on his funerary items with multiple spellings.[bd] The eleven variant spellings of this god—Yaa, Ya, Yiya, Yayi, Yu, Yuyu, Yuya, Yaya, Yiay, Yia, and Yuy[be]—on Yuya's afterlife items indicate the foreign origin of his god. Ancient Egyptian scribes appear to have been so unfamiliar with this foreign god that they used all possible spellings of his name to ensure that the correct one was included. Clearly, this was an important element of Egyptian funerary procedures, which means they probably did not want to make a mistake on something so essential.

Since the actual name of this man was unknown, Egyptologists turned to the writings on the funerary items for inspiration. The name "Yuya" was selected by Egyptologists from the long list of variations of the names for this man's god. Thus the mummy became known as "Yuya" from that point on.

The name of a non-Egyptian god on Yuya's funerary items is quite extraordinary. Only a truly exceptional individual would have been granted permission by pharaoh to worship his own non-Egyptian god and have this god's name written on the items chosen to accompany him into the afterlife. According to Egyptologist Ahmed Osman, no foreign (i.e. non-Egyptian) gods have ever been found on Egyptian funerary items except in the case of Yuya.[229] This fact would appear to make Yuya a one-of-a-kind Egyptian vizier, just like Joseph of Egypt.

---

bd   The ancient Egyptian scribes were most likely unsure of how to spell the Hebrew word "Yah" or "Yahweh" in Egyptian.

be   These multiple spellings of Yuya's God could open the door for Yuya's actual name to be Yusef—Joseph.

-3-
Yuya and Joseph of Egypt both Served two Pharaohs

Records indicate that Yuya served two consecutive pharaohs, which is also true of Joseph of Egypt. The Bible mentions a second pharaoh who began his reign while Joseph was vizier of Egypt: "Now there arose up a new king over Egypt, which knew not Joseph. . . . And the Egyptians made the children of Israel to serve with rigour: and they made their lives bitter with hard bondage" (Exodus 1:8, 13-14 KJV). Egyptologist Ahmed Osman believes that the two pharaohs were Tuthmosis IV and his son, Amenhotep III. Amenhotep III's name is on at least three items found in Yuya's tomb: a coffer, Sitamun's kohl-tube of bright blue glazed faience (sintered-quartz ceramic probably from Sitamun's palace in Thebes), and a vase.[230]

Both Yuya and Joseph of Egypt are said to have been given gold necklaces and rings by pharaoh. Genesis Chapter 41 mentions these items:

> And Pharaoh said unto Joseph, See, I have set thee over all the land of Egypt. And Pharaoh took off his ring from his hand, and put it upon Joseph's hand, and arrayed him in vestures of fine linen, and put a gold chain about his neck; And he made him to ride in the second chariot which he had; and they cried before him, Bow the knee: and he made him ruler over all the land of Egypt (Genesis 41:41-43).

A gold chain was discovered in Yuya's tomb. Based on details from descriptions of the tomb by British Egyptologist Percy E. Newberry, Yuya was buried with this gold chain around his neck.²³¹ This would seem to indicate that this chain was of special significance or importance to Yuya. Could this be the same chain Pharaoh gave to Joseph? We may never know for certain, but it seems likely.

Although a gold chain was found in the tomb, a gold ring was not. One explanation for the absence of a gold ring is that it may have been passed on to another Egyptian royal by Joseph of Egypt. Another explanation may be that it was taken from the tomb by bands of grave robbers who looted Yuya's tomb multiple times before it was officially discovered in 1905. According to Ahmed Osman, Yuya was either given the title on his funeral papers "Bearer of the Ring of the King of Lower Egypt" or "Seal-Bearer of the King of Lower Egypt." Both references support the idea of the ring's existence, which means it was probably stolen.²³²

-4-

Joseph of Egypt was Handsome (Genesis 39:6)

According to the Bible and the Koran, Joseph of Egypt was a handsome man. The Revised Standard Version of the Bible states, "And Joseph [of Egypt] was handsome and good looking" (RSV Genesis 39:6). Muhammad, the founder of Islam, admired Joseph of Egypt so much that the entire twelfth chapter of the Koran was devoted to his beauty, piety, and admirable personality traits.²³³

With regards to Yuya's handsomeness, photographs of his mummy demonstrate just how attractive he was. The images of

Yuya in this section indicate that he had a fairly symmetrical face, high cheekbones, a strong chin, an angular jawline, and a commanding presence—all traits many individuals, especially women, naturally find attractive.[234] It is no wonder that Potiphar's wife threw herself at him (See Genesis 39:6-20).

Yuya of Egypt[235]

Featured above is a photograph of the head of Yuya of Egypt's mummy. In the early 1900s, Sir Grafton Elliot Smith, an Australian-British anatomist described Yuya's mummy as one of the finest examples of embalming practices of the 18th Dynasty.[bf] Because of the fine embalming work performed by the Egyptians, Yuya's attractiveness and living image have been preserved, even thousands of years after his death. We can observe Yuya's striking physical features above including a high nasal bridge, a long and thin nose, a long and narrow face, high cheekbones, and a long-headed (dolichocephalic) skull—many features that are considered attractive to the general population.[236]

---

bf   Yuya is said to have lived over 3,400 years ago (c. approximately 1390 BC).

Yuya of Egypt[237] & **Rhys Ifans,** Welsh actor and musician (b. 1967)[238]

Photographs of Yuya's mummy and Rhys Ifans are featured above. Due to the similar physical characteristics of these two individuals, they are look-alikes. Although Yuya was embalmed, it is apparent from this photograph that he possessed similar facial proportionality to Ifans. Their eye sockets, ears, foreheads, rectangular-shaped faces, long and thin noses, mouths, and jawlines are situated in very similar places and almost perfectly match up. Yuya's cheekbones are slightly higher and his chin slightly wider, but other than that, these two men strikingly resemble one another. Aside from the features that directly contribute to their attractiveness, it is worth noting that both men also share features such as high nasal bridges, long and thin noses, long and narrow faces, high cheekbones, and long-headed (dolichocephalic) skulls.[239]

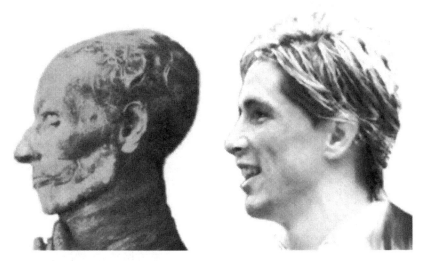

Yuya of Egypt (1390 BC)
& **Fernando Torres,** a Spaniard with Sephardic Jewish genetics (b. 1984)[240]

Yuya (featured above to the left) has a similar profile to Fernando Torres—a world-renowned Spanish soccer star (featured above to the right). Torres is widely-regarded as a handsome man and shares many physical similarities with Yuya. Fernando Torres possesses Sephardic Jewish ancestry, which is why he has many common Semitic physical features.[bg] Both men possess high nasal bridges, long and thin noses, and long and narrow faces. Their faces are rectangular in shape, and they have high cheekbones, and long-headed (dolichocephalic) skulls, although Torres has a shorter head than Yuya. Because of the phenotypic similarities between these two handsome men, it is apparent they share similar Semitic stock.[241]

---

bg The Sephardi Jews are "the Jews of Spain." Their ancestry is Middle-Eastern.

-5-
Yuya and Joseph of Egypt Shared Similar Titles

A way to refer to "**Yuya**" in Egyptian (found within his funeral papyrus)

Funerary texts made of papyrus in Yuya's tomb contain a number of titles attributed to Yuya. Many of the titles match the ones given to Joseph of Egypt.[242] "Great of the great ones" is a title given to Yuya. He is also called: "The great favorite of the Sovereign" and "The favorite of the great god."[243] In the Bible, Pharaoh said of Joseph of Egypt, "Thou shalt be over my house, and according to thy word shall all my people be ruled: only in the throne will I be greater than thou."[244]

Yuya was known as an "Overseer of the Cattle of Min, Lord of Akmin," and "Overseer of the cattle of Amun."[245] This is yet another specific similarity between Yuya and Joseph of Egypt, since, in the Bible, Joseph is also mentioned as one of the "rulers over . . . cattle" (Genesis 47:6).

The title most commonly attributed to Yuya appears over 20 times on his funeral papyrus. This title is "The holy father of the Lord of the Two Lands [Upper and Lower Egypt]."[246] In addition, two similar titles are given to him in the funeral papyrus: "The divine father who loves *his* Lord"[247] and "Praised of *his* God."[248] These distinct titles are important because the word "his" is used in conjunction with God and Lord, which implies that *his* God differed from common Egyptian gods (Egyptian gods included the Pharaoh himself).

Yuya is also called 🔣 "Seal-bearer"[249] or "Chancellor," which denotes a high ranking office, most likely second to a pharaoh and was virtually the same title Joseph of Egypt was given by Pharaoh. The title "Chancellor" could also be interpreted as "vizier," which in ancient Egypt was the highest official serving the pharaoh. The pharaohs of Egypt primarily chose their viziers based on the merits of great loyalty and talent. Joseph of Egypt was both loyal and talented.[250]

Twice was Yuya given the title 🔣 "Prince," which indicates his rank at court. Yuya was also known as 🔣 "The beloved priest," 🔣 "the only friend" ("sole friend" or "unique friend" according to Egyptologist Ahmed Osman);[251] and 🔣 "The chief of the Rekhit [privileged caste]."[252]

The 41st chapter of Genesis discusses Joseph's relationship to Pharaoh: "Pharaoh said unto Joseph . . . I have set thee over all the land of Egypt [Upper and Lower Egypt]. . . . Bow the knee: and he made him ruler over all the land of Egypt" (Genesis 41:41-43). Yuya and Joseph were both Lords over all of Egypt.

Also, in Genesis, Joseph of Egypt is called "A father of pharaoh" (Genesis 45:8). Ahmed Osman claims that Yuya was the only Egyptian official to have ever been given this title (with the exception of Yuya's son, Ay, who held the similar title *Itnetjer*, or "father of the god").[253]

Yuya's chariot found in his tomb

A small, intact chariot was found in Yuya's tomb (featured above). This rare discovery of an intact chariot in a tomb is important since it places Yuya in a time period after the Hyksos—a Semitic and Canaanite group to which Asenath, Joseph of Egypt's wife, possibly belonged—introduced the chariot to Egypt. According to Egyptologist Ahmed Osman, Yuya was known as the "Master of the Horse" and "Deputy of His Majesty in the Chariotry."[254] Joseph of Egypt's chariot is mentioned in the Bible on three occasions. In the 41st chapter of Genesis it reads, "And he [Pharaoh] made him [Joseph of Egypt] to ride in the second chariot [.] . . . [A]nd he made him ruler over all the land of Egypt" (Genesis 41:43).[255]

Joseph of Egypt and Yuya were both known for their wisdom. In the Bible, Pharaoh said to Joseph of Egypt: "There is nobody as discreet and wise as you" (Genesis 41:39). A similar distinguished title was given to Yuya: "The only wise, who loves *his* God."[256] According to Ahmed Osman, two more titles were given to Yuya: "The Wise One" and "He whom the King made Great and Wise, whom the King has made his Double."[257]

The Bible states that Joseph of Egypt lived to be 110 years

old. However, many Bible literalists believe that he may not have lived that long. A statue of Pharaoh Amenhotep, son of Hapu, as an old man, bears an inscription which states that Amenhotep was 80 years old when the statue was made, and that "Amenhotep hopes to live to the age of 110."[258] Amenhotep, a pharaoh and a known Egyptian magician in Yuya's time, wanted to be known as a wise man, and so he aspired to live to the age of 110. This age appears to be symbolic rather than literal. If it is Yuya's true age at his death, he aged very well.

In 1865, British scholar Charles W. Goodwin suggested that the age which the biblical narrator assigned to Joseph at the time of his death (110) reflected the Egyptian tradition of using the age of 110 to denote symbolic old age and thus, wisdom. According to Egyptologist Ahmed Osman, this idea has become increasingly accepted by Egyptologists. Egyptologists Gustave Lefebvre and Josef M. Janssen were able to show from Egyptian texts that at least twenty-seven Egyptian persons were said to have reached the age of 110 years.[259]

-6-

The DNA of Yuya and North American Indians

Yuya's DNA matches the autosomal[bh] ancestry of groups of North American Indians living in the Great Lakes region and on the Great Plains. If Yuya and Joseph of Egypt are the same person, then this corresponds with the claims of Joseph Smith in 1833 concerning the ancestry of North American Indians: "The Book of

---

bh Autosomal refers to any of the chromosomes other than the sex-determining chromosomes (i.e., the X and Y) or the genes on these chromosomes.

Mormon is a record of the forefathers of our [United States] ... tribes of Indians[.] ... By it, we learn that our ... Indians, are descendants from that Joseph [who] was sold into Egypt."²⁶⁰

For reasons unknown, Yuya's story, historical importance, and DNA have been downplayed, and possibly even suppressed, by modern Egyptologists. This kind of information is often purposefully hidden when it does not corroborate the Afrocentrism of Egyptian Egyptologists. Zahi Hawass, an Egyptian Egyptologist and former Secretary General of Egypt's Supreme Council of Antiquities in charge of the study of some Amarna mummies, appears to be at least one example of this practice.²⁶¹

For example, in 2010 he never published the findings of King Tut's paternal DNA, including Tut's genetic haplogroup,[bi] even though the data had been collected and many groups around the world were clamoring for the knowledge.[bj] Far from an official release of this data, this information was kept behind closed doors until it was inadvertently leaked. The Discovery Channel aired a program called "King Tut Unwrapped: Royal Blood"²⁶² to show the public how the Amarna mummies were related. On multiple occasions, the camera panned over specific data from King Tut's and also Yuya's DNA test results, and geneticists from Switzerland were watching. A Swiss team was actually able to decipher the genetic haplogroup of King Tut from the data aired on the TV special.²⁶³

---

bi   In molecular evolution, a haplogroup (from the Greek: ἁπλούς, haploûs, "onefold, single, simple") is a group of similar haplotypes that share a common ancestor having the same single nucleotide polymorphism (SNP) mutation in all haplotypes. That is, haplogroups are sets of similar combinations of closely linked DNA sequences on one chromosome that are often inherited together.

bj   DNA testing company iGENEA put pressure on researchers who studied Tutankhamun's DNA to publish Tutankhamun's full DNA report to confirm his Y-DNA results. They refused to respond. Genzlinger, Neil (19 Feb. 2010). "CSI: Egypt, Complete With DNA Tests of Mummies". The New York Times. Retrieved 6 March 2013.

Similar to Tut's DNA leak, other Amarna mummies including Yuya, Tuya, Amenhotep III, and Akhenaten all had their DNA tested, and portions of their DNA test results were also inadvertently released.[264]

In 2012, DNA Tribes—a genetic ancestry analysis company—performed an 8 marker Short Tandem Repeat (STR) profile[bk] of the Egyptian mummy Yuya, and have suggested that, based on the information they had ascertained, Yuya's DNA closely matches the DNA of indigenous North Americans.[265]

Map of the American Continent with highlighted circular focal points[bl]

The map provided above indicates the American Indian groups to which Yuya is most genetically related. The biggest

---

bk   When only 8 genetic markers are used by genetic ancestry analysis companies, it does not provide definitive information about the ancestral origin of an individual. The results from this particular genetic ancestry analysis of Yuya only suggests that he was closely related to indigenous Americans (about 37 genetic markers are necessary to obtain definitive results).

bl   A map that includes the US with highlighted circular focal points for DNA Tribes Match Likelihood Index (MLI) of Yuya (upper: North/Northeastern Amerindian region; lower: Athabaskan region of North America). Note: autosomal DNA is discussed further in *Chapter 10: Genetic Markers of Semites, East Asians, and Siberians.*

circles on the map are located on the American continent and represent North American Indian tribes. Each circle on the map represents an approximate Match Likelihood Index (MLI)—an indication of how strong Yuya's genetic relation is to specific groups of American Indians. The top circle belongs to the Great Lakes region and the other commensurate circle belongs to the Great Plains region. Tiny circles are found on the map indicating a very minor connection to indigenous peoples of Mexico, Guatemala, and the western coasts of South America.

DNA Tribes found that Yuya has an approximate MLI of 14.01 for the natives in the area corresponding with the Great Lakes region, and 13.97 for the natives in the Great Plains region.1 Great Lakes region Indians include the Delaware (Lenni Lenape), Shawnee, Sauk & Fox, Huron, and Iroquois—all tribes visited by early Latter-day Saint missionaries on the first missions to the Lamanites. Great Plains region Indians also include the Blackfoot, Cheyenne, Crow, Kaw, Lakota Sioux, Mandan, Plains Ojibwe, and Yankton Sioux. Since the MLI for the indigenous Mesoamericans corresponding with Central America on this map is approximately 0.5, to find out the indigenous American group to which Yuya of Egypt was most related, we divide 14.01 by 0.5 to get approximately 28. This means that natives from the Great Lakes region of North America are 28 times more likely to be related to Yuya of Egypt than indigenous Mesoamericans.

Yuya and Kihue

Yuya of Egypt looks very similar to select North American Indians, including a full-blooded Huron of the Great Lakes region named Kihue.[bm]

A **Siberian** (Nganasan) man (1800s)
Yuya, vizier of Egypt (who lived around 1390 BC)
& **Kihue**, a full-blooded Huron (1930 AD)[266]

Featured above are a Nganasan man from Siberia, Yuya, and Kihue, a full-blooded Huron[bn] from Ohio near the Great Lakes. Yuya and Kihue both share similar Semitic facial features which are virtually non-existent among East Asian and Siberian populations. Yuya and Kihue both have similar long and aquiline noses, cheekbones, orbitals (eye sockets), mouths, head shapes,[bo] and smile lines from flexing the zygomatics (cheek muscles),

---

bm The Huron (also known as the Wyandot) traditionally spoke the Wyandot language, which stems from the Iroquoian language. The Huron were one of the first tribes to whom the Prophet Joseph Smith sent missionaries to preach the gospel of Jesus Christ, which means they belonged to the Lamanite remnant

bn "I am a Wyandot [Huron] Indian, and the sole surviving full-blooded member of that tribe. My father died Sept. 9, 1871, aged 100 years. My mother died March 21, 1872, aged 106 years. Both are buried in the Indian cemetery at Upper Sandusky [Ohio]." Kihue (Bill Moose Crowfoot) brief history, as told to the editor of the Worthington News, Worthington, Ohio.

bo Long (dolichocephalic) heads.

whereas the Siberian man has a flat face and a low nasal bridge. Kihue bears little resemblance to the average Far Easterner (like the Siberian man in this section), but looks like he could be a direct relative of Joseph of Egypt and belong to the Lamanite remnant.

On January 6, 1836, in a meeting of the Latter-day Saint High Council at Kirtland, Ohio—the same state to which Kihue belonged—the Prophet Joseph Smith connected North American Indian tribes such as the Huron to the Lamanite remnant.[267] Thus it appears to be no coincidence that one of the first Latter-day Saint missions was to Kihue's tribe, the Huron, and that Kihue looks so similar to Yuya of Egypt.

Was Yuya Really Joseph of Egypt?

Could Joseph the Patriarch and Yuya be one and the same person?
—**Ahmed Osman, Egyptian historian and scholar**[268]

Based on the seven main points in this chapter, Yuya fills the shoes of Joseph of Egypt nicely.[269] If there was an actual Semitic man from the Bible who became vizier over all Egypt, he was Yuya. The odds of discovering another attractive Semitic male like Yuya, one who worked his way through the upper ranks of Egypt to be so highly respected by the pharaohs of his time that he was buried in the Valley of the Kings, are astronomical.

In and of itself, the idea that Yuya is an ancient Semitic-looking man who resembles members of the seed of Joseph (such as the many indigenous North Americans mentioned earlier) is significant. Furthermore, the fact that Yuya of Egypt appears to possesses genetic markers that match indigenous North Americans from the Great Lakes region and the Great Plains is astounding. Even if Yuya is not Joseph of Egypt, but is simply a Semite or a man of Hebrew descent, the name similarities of the god they both worshiped, the titles they both held, and the artifacts discovered in Yuya's tomb are enough to make one at least entertain the possibility. Because it is plausible that Yuya is Joseph of Egypt, we may finally be able to trace the literal genetic lineage of Joseph of Egypt!

5

Locating the Seed of Joseph

Moroni said unto them: Behold . . . we are a remnant of the seed of Joseph, whose coat was rent by his brethren into many pieces[.] . . . —**Moroni, Book of Mormon Prophet (Alma 46:23)**

Behold, our father Jacob also testified concerning a remnant of the seed of Joseph. And behold, are not we a remnant of the seed of Joseph? And these things which testify of us, are they not written upon the plates of brass which our father Lehi brought out of Jerusalem? —**Nephi, son of Nephi, Book of Mormon Prophet (3 Nephi 10:17)**

Wherefore, Joseph [of Egypt] truly saw our day. And he obtained a promise of the Lord, that out of the fruit of his loins the Lord God would rise up a righteous branch unto the house of Israel; not the Messiah, but a branch which was to be broken off, nevertheless, to be remembered in the covenants of the Lord that the Messiah should be made manifest unto them in the latter days, in the spirit of power, unto the bringing of them out of darkness unto light—yea, out of hidden darkness and out of captivity unto freedom. —**Lehi, Book of Mormon Prophet (2 Nephi 3:5)**

The great majority of those who become members of the Church are literal descendants of Abraham through Ephraim, son of Joseph. —**Joseph Fielding Smith, Latter-day Saint Church Historian and President (1923)**[270]

Joseph of Egypt is an essential figure in the Book of Mormon and a key individual to locating the seed of Joseph of Egypt which includes the main Lamanite remnant. BH Roberts, a Latter-day Saint General Authority and historian understood the great significance of Joseph of Egypt to Latter-day Saint doctrine and teachings when he stated in 1909, "The Book of Mormon throughout is true to this Josephic idea; it is impregnated with it. Joseph [of Egypt] is the central figure throughout."[bp] Latter-day Saint apostle Parley P. Pratt also apparently understood Joseph of Egypt's significance to Church teachings when he connected indigenous North Americans to Book of Mormon peoples in 1851: "[T]he Red Men of America . . . are descended from . . . the tribe of Joseph . . . [of] Egypt."[271]

As prophesied in scripture, the seed of Joseph of Egypt is found all over the world. Genesis speaks of the numerousness of Joseph of Egypt's future descendants when it says that his genetic lineage would spread out beyond his homeland to many parts of the globe (See Genesis 49:22). Fortunately methods exist to help us locate the scattered seed of Joseph of Egypt. Physical trait recognition is one of the ways to recognize Joseph of Egypt's literal (rather than adopted) lineage.

Researchers around the world are revisiting the idea that our faces can tell us about our ancestry. What is emerging is a scientifically-based "new physiognomy" that allows people to be identified by their physical traits.[272] This new form of physiognomy can be used to effectively determine a number of genetic traits, and also the predominant ethnicity of individuals.

---

bp   BH Roberts, *New Witnesses for God,* 3 Vol. [1909], 3:106. Joseph was the great-grandson of Abraham the Great Hebrew Patriarch of the Old Testament. An Israelite by birth, Joseph is Hebrew through Abraham's son Isaac and grandson Jacob. Since all Hebrews are Semites, anyone with Hebrew blood is related to Shem—a son of the ark-building Noah. Joseph of Egypt is genetically related to Shem and is therefore a literal Semite.

Physical trait recognition involves the identification of common physical traits and the categorization of individuals into helpful categories based upon trait patterns. For instance, as previously mentioned, high nasal bridges and long aquiline noses are common physical traits among Semites and Israelites. Virtually the only way other ethnicities can possess Semitic and Israelitish physical traits is if they admix with individuals who belong to these lineages. Note that these Semitic traits are very uncommon to other ethnicities.

Physical resemblance can be used to determine whether two individuals belong to the same genetic lineage via comparison of physical traits. Identical twins who have been reared apart, but share many temperament and character similarities, are great examples of how genetic lineages are able to be identified by appearance alone. This same principle applies to individuals with look-alikes (doppelgängers). Strictly speaking, the more two people look alike physically, the greater chance they will share genetic similarities.

Physical trait recognition and physical resemblance have been used by individuals in helpful and less-than-helpful ways for mankind. People with certain skin colors, for instance, have been discriminated against for millennia. Black, yellow, red, and white skin tones have all been turned into racial slurs. However, a lot of good can also come from the proper utilization of physical trait recognition.[273]

As a case in point, Dr. Peter Hammond of the University of London is currently using facial recognition technology to detect rare genetic disorders such as Williams Syndrome, Smith-Magenis Syndrome, Fragile X Syndrome, and Jacobsen Syndrome.[274] Hammond's technology helps doctors correctly diagnose certain rare genetic disorders by appearance alone. Their computer scans can detect certain physical trait patterns often missed by the naked eye. With the help of advanced computer technology, and the verifiable connection between appearance and genetics, Dr. Hammond and his team do much good in the world.

Morphological characteristics (physical traits) and genetics are currently being used by forensic scientists to determine the races of deceased individuals.[275] Forensic anthropologists are regularly presented with human "material ranging from bits of bone, the species of which a medical examiner or coroner is unable to identify, to whole, obviously human skeletons in various stages of decomposition."[276] Many of these individuals uncannily determine the race of deceased persons with a high degree of accuracy using the strong correlation between appearance and genetics. Since they can determine the race, and even the general appearance of some individuals who do not even have faces when discovered, one cannot help but wonder what can be inferred from individuals who have faces (like the example of Yuya of Egypt from the previous chapter). A person's ethnicity and genetic lineage can be determined much more easily from a whole face rather than mere charred remains.

Hole in the Sky the younger, an Ojibwe (1860)²⁷⁷ & another **Ojibwe** man²⁷⁸

When morphological traits of individuals from within the same American Indian tribe are observed, interesting patterns emerge. Featured above are two men from the Ojibwe tribe of North America. Hole in the Sky (featured on the left) clearly possessed common Caucasian and Semitic physical traits,²⁷⁹ whereas the traits of the Ojibwe man on the right were manifestly more East Asian in appearance.[bq]

---

bq The Ojibwe man on the right lacks facial angularity and a high nasal bridge, which are

Hole in the Sky the younger, an **Ojibwe**, (1860),[280] a **Caucasian** man, an **Ojibwe** man[281] & an **East Asian** man

Featured above on the far left is Hole in the Sky, the same Ojibwe Indian that was previously highlighted. He possessed facial characteristics that more closely resembled the lineaments of Yuya of Egypt's face. Hole in the Sky also shares similar facial characteristics of the Caucasian face featured second from the left. Both of these faces possess long and thin noses, high nasal bridges, and rectangular-shaped faces—traits commonly found among Caucasian and Semitic groups. Second from the right is a photograph of an East Asian-looking Ojibwe man (the same man highlighted on the previous page). He possessed facial traits most common with Far Easterners like the East Asian face featured on the far right. Both of these men have epicanthic eye folds (hooded eyes), flat noses with low nasal bridges, and round faces—all physical characteristics common among East Asians and Siberians.

---

common among Joseph of Egypt's lineage. This particular Ojibwe man on the right has a very round face.

Joseph Smith's Family & Joseph of Egypt

Joseph Smith III, Joseph Smith, Jr.'s son (1832-1914) & a **Samaritan** priest

The Prophet Joseph Smith was known as a pure Ephraimite,[282] which would make his son, Joseph Smith III, an Ephraimite as well. Joseph Smith III and a Samaritan priest are strikingly similar in appearance, as we can see from the photographs above. They are clearly doppelgängers. Samaritans—an ancient Semitic group that inhabited Samaria in biblical times—have long claimed that groups of their people belong to the lineages of Ephraim and Manasseh.[283] Based on how similar this Samaritan looks to Joseph Smith's son, this claim would appear to be true. It seems from the appearance of the two men featured above that they belong to the literal seed of Joseph of Egypt.[br]

---

br   According to revelation received through the prophet Joseph Smith, the order of the priesthood was confirmed to be handed down from father to son, and rightly belongs to the literal descendants of the chosen seed, which is currently Joseph of Egypt's seed, to whom the promises were made (D&C 107:40). This is why the majority of white LDS members are said to literally belong to the tribe of Ephraim and occasionally the tribe of Manasseh. Lehi from the Book of Mormon "knew that he was a descendant of Joseph; yea, even that Joseph who was the son of Jacob, who was sold into Egypt" (1 Nephi 5:14).

Hyrum Smith (1844 death mask)²⁸⁴ & **Yuya** mummy (1300 BC)²⁸⁵

On December 9, 1834, Patriarch Joseph Smith, Sr., gave a patriarchal blessing to his son, Hyrum.^bs In the blessing, he declared that Hyrum was a "true descendant" of the ancient Joseph of Egypt and that Hyrum's "posterity shall be numbered with the house of Ephraim."²⁸⁶ Like his brother Joseph Smith, Jr., Hyrum was a literal descendant of Joseph of Egypt. Because of the genetic connection between Hyrum and Joseph of Egypt, it appears to be no coincidence that Hyrum looked similar to Yuya (Joseph of Egypt). When Hyrum Smith's death mask is juxtaposed with the face of Yuya's mummy (both featured above), the

---

bs Latter-day Saint scholar Daniel H. Ludlow has remarked concerning direct lineages discussed in patriarchal blessings, "Lineages declared in patriarchal blessings are almost always statements of actual blood lines; they are not simply tribal identifications by assignment. . . ." Daniel H. Ludlow, "Of the House of Israel," Ensign, Jan. 1991, pp. 52, 54–55.

physical similarities are astounding. If Yuya's nose had not been stuffed with material in the embalming process after his death, the two men would have resembled each other even more closely.

Both Hyrum and Yuya of Egypt were handsome men who possessed common Semitic physical features: high nasal bridges, long and thin noses, and long and narrow faces. Their facial symmetries were comparable, and they both had similar foreheads, high cheekbones, angular jawlines, and rectangular-shaped faces.

Although most comparisons are made between the Prophet Joseph Smith and Joseph of Egypt, Hyrum Smith was very similar in disposition to Joseph of Egypt—the two shared many of the same character traits and temperament characteristics.[287] A genotype-phenotype correlation existed between Hyrum Smith and Joseph of Egypt, as it did with his brother Joseph and Joseph of Egypt. Hyrum and Joseph Smith were two men who came from the seed of Joseph of Egypt and brought forth much righteousness. Genetic and spiritual traits found in the branches of Joseph of Egypt's family tree increases the likelihood of having righteous men and women in their family:

> Joseph [of Egypt] . . . obtained a promise of the Lord, that out of the fruit of his loins the Lord God would raise up a righteous branch unto the house of Israel; not the Messiah, but a branch which was to be broken off. . . . And his name shall be called after me [Joseph]; and it shall be after the name of his father. And he shall be like unto me (2 Nephi 3: 5, 15).

Émile Meyerson, Joseph F. Smith & Paul Natorp (1903)

Almost all, if not all, members of the Joseph Smith, Sr., family shared physical commonalities with individuals of Semitic and Israelitish bloodlines, especially from the genetic lineages of Joseph of Egypt and Judah. The Smith family was known for their high nasal bridges and long and prominent noses. One can simply view the profiles of members of the immediate family and also the extended family to see just how prominent their high nasal bridges were.

Hyrum Smith's son, Joseph F. Smith (featured above in the middle), for example, possessed a high nasal bridge and resembles others with the same features. He looks like Émile Meyerson (featured above on the left), a Polish-born French epistemologist, chemist, and philosopher of science with Jewish genetics, and Paul Gerhard Natorp (featured above on the right), a German philosopher and educationalist. Even without beards, each of these men possessed physical similarities. All three men had high nasal bridges, long noses, and long rectangular-shaped faces, which are common physical traits among Semites.

John Smith (Joseph Smith, Jr.'s uncle), Irish genetics (1781-1854)
& **Irwin Shaw**, Jewish genetics (1913-1984)[288]

John Smith (featured above on the left) was a brother of Joseph Smith, Sr., who possessed a number of common Semitic physical characteristics. In fact, John's profile is quite similar to Irwin Shaw (featured above on the right), a Jewish-American playwright, screenwriter, novelist, and short-story author.[289] Both John Smith and Irwin Shaw had extremely high nasal bridges, long and rectangular-shaped faces, long and prominent noses, and long-headed (dolichocephalic) skulls. A number of members of the Smith family shared similar physical characteristics with John since these traits appear to have belonged to the genetic lineage of the Smith's.

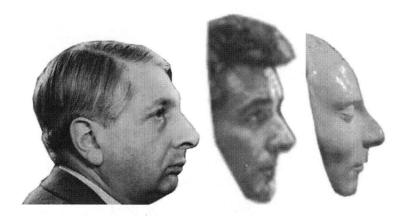

Giorgio de Chirico, an Italian (1888-1978),
Leonard Bernstein, a Jew (1918-1990)
& **Joseph Smith, Jr.** (death mask 1844)

Similar to other members of his family, the Prophet Joseph Smith physically resembled Semites and Israelites. Featured above (from left to right) are Giorgio de Chirico, Leonard Bernstein, and the Prophet Joseph Smith. Giorgio de Chirico was a famous Italian painter, Leonard Bernstein was a world-renowned music composer with Jewish genetics, and Joseph Smith was, of course, an American who founded The Church of Jesus Christ of Latter-day Saints.[290] As one can see from these pictures above, all three men possessed high nasal bridges and long noses. They all have the prototypical "look" of Israelite males.

Other Descendants of Joseph of Egypt

Like many members and relatives of the Smith family, President Brigham Young was certain that he belonged to the literal tribe of Joseph of Egypt. On April 8, 1855, President Young delivered a discourse in the Great Salt Lake City Tabernacle. In his speech, he discussed his own Ephraimite bloodline and the gathering of Israel:

> It is Ephraim that I have been searching for all the days of my preaching and that is the blood that ran in my veins when I embraced the gospel. If there are any of the other tribes of Israel mixed with the Gentiles we are also searching for them. . . . We want the blood of Jacob, and that of his father Isaac and Abraham, which runs in the veins of the people. There is a particle of it here, and another there, blessing the nations as predicted.[291]

Brigham Young appears to have understood the great importance of the blood of Ephraim in the latter days. Ephraim was given the birthright by his father, Joseph of Egypt, and the scriptures prophesy that in the last days his descendants would have the privilege and responsibility to bear the priesthood, take the message of the restored gospel of Jesus Christ to the world, and raise an ensign to gather scattered Israel.[292] Although the tribes of Ephraim and Judah are meant to play important leadership roles for covenant Israel in the last days, it is ultimately the job of the descendants of Ephraim to unite all the tribes of Israel.

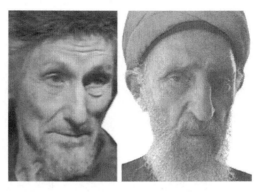

John C. Lilly, an American[293] & a **Samaritan** High Priest[294]

Individuals outside of the LDS Church also appear to have the blood of Israel coursing their veins. These individuals often come from the Middle East, Europe, and the United States of America. John Cunningham Lilly (featured above), an American philosopher, writer, and inventor, has the prototypical "look" of an Israelite male. Lilly has British genetics but resembles a Samaritan High Priest (featured above). Since it is well-known that Samaritans claim Israelite lineages of Ephraim, Manasseh, and Levi, it is likely that Lilly also possesses the genetic lineage of an Israelite.

Several western Christian groups, especially the Church of God in Christ, claim that the United Kingdom primarily descended from Ephraim. Latter-day Saints also claim that a significant portion of its members descended from the tribe of Ephraim. Ephraimites, according to the LDS Church, have been assigned to restore the lost tribes in the latter days, as prophesied by Isaiah. It is also prophesied that the tribes of both Ephraim and Judah are to play important leadership roles for covenant Israel in the days before Jesus Christ returns to the earth.

Amran Ishak, a **Samaritan** High Priest (1960)[295]
& **Seyid Riza,** a Kurdish religious leader[296]

Samaritans and Kurds often look similar. By no coincidence, these two Middle Eastern groups share similar Semitic genetic lineages. This correlation is a great example of how the genotype-phenotype connection works. The more two individuals possess similar physical traits, the greater the chance they come from similar races, ethnicities, and genetic lineages, with identical twins as the clearest example of this principle. Amran Ishak (featured above on the left) was a Samaritan High Priest and Seyid Riza (featured above on the right) was a highly revered Kurdish revolutionary. These two highly conscientious men have the common "look" of Israelite males. They both have the tell-tale facial features: long and aquiline noses, high nasal bridges, and long and narrow rectangular-shaped faces. Take away the beards of these men and add headdresses and they would look similar to certain North American Indians of the Great Lakes and Great Plains regions.

A **Dogrib** Indian of North America
& **Yitzhaq ben Amram ben Shalma ben Tabia**, a **Samaritan** High Priest (1920)[297]

Featured above is a rare photograph of a Dogrib Indian of North America with a full beard and also a photograph of a bearded Samaritan High Priest. This Dogrib Indian had the prototypical "look" of an Israelite, especially one from the House of Joseph of Egypt, as did the Samaritan. It is uncanny how physically similar this Dogrib Indian is to Yitzhaq ben Amram ben Shalma ben Tabia. It is easy to observe, once again, the long and aquiline noses, high nasal bridges, and long and narrow rectangular-shaped faces of these two individuals.

Paternal Dogrib Indian genetic markers are often different from most indigenous Americans from across the continent. Dogrib DNA most closely matches Israelite genetic markers, especially from the tribe of Joseph of Egypt. It should come as no surprise, then, that this Dogrib Indian looks so much like other Israelites.

Yuya of Egypt & an **Ojibwe** man (1900)

Featured above is an example of the common physical traits of North American Indians and Yuya of Egypt (Joseph of Egypt). This Ojibwe man in the photograph featured above could almost be his twin: he possessed a similar brow, high cheekbones, high nasal bridge, long and narrow nose, and rectangular-shaped face. Many of the Ojibwe differ in appearance from the majority of indigenous Americans who more often possess East Asian and Siberian physical characteristics, particularly those living in Mesoamerican countries.

Gabriel Cousens, a Jewish Rabbi,[298]
Yuya, a vizier of Egypt[299]
& **Pine Bird**, an Oglala Sioux, (1907)[300]

Featured above are photographs of three men who share similar Semitic-looking facial features: Gabriel Cousens (an ordained Jewish Rabbi on the left), Yuya (a former vizier of Egypt who is featured in the middle), and Pine Bird (an Oglala Sioux who is featured on the right). By appearance alone, these men all appear to belong to similar Semitic genetic groupings. All of them possess long and aquiline noses, high nasal bridges, and long and narrow rectangular-shaped faces—physical characteristics most commonly found among the houses of Judah and Joseph of Egypt. The physical characteristics of these three men are rarely found in East Asia and Siberia, which implies that the genetic markers possessed by them do not match the DNA of Far Easterners.

**Yuya of Egypt
& Mico Chlucco, King of the Seminoles** (1792)[301]

Similar to the Dogrib, Sioux, and other North American Indian tribes, the Seminoles who originated in Florida resemble Semities and Israelites.[bt] Yuya of Egypt (featured above to the left) and Mico Chlucco, a King of the Seminoles (featured above to the right) both have the "look" of prototypical Israelite males. If this depiction of Mico Chlucco is true to form, then these two men shared many physical similarities and likely possessed Semitic blood from the House of Israel through Joseph of Egypt's lineage.

---

bt The Seminoles, who were once a powerful nation living in many parts of Florida, currently live in Oklahoma. Only a small percentage of Seminoles remain the "Sunshine State" (Florida). The Seminole tribe is related to Native American tribes including the Choctaw, Miccosukee, and the Muscogee (Creek Indians). They are the only tribe in American who never signed a peace treaty.

Red Cloud, an Oglala Sioux[302]
& **Omar Mukhtar** (1858-1931), a Libyan leader[303]

Many Sioux males possess the "look" of Israelite males. Featured above is Chief Red Cloud of the Oglala Sioux, and Omar Mukhtar, a former Libyan leader. Although Libya is a North African country and does not belong to the Middle East, common paternal Libyan DNA resembles Middle Eastern and Semitic DNA, which means that at some point in the genetic record, Semites made their way to Libya.[304] Both men featured above have long and aquiline noses, high nasal bridges, and long and narrow rectangular-shaped faces—common traits of Semites. These men have very masculine faces and bodies. This is common among certain groups of Israelites. This physiology is prominent in descriptions of the Israelite men in the Book of Mormon, and can be observed among select North American Indian tribes. It is uncommon to find ultra-masculine faces[bu] and bodies among many indigenous Americans, especially indigenous Mesoamericans. Many indigenous Mesoamericans are not often large of stature but have neotenous (youthful) East Asian physical features.[bv]

Jack Red Cloud, a Lakota Sioux (1862-1928)[305]
& a **Bedouin** Arab[306]

Other members of the Sioux Indian tribe have the "look" of prototypical Israelites and other Semites. Lakota Sioux Chief Jack Red Cloud (featured above on the left) is a good example of a Sioux man who possessed common traits of an Israelite. He looked a lot like a Bedouin Arab (featured above on the right). Both men

---

bu   Ultra-masculine faces (faces with masculine facial dimorphism) are more rectangular-shaped with strongly-defined Jawlines and often include heavier brow ridges, smaller eyes, long noses with high nasal bridges, and thinner lips.
bv   Neotenous facial features include big eyes, roundness of the face (less angular jawline), smaller chin, and fuller lips.

possessed common Semitic physical features including prominent long and thin noses, high nasal bridges, and long and narrow faces.

According to Latter-day Saint scholar Hugh Nibley, admixture between Arabs and the tribe of Manasseh[bw] occurred frequently, so it is possible that the Arab featured in this section possessed the blood of Manasseh (the blood of the Lamanites).

---

bw   Concerning the tribe of Manasseh, Latter-day scholar Hugh Nibley said, "[O]f all the tribes of Israel, Manasseh was the one which lived farthest out in the desert, came into the most frequent contact with the Arabs, intermarried with them most frequently, and at the same time had the closest traditional bonds with Egypt." Hugh W. Nibley, "Lehi and the Arabs."

6

The Physical Traits of Semites and Israelites

The children of Shem; Elam, and Asshur, and Arphaxad, and Lud, and Aram. —**Genesis 10:22 (KJV)**

There is abundant evidence . . . that physical features played an important role in determining Jews well before the closing decades of the nineteenth century. . . . Jews were defined by physical characteristics well before the appearance of anti-Semitism. —**Joel Carmichael, historian (1992)**[307]

Although what Lehi and his DNA looked like are unknown, it is possible to ascertain general information about his genetics and appearance.[bx] The Bible, Book of Mormon, ancient artifacts from the Old and New World, Joseph Smith, and other early brethren left clues that allow us to learn more about Lehi and

---

bx A Latter-day Saint scholar has claimed that it is impossible to know anything about Lehi's genetic markers and appearance. "We do not know the genetic composition of the Jews at the time of the diaspora when the Lehite and Mulekite parties left the land of Israel." Arthur E Mourant et al. 1998 "the Jews in Palestine", "the Genetics of the Jews."

his traveling caravan. In conjunction with these clues, the use of research and reasoning allow us to narrow down the search for the physical traits and genetic markers of Lehi and company. For example, it is known that Lehi and his traveling group were mostly, if not all, Hebrews, Semites, and Israelites. According to the Book of Mormon, none of them were from the Far East, Sub-Saharan Africa, or Australia. Instead, they all spoke, wrote, acted like, and looked like Semitic peoples because they were all Middle Easterners.[308]

The Book of Mormon is all about ancient Semites and Israelites, with few exceptions. Although written in "reformed Egyptian," the Book of Mormon reads like a Semitic text with its many Hebraisms.[309] This makes sense because members of the House of Israel are Hebrews—an isolated Middle Eastern group who are practically the only people on the planet to use the Hebrew language.[310]

Lehi and his caravan may have admixed with other ethnicities along their journey to the New World, and even with natives already located in the Americas, but they still possessed the genetic lineages of Semites.[311]

Common physical traits of ancient Semites and Israelites show up on coins, reliefs, effigies, and mummies from the Middle East. The following are examples that allow us to better see and understand the physical traits of Semitic people during Book of Mormon times. Note that these same traits have proven dominant and are still obvious and present in modern-day Semitic peoples.

The Physical Traits of Ancient Semites

Two **Israelite effigies** discovered at a site in Israel (600 BC-750 BC)[312]

At about the time Lehi and his caravan left Jerusalem for the New World, Israelites were making effigies to capture the essence of what their people looked like. Two ancient Israelite figurines (featured above) provide information about the prototypical physical features of ancient Semites close to Lehi's day.[by] It is possible that members of Lehi's caravan shared common physical traits of these ancient Israelite effigies.

As could be expected, the effigies featured above possess long and prominent noses, high nasal bridges, and angular face shapes. Many of these prototypical Semitic physical traits are uncommon among indigenous Siberians and East Asians,[bz] but extremely common among Semites and North American Indians from the Great Lakes region and the Great Plains. Both of these figurines possess the prototypical "look" of Israelites.

---

by   Any good geneticist will tell you that phenotype is highly correlated with genotype (genetics). This is why individuals who look the same or similar, like identical twins, will often share similar, or the same, genetic traits.

bz   Siberians, and other North Asians, typically possess low nasal bridges and flat faces instead of large noses, high nasal bridges, and long faces.

Levantine workmen loading Phoenician boats (8th century BC), an Assyrian relief[913]

An Assyrian relief (featured above)[ca] depicts workmen of the Levant dating back to approximately 100-200 years before the Book of Mormon begins.[cb] These workmen all possessed high nasal bridges, long noses, long faces, and long-headed (dolichocephalic) skulls. Doubtless, many members of Lehi's caravan, including Lehi himself, shared similar visages of these Levantine workers because Lehi and his traveling group all came from the Levant. Interestingly, if feathers were placed on the heads of these men, they could pass for certain North American Indians, especially ones living near the Great Lakes region and on

---

ca   A relief in this instance is a carving in which the design stands out from the surface, to a greater extent.

cb   The Levant region includes Israel, Palestinian territories, Jordan, Syria, and Lebanon.

the Great Plains.^cc Although the overall appearance of these featured Levantine workers is common among natives of North America, it is much less common among indigenous Mesoamericans and South Americans.

Ancient **Nabataean Kings: Rabel II** (70-106 AD) & **King Aretas IV** (9 BC-40 AD)[314]

The two coins featured above belong to a Semitic group known as the Nabataeans.[315] Nabataeans rose in power beginning in the 4th century BC after the fall of the kingdom of Judea. King Rabel II is featured on the coin to the left, and the coin on the right features King Aretas IV.[316] The high nasal bridges, long and aquiline noses, and long faces displayed on these coins support a Semitic origin of these people. It is possible that many, or perhaps all, of the men who belonged to Lehi's caravan had visages similar to the Semitic Nabataeans. The two men on the coins featured above could pass for certain North American Indians of the Great Lakes region and the Great Plains.

---

cc   See *Appendix 3: Egyptians, American Indians, and Feathers* of this book to learn more about feathers and the Old World.

Artabanus II, Parthian coin (128 BC-124 BC)[317]

Other groups of Semites made coins and effigies to depict the common appearance of their peoples. Featured above is a Parthian coin displaying the profile of King Artabanus II (128 BC - 124 BC). Parthia, the homeland of Artabanus II, is a region of north-eastern Iran. The Parthian language, which included a Semitic alphabet, partially preserved the Parthians' Semitic origins. As we can see, King Artabanus II possessed prototypical Semitic physical features, such as a long nose and a very high nasal bridge.

The men of Lehi's caravan may have possessed visages and characteristics similar to those of the Semitic Parthians, since these are the same kinds of physical traits common to select North American Indians of the Great Lakes region and the Great Plains. Once again, the phenotype-genotype connection makes it possible to prognosticate what Lehi and his caravan looked like all those years ago.

A **Sabaean** Alabaster head (2 views), Southern Arabia (100 BC-100 AD)[318]

Featured above is an ancient alabaster effigy[cd] of a Sabaean man from Southern Arabia that dates back to between 100 BC - 100 AD. This ancient effigy possesses prototypical physical features for individuals from Yemen: a high nasal bridge, a long and thin nose, and a long and narrow face. The men of Lehi's caravan may have shared common physical traits with this ancient Sabaean effigy since so many Semitic groups possess these recognizable physical traits.

The Sabaeans were a Semitic-speaking people of ancient Saba in Yemen—a country that currently occupies the southwestern to southern end of the Arabian Peninsula. With the exception of facial hair, the appearance of the Sabaean effigy featured above is common among indigenous North American Indians of the Great Plains and Great Lakes regions, but less common among other indigenous Americans.

---

cd   An effigy is a representation of a specific person in the form of sculpture or other mediums.

Physical Traits of Modern Era Semites

Otto Frank, Anne Frank's father, **Jewish** genetics (1933)[319]

Featured above is a photograph of Otto Frank,[320] a Jewish man known for being the father of Anne Frank—the famous Jewish girl who kept a record of her time spent hiding from the Nazis (what is known today as *The Diary of Anne Frank*).[321] As might be expected, Semites of the modern era such as Otto Frank share similar physical characteristics with ancient Semites. From this photograph of Otto, his prototypical Semitic features are recognizable. He possessed a high nasal bridge, a long and thin nose, and a long and rectangular-shaped face.

A **Mandanean,** Semitic genetics
& **Abraham Lincoln,** English and Dutch ancestry[322]

A Mandanean man and Abraham Lincoln are featured above. Mandaeans are Semites who fled from the Jordan Valley in the Middle East in about 70 AD and apparently ensconced in northern Mesopotamia. Many Mandaeans settled in modern-day southern Iraq near the Tigris and the Euphrates rivers. Both the Mandaean man and Abraham Lincoln have high nasal bridges, long and thin noses, long and narrow faces, and long-headed (dolichocephalic) skulls.

In an attempt to argue that Lincoln was of Semitic stock, Jewish Union Rabbi Isaac Wise gave an address following Lincoln's death, which was published April 20, 1865, in the Cincinnati Commercial:

> Abraham Lincoln believed himself to be bone of our bone and flesh of our flesh [Semitic]. He supposed himself to be of Hebrew parentage, he said so in my presence, and indeed he possessed the common features of the Hebrew race both in countenance and features.[323]

Abraham Lincoln's highly recognizable physical features all denote he was indeed of Semitic stock.

Abraham Lincoln (English and Dutch ancestry)[324]
& **Daniel Day-Lewis** (Jewish, Irish, and English ancestry)[325]

Featured above is Abraham Lincoln, former US President, and also Daniel Day-Lewis, an American actor with Jewish and European genetics.[326] Because of his acting ability and resemblance to Abraham Lincoln, Daniel Day-Lewis played him in Steven Spielberg's 2012 film *Lincoln*. These two men undoubtedly belong(ed) to Semitic lineages.

A **Jewish** woman & an **Etruscan** woman (768 BC-264 BC)

Featured above is a picture of a Jewish woman and also a painting of an Etruscan woman that was discovered on an ancient Etruscan wall. The Etruscans belonged to an ancient Italian civilization that were conquered by the Romans. Because both women possess prototypical Semitic physical features—high nasal bridges, long and thin noses, and long and narrow faces—it is possible the Etruscans were originally Semites.

Al Pacino, Italian genetics (b. 1940)[327]
& **Leonard Cohen**, a Jew (b. 1934)[328]

Featured above is Al Pacino, an Italian-American actor with Sicilian ancestry, and Leonard Cohen, a Canadian music artist with Jewish genetics.[329] These two men could pass for brothers, or even twins. Italians—especially Sicilian Italians—and Semites share similar genetic lineages. This fact indicates common Semitic origin.

Leonard Cohen, Jewish genetics (b. 1934)[330]
& **Al Pacino**, Italian genetics (b. 1940)[331]

From another view of these two men, we can see just how correlative Sicilian genetics are to the DNA strands of many Jews. This connection between Sicilians and Jews further corroborates the strong correlation between DNA and appearance.³³² Both men possess high nasal bridges, long and thin noses, and long and rectangular-shaped faces: the prototypical "look" of Israelite males.

Since Cohen's last name means "priest" in Hebrew, he most likely belongs to the Cohen Modal Haplotype (CMH)—the priestly lineage of Jews that might be directly connected to Aaron, the brother of Moses of the Bible and Torah. Although the exact paternal lineages of Pacino and Cohen are unavailable to the public, both men clearly belong to Semitic lineages.

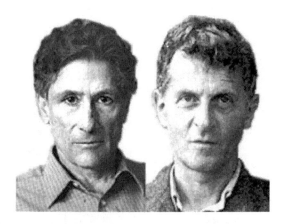

Edward Wadie Saïd, a Palestinian Arab³³³
& **Ludwig Wittgenstein**, a Jew³³⁴

Edward Wadie Saïd, a Palestinian American literary theorist with Semitic genetics, and Ludwig Wittgenstein, an Austrian-British philosopher with Jewish genetics, are featured above. Saïd's mother, Hilda, who was born in Nazareth, had a Palestinian father and a Lebanese mother. Saïd's father was a Palestinian. Wittgenstein's paternal great-grandfather was Moses Meier, a Jewish land agent. Both Saïd and Wittgenstein possessed high nasal bridges, long and thin noses, and long and rectangular-shaped faces (prototypical traits of Semites). Since Palestinian Arabs and Jews share similar phenotypes and genotypes, the two groups are connected.

Rev. Joseph Vasilon, a **Greek Orthodox priest** (c. 1910)[335] & a **Jewish rabbi** (Early 1900s)

A Greek Orthodox priest and a Jewish rabbi are featured above. Both religious men possessed high nasal bridges, long and thin noses, and long and rectangular-shaped faces. Many Greeks, especially Greek Cypriots, resemble Jews and other Semites. This fact supports the genotype-phenotype connection, because not only do many Greeks and Jews share similar visages, they also share similar genetic lineages.[336] Individuals from the Middle East traveled to places like Greece to ensconce there. When Middle Easterners established settlements in areas outside of their homelands (like Greece), they brought their cultures and genetic markers with them.

Samuel Beckett, Irish genetics
& **J. Robert Oppenheimer,** Ashkenazi Jewish genetics[337]

Studies performed by evolutionary biologist Lara M. Cassidy and her team from Trinity College Dublin in Ireland indicate that a large number of Irishmen are related to ancient Middle Easterners.[338] This is one reason why the Irish often possess the common appearance of Semites and their accompanying prototypical Semitic genetic markers.[339] Samuel Beckett (featured above to the left), an Irish avant-garde novelist, playwright, theater director, and poet, appears to have had Semitic genetics because he so closely resembled J. Robert Oppenheimer (featured above to the right), a Jewish-American theoretical physicist.

Both men featured above had similar large aquiline noses, high nasal bridges, and long rectangular-shaped faces, which of course are most commonly found among individuals with Semitic ancestry. Beckett and Oppenheimer also possessed similar mouths, cheekbones, thin lips, tight skin, glasses, jawlines, hairlines, necks, and other physical features.

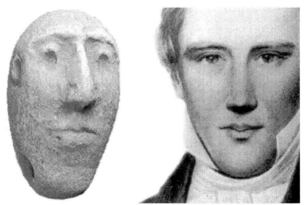

A **Celtic stone head amulet** (300 BC-1 BC)³⁴⁰
& **Joseph Smith, Jr.**³⁴¹

The Prophet Joseph Smith (featured above to the right) shared common facial features with an ancient Celtic effigy from Ireland that dates back to around 300 BC-1 BC (featured above to the left). Both faces show prototypical Semitic physical traits: long and thin noses, high nasal bridges, and long and narrow faces. It is no coincidence that Joseph Smith looks like an ancient Celtic effigy since he possessed Irish blood on his father's side.³⁴²

According to a previously mentioned study, Irishmen are mostly Semitic, or at least Middle Eastern, in origin.³⁴³ A 2013 study conducted by scientists including Mark A. Jobling at the University of Leicester used DNA tests to discover that 85% of Irishmen are descendants of farming people from the Middle East and Anatolia.³⁴⁴ This finding supports the idea that a number of people from the UK have Middle Eastern ancestry, including the Prophet Joseph Smith.ᶜᵉ

---

ce   See *Chapter 10: Genetic Markers of Semites, East Asians, and Siberians*.

From this chapter, we have learned that even though the exact appearance and DNA of Lehi's family and traveling caravan are unknown, we have a general idea of what Semitic peoples looked like in ancient times because ancient figurines, sculptures, effigies, and reliefs that depict ancient Semites exist. Another takeaway from this chapter is that a consistency exists between the appearance of ancient and modern Semites. Hence we have a reliable way to determine what the main Lamanite remnant (Lehi's descendants) should generally look like.

The next chapter, *Chapter 7: American Indian Depopulation*, discusses the idea that Semitic traits were probably more common among American Indians prior to the massive depopulation of indigenous Americans. This massive depopulation appears to have made it much more difficult (but not impossible) to recognize the main Lamanite remnant in our modern era.

7

American Indian Depopulation

Where today are the Pequot? Where are the Narragansett, the Mohican, the Pocanet, and other powerful tribes of our people? They have vanished before the avarice and oppression of the white man, as snow before a summer sun. Will we let ourselves be destroyed in our turn without a struggle, give up our homes, our country bequeathed to us by the Great Spirit, the graves of our dead and everything that is dear to us? I know you will cry with me, Never! NEVER! ... The annihilation of our race is at hand unless we unite in one common cause against the common foe. ... Sleep not longer, O Choctaws and Chickasaws. ... Will not the bones of our dead be plowed up, and their graves turned into plowed fields? —**Tecumseh, Shawnee Chief (1811)**[345]

Nearly all scholars now believe that widespread epidemic disease, to which the natives had no prior exposure or resistance, was the overwhelming cause of the massive population decline of the Native Americans. —**Noble David Cook, American historian (1998)**[346]

Hitler's concept of concentration camps as well as the practicality of genocide owed much, so he claimed, to his studies of English and United States history. He admired the camps for Boer prisoners in South Africa and for the [North American] Indians in the wild west; and often praised to his inner circle the efficiency of America's extermination – by starvation and uneven combat – of the red savages who could not be tamed by captivity. —**John Toland, American historian (1991)**[347]

J oseph of Egypt's descendants from among the Lamanite remnant are still alive today. However, currently it appears they are much more difficult to locate than in previous centuries. Entire populations of Amerindians have been nearly wiped off the face of the earth after Europeans set foot on the American continent.[348]

North American Indian depopulation occurred mostly during the first European contact period, which apparently led to a population decline from millions to an estimated 530,000 by the year 1900.[349] This widespread depopulation nearly annihilated a number of indigenous North American genetic lineages (many of which might have been Semitic in origin and belonged to the Lamanite remnant).[cf] In 1873, Elder Wilford Woodruff remarked on this very fact:

> The Lamanites will blossom as the rose on the mountains. . . . [T]hough I believe this, when I see the power of the nation destroying them from the face of the earth, the fulfillment of that prophecy is perhaps harder for me to believe than any revelation of God that I ever read. It looks as though there would not be enough left to receive the Gospel; but notwithstanding this dark picture, every word that God has ever said of them will have its fulfillment, and they, by and by, will receive the Gospel. It will be a day of God's power among them, and a nation will be born in a day. Their chiefs will be filled with the power of God and receive the Gospel . . . and we shall help them. They are branches of the house of Israel[.][350]

The collective population of indigenous North, Central, and South Americans "was reduced from over 44 million persons down to 2 or 3 million in fewer than 100 years and was eventually conquered by a small group of Europeans."[351] This prolonged widespread decimation drastically altered the demographics and genetic composition of indigenous Americans. As Michael H. Crawford, a British ancient historian, remarked in 2001 concerning the subject, "Amerindian populations that survived the initial contact passed through a tight selective 'bottleneck' that must have altered their genetic constitution."[352]

---

cf   The "seed" of Joseph of Egypt among the Lamanite remnant appears to be much less represented in modern times due to the massive depopulation of indigenous Americans.

Before millions of indigenous Americans were annihilated by deadly European diseases, warfare, and other factors, a wider variety of physical traits (i.e. eye colors, hair colors, skin tones) and DNA could have existed among American Indians; much more than one will observe today. Had these things not played such devastating roles in the New World, Joseph of Egypt's descendants among the Lamanite remnant probably would have been much more prevalent than they are presently. Still, it is uncertain whether this devastation could have been avoided since it appears inevitable that the white man was drawn to the New World and therefore would carry with him deadly diseases.

Examples of extinct North American Indian tribes include the Ababco, Ais, Agua Dulce, Alabama, Apalachicola,[353] Calusa, Chakchiuma,[354] Chatot, Chiaha, Chisca, Coree (Core),[355] Doegs, Griga, Guale, Hachaath, Hitchiti,[356] Ibitoupa, Karankowa, Koasati,[357] Koroa, Mandan, Mobile, Muklasa, Nacisi, Neutrals, Nynee, Oconee,[358] Osochi, Pascagoula, Pensacola, Pohoy, Potano, Powhatan, Saturiwa, Sawokli, Tamathli, Tawasa, Tekesta,[359] Tuskegee,[360] Utina, Westo, Yamasee, Yazoo, Yui, and Yustaga.

Deadly European Diseases

The populations of the Americas were drastically reduced by disease, warfare, and slavery until the extinction of some groups and the attainment of a population nadir [low point] towards the end of the nineteenth century. This population reduction has forever altered the genetics of the surviving groups, thus complicating any attempts at reconstructing the pre-Columbian genetic structure of most New World groups. —**Michael H. Crawford, British ancient historian (1998)**[361]

The three Old World diseases brought to the Americas that caused the most damage to American Indian populations were smallpox, typhus, and measles.[362] One unique tribe that was nearly decimated by smallpox—the Mandan—might have provided insight into the people of the Book of Mormon had the remaining survivors' DNA not virtually disappeared due to assimilation into other tribes. The Mandan, a once great and noble people who roamed the banks of the Missouri river and wandered across the Great Plains of North America, were often unknown to European American settlers. A great majority of Americans of the 19th century were completely unaware of the fact that significant amounts of Mandan Indians possessed blue eyes, light hair, and fair skin—traits not inherited due to any known recent European admixture.

The devastating effects of smallpox began for the Mandan in June of 1837, but destroyed the tribe in 1838. A ship known as "St. Peter's" arrived on that date at Fort Clark, which was sixty miles north of the current location of Bismarck, North Dakota. Knowing men were aboard the boat with smallpox, Francis A. Chardon and others of the American Fur Company tried to keep the Mandans away from the boat, but proved unsuccessful. Because of the exposure to smallpox, many Mandan contracted the deadly European disease.

Two Mandan villages (the same ones that had provided aid to Lewis and Clark during the winter of 1804-1805) were devastated. Only handfuls of the Mandan survived out of a population of sixteen hundred to eighteen hundred.[363] Brigadier General W. H. Jackson described the chilling and tragic fate of the once-great Mandan:

> In the spring of 1838[,] that dreaded scourge of the Indians, small-pox, made its appearance among the Mandans, brought among them by the employees of the fir [fur] company. All the tribes along the river suffered more or less, but none approached so near extinction as the Mandans. When the disease had abated, and

when the remnant of this once powerful nation had recovered sufficiently to remove the decaying bodies from their cabins, the total number of grown men was twenty-three, of women forty, and of young persons sixty or seventy. These were all that were left of the 1,800 souls that composed the nation prior to that ... disease.[364]

On July 30, 1837, Mandan Chief Máh-to-tóh-pa, aka "The Four Bears," gave a speech after he had been fermenting with smallpox for a time.[365] Virtually all of his tribe had been obliterated by the disease that year. Before he passed away, he vehemently declared to his suffering people:

> I have never called a white man a dog, but today, I do pronounce them to be a set of black-hearted dogs. They have deceived me. Them that I always considered as brothers have turned out to be my worst enemies. . . . today I am wounded, and by who? By those same white dogs that I have always considered and treated as brothers. I do not fear death, my friends ... but to die with my face rotten, that even the wolves will shrink at horror at seeing me ... your wives, children, brothers, sisters, friends, and ... all that you hold dear—are all dead, or dying, with their faces all rotten, caused by those dogs the whites.[366]

That same year, many Mandans committed suicide after losing friends and family members.[367] The handful of surviving Mandan fled to neighboring camps or else attempted survival alone in the wilderness but to no avail. Some starved to death after prolonged periods of grief and depression and refusal to eat. The will to live quickly fled after they saw little to no hope of recovery.

The practices of Mandan shamans and holy men were rendered useless against deadly European diseases. Their collective lack of knowledge about how to treat foreign fatal diseases even led to malpractice. Frequently, the shamans and holy men made conditions worse for smallpox sufferers by placing them in sweat lodges to recover. This proved only to compound the problem by inflaming the open sores.

Other Indian tribes suffered the effects of deadly diseases. Clinton A. Weslager, author of *The Delaware Indians: A History* (1972), said concerning the Delaware (Lenni Lenape) Indians:

> At a conference held at Burlington, one of New Jersey's early settlers, Thomas Budd, was told by a Delaware [Lenni Lenape] Indian that three separate smallpox epidemics had ravaged the native villages. The Indian said the first epidemic occurred in his grandfather's time, a second in his father's time, and a third, which he himself well remembered, spread through the Indian settlements. With each recurrence of the disease bringing disastrous results, the Indians believed the whites wanted to get rid of them and deliberately infected them by selling them matchcoats that had been exposed to smallpox germs.[368]

The devastating effects of smallpox, and other deadly diseases of that time period,[369] date back to even earlier eras. On May 2, 1792, Commander Vancouver[370] described signs of disease and devastation at a nearby Indian village:

> The houses ... of the Indians . . . had . . . fallen into decay; their inside, as well as a small surrounding space that appeared to have been formerly occupied, were overrun with weeds; amongst which were found several human sculls [sic], and other bones, promiscuously scattered about.[371]

In about 1630, smallpox, measles, influenza, whooping cough, and diphtheria spread from the Europeans to the Iroquois.[372] In 1755, a smallpox epidemic killed 50% of the Nez Percé population. Nearly 70% of the Wampanaog Indians were destroyed by fevers and illness brought by foreigners in 1764. By the 1860s, smallpox epidemics were predominantly responsible for the end of tribal warfare, due to the drastic decrease in population.[373]

Warfare

Papoose Conewa, "**Passaconaway**," a Pennacook (b. about 1565)

Chief Passaconaway (featured above), otherwise known as "Child of the Bear,"[374] was a holy man and chieftain of the Pennacook Indians of North America.[cg] He engaged in war with the English and

---

cg   Legend has it that Passaconaway was a giant of a man; a genius with magical powers. He purportedly could make water burn, and get trees and rocks to dance. According to folklore, he was able to cause desiccated leaves to become lush green. Out of dead snake skin, it is said he could make a living one. Some even said he could become invisible and create thunderstorms at will. Passaconaway. http://en.wikipedia.org/wiki/Passaconaway

died in battle alongside others of his tribe. In a 1660 farewell speech prior to his death, Passaconaway lamented the loss of his men from battles with the English,

> The English came, they seized our lands; I sat me down at Pennacook. They followed upon my footsteps; I made war upon them, but they fought with fire and thunder; my young men were swept down before me, when no one was near them. I tried sorcery against them, but they still increased and prevailed over me and mine, and I gave place to them and retired to my beautiful island of Natticook. . . . I who have had communion with the Great Spirit dreaming and awake—I am powerless before the Pale Faces.[375]

Passaconaway and his people were felled like trees, just like many other tribes who fought Europeans wielding "fire and thunder." They were no match against the firepower of the Europeans.

Warfare decimated other tribes, including groups within the Iroquois Confederacy. Historical artist George Catlin once remarked that the Iroquois were

> [o]ne of the most numerous and powerful tribes that ever existed in the Northern regions of our country [US] . . . [but] by their continual wars with the French, English, and Indians, and dissipation and disease, they have been almost entirely annihilated. The few remnants of them have long since merged into other tribes, and been mostly lost sight of. . . . *Not-o-way* [an Iroquois] was an excellent man, and was handsomely dressed . . . He told me, however, that he had always learned that the Iroquois had conquered nearly all the world; but the Great Spirit being offended at the great slaughters by his favourite people, resolved to punish them; and he sent a dreadful disease amongst them that carried the most of them off, and all the rest that could be found, were killed by their enemies.[376]

During wartime, a reprehensible tactic to rid the world of American Indians appears to have found its way into the New World for the first time during the 1763 siege of Fort Pitt. During the height of Ottowan Chief Pontiac's[377] uprising against unfair British requirements, his people raised hatchets to the British. In an attempt to break the Ottowan siege, a number of British militants, including British general Jeffrey Amherst, provided blankets for natives taken from disease-infected corpses. He wanted them to suffer and die miserable deaths from smallpox. In 1763, in Pennsylvania, British general Jeffrey Amherst gave the order: "[Kill] the Indians by means of blankets. . . . [T]ry every other method that can serve to extirpate this execrable race."[378]

William Trent, a prominent merchant associated with William Penn in the founding of Pennsylvania, wrote in his journal on June 24th, 1763, concerning some biological warfare used against the North American Indians. Trent's journal reads:

> The Turtles Heart [,] a principal Warrior of the Delawares [,] and Mamaltee [,] a Chief [,] came within a small distance of the Fort Mr. McKee [.] [We] went out to them and they made a Speech. . . . [W]e gave them two Blankets and an Handkerchief out of the Small Pox Hospital. I hope it will have the desired effect.[379]

When Trent stated, "I hope it will have the desired effect," he implied that he wanted the Indians to die painful and humiliating deaths via smallpox. Death by smallpox is an ignominious way to die due to what it does to the skin on the face and body. It can turn beautiful people into virtually unrecognizable creatures. All one has to do is look up images for the word "smallpox" to view the deleterious effect it has on the skin of those infected.

Some Europeans in the New World had the audacity to claim God's blessing upon their baleful acts, which included germ warfare.[ch] Their hubris and deluded thinking led one European man in North America to state: "It has pleased Our Lord to give the said people [North American Indians] a pestilence of smallpox that does not cease...."[380] These Europeans appear to have viewed the scourge as "a sign from God" that the American Indians had incurred God's wrath and therefore deserved punishment.

The loss of many individuals within the North American Indian population reduced the collective knowledge of the history and ceremony of the various tribes. One early 18th century Charleston, North Carolina Indian stated, "[T]hey [the North American Indians who were highly affected by epidemics and warfare] keep their festivals and can tell but little of the reasons: their Old Men are dead."[381] This loss of collective knowledge of history and ceremony of the various tribes increased the difficulty of acquiring knowledge about early American Indian cultures, languages, and history, because once their oral history was lost, traditions and tribal identity faded.

The Indian Removal Act of 1830

After the injuries we have done them [the North American Indians], they cannot love us, which leaves us no alternative but that of fear to keep them from attacking us. But justice is what we should never lose sight of, and in time it may recover their esteem.
—**Thomas Jefferson, 3rd US President (1786)**[382]

On December 6, 1830, in a message to US Congress, President Andrew Jackson called for the relocation of select American Indian tribes to lands west of the Mississippi River. This act created a

---

ch As a result of pernicious germ warfare, many European explorers and traders received death threats from rancored victims and relatives of the deceased.

dividing line that was known among Latter-day Saints as "the borders of the Lamanites" (D&C 28:9). Jackson created this act to open new land for settlement by United States citizens, but he mostly wanted to mine the gold and other resources located on Cherokee lands and the lands of other tribes. Because of the violent responses of North American Indians to the Indian Removal Act of 1830, thousands died.

The removal process included inhumane treatment of indigenous North American tribes. Men, women, and children were forced to walk many miles all day long, often with very little rest. Many Indians perished due to famine, disease, and harsh traveling conditions while making their way to their assigned reservations.

Even before the Indian Removal Act of 1830, indigenous North American Indians dealt with the pain and death associated with forced relocation. Non-on-da-gon, a leader of the Delaware (Lenni Lenape) tribe, dealt with consistent relocation. Sadly, by the 17th century following approximately 200 forced relocations, the Delaware tribe had lost nearly ninety percent of its population. Historical artist, George Catlin wrote a sympathetic account of the Delaware who had been relocated to the Ohio River Valley, the Illinois River region, and finally across the Missouri River to Fort Leavenworth (in present-day Kansas) from their native lands of Eastern Pennsylvania, New Jersey, and Delaware. Catlin Lamented:

> In every move the poor fellows have made, they have been thrust against their wills from the graves of their fathers and their children; and planted as they now are, on the borders of new enemies, where their first occupation has been to take up their weapons in self-defense, and fight for the ground they have been planted on.[383]

Throughout history, the Cherokee have been a powerful North American Indian tribe. However, the US government broke many of their spirits when they enforced the Indian Removal Act of 1830. The Cherokee were forcibly removed from their homelands and forced to walk approximately 2,200 miles to a reservation set aside for them by the US government. This walk, known today as the "Trail of Tears," was responsible for the death of 4,000-8,000 of the approximately 16,000 Cherokee who were removed between 1836 and 1839. Because of the common Semitic-looking appearance of the Cherokee, many of them who died on the trail may have belonged to the seed of Joseph of Egypt.

Before the "Trail of Tears," Andrew Jackson stated in his *Sixth Annual Message to Congress*, December 1, 1834:

> I regret that the Cherokees east of the Mississippi have not yet determined as a community to remove. How long the personal causes which have heretofore retarded that ultimately inevitable measure will continue to operate I am unable to conjecture. It is certain, however, that delay will bring with it accumulated evils which will render their condition more and more unpleasant. The experience of every year adds to the conviction that emigration, and that alone, can preserve from destruction the remnant of the tribes yet living amongst us.[384]

Years before Jackson's time in office, President Thomas Jefferson attempted to create peaceful interactions with North American Indians. Contrary to what some might believe, Jefferson originally tried very hard to work with indigenous Americans so they could become part of American society.

President Thomas Jefferson was quite interested in North

American Indian tribes. According to Stephen E. Ambrose, author of *Undaunted Courage: Meriwether Lewis, Thomas Jefferson and the Opening of the American West*, "Jefferson had a passion for Indian language, believing he would be able to trace the Indian's origins by discovering the basis of their language."[385] Due to his interest in North American Indian tribes, Jefferson even had the Lewis and Clark Expedition report to him about them while he was serving as US Commander-in-Chief.

In Thomas Jefferson's letters, his true feelings towards the American Indians often manifested themselves. Jefferson wrote a letter in 1803 to William H. Harrison[ci] about North American Indians, that said,

> Our system is to live in perpetual peace with the Indians, to cultivate an affectionate attachment from them by everything just and liberal which we can do for them within the bounds of reason, and by giving them effectual protection against wrongs from our own people.[386]

Until indigenous North Americans rejected Jefferson's proposals, he had always wanted them to keep their homelands.[387] Jefferson first planned to guide North American Indian tribes on the road to civilization, which included setting up agriculturally-based societies among them.[388] The policy he proposed for dealing with North American Indians on United States soil was also "to let our settlements and theirs meet and blend together, to intermix, and become one people."[389] Jefferson's hopes of unity with his Indian brothers and sisters dissipated quickly after North American Indian

---

ci William H. Harrison would later become the 9th US President.

tribes did not agree with his vision for them. Many tribes were resistant to any form of assimilation. Because of the recalcitrance of many North American Indians to assimilate into the United States culture at the time,[390] Thomas Jefferson proposed that the North American Indians should emigrate across the Mississippi River and maintain a separate society, an idea that manifested itself by the Louisiana Purchase of 1803.[391]

The Prophet Joseph Smith's personal opposition to forced Indian removal mirrored President Thomas Jefferson's original plan he had for the Native Americans. However, unlike President Jefferson, Joseph desired to extend the United States "from the east to the west sea," but only if the North American Indians gave their consent.[392] How this could have transpired in the United States in a smooth manner is difficult to say, largely due to the fact that some tribes were extremely hostile towards Europeans and distrustful of them. However, if anyone could have succeeded, it probably would have been Joseph Smith. He was known among certain North American Indian tribes for his strength, kindness, and diplomacy.

Joseph Smith believed that the North American Indians deserved better treatment than they had received at the hands of the US government, US citizens, and early European settlers. It is probably one of the main reasons why he treated Native Americans so well. On January 25, 1842, Joseph Smith told Latter-day Saint John C. Bennett and others that: "[T]he Indians have greater cause to complain of the treatment of the whites, than the negroes."[393]

Looking Glass (Allalimya Takanin), a Nez Percé chief (1871 or 1877)[394]

Other tribes, including the Nez Percé, fought against the US Government's decision to forcibly remove Indians from their homelands. Featured above is "Looking Glass" (1832-1877), the principal war chief of the Nez Percé. American history buffs have doubtless heard of the Nez Percé chief, Chief Joseph, who famously said, "I will fight no more forever." However, most people have probably never heard of "Looking Glass,"[395] who actually outranked Chief Joseph in the Nez Percé War (1877).

When war broke out against the Nez Percé in 1877, young Looking Glass was elected head of the anti-white faction as was his father before him (Looking Glass, Sr.). Looking Glass, Jr.

> continued the tradition of passive resistance against land sales while preaching peaceful coexistence with surrounding settlers. . . . He was joined in his resistance by another notable leader, Chief Joseph of the Wallowa band.[396]

Chief Looking Glass and Chief Joseph[397] worked together after their bands joined forces. In 1877, with Chief Joseph by his side, Looking Glass, Jr., led the Nez Percés from their reservation to Montana. Because of his soured encounters with certain US political and military leaders, Looking Glass declared, "I will never surrender to a deceitful white chief."[398]

White troops sent by President Ulysses S. Grant but led by Colonel Nelson Miles caught up to the Nez Percé near the Bear Paw Mountains of North-Central Montana. There they laid siege to the Indian camp. Looking Glass, anticipating that an approaching rider outside of their camp was a messenger from the Sioux leader, Sitting Bull, sprang up on the rocks to get a better view of the approaching rider. The moment he perched up to get a better look, he was cut down by a Cheyenne scout sniper's bullet. Because of the death of their principal War Chief, Looking Glass, Nez Percé defenders became disheartened and gave up their resistance. Chief Joseph surrendered to Colonel Miles that same day.

8

The Lamanite Remnant

I was . . . informed concerning the aboriginal inhabitants of this country [.] . . . In this important and interesting book [the Book of Mormon] . . . [w]e are informed . . . that America in ancient times [had] been inhabited by two distinct races of people. . . . The second race came directly from the city of Jerusalem, about six hundred years before Christ. They were principally Israelites, of the descendants of Joseph. . . . The principal nation of the second race fell in battle towards the close of the fourth century. The remnant are the Indians that now inhabit this country. —**Joseph Smith, Jr., Latter-day Saint Prophet (1842)**[399]

I would speak somewhat unto the remnant of this people [the Lamanites] who are spared, if it so be that God may give unto them my words, that they may know of the things of their fathers; yea, I speak unto you, ye remnant of the house of Israel; and these are the words which I speak: Know ye that ye are of the house of Israel. —**Mormon, Book of Mormon prophet, Mormon 7:1-2 (about 385 AD)**

The American [Indian] Race is marked by . . . long, black, lank hair . . . The cheekbones high, the nose large and aquiline. —**Samuel George Morton, natural scientist, *Crania Americana* (1839)**[400]

Who are the Lamanites? The term "Lamanite" is defined in many ways throughout the Book of Mormon.[401] However, in this book, "Lamanite" is used to refer to any member of Lehi's caravan or any group with the specific Middle Eastern genetics of Lehi's family.[402] Specifically, members of these lineages possess actual Lamanite DNA. A definition from the *Encyclopedia of Mormonism* perfectly encapsulates this book's primary usage of the term "Lamanite:" "[A] people of Israelite origin from the Book of Mormon."[403]

The Book of Mormon states that God led a remnant of Joseph of Egypt's seed out of Jerusalem so that this genetic branch would not be destroyed (See Ether 13:7).[404] According to the Book of Mormon, the Lamanite remnant belongs to Joseph's bloodline, so, they must be somewhere in the Americas. DNA results from genetic tests performed on indigenous Americans have yet to provide irrefutable proof about the existence of ancient Semitic groups in the Americas, and this lack of concrete evidence makes it difficult for Latter-day Saints to discuss the whereabouts of the Lamanite people. Many are confused, including a young Peruvian attendee of a 2003 DNA and Book of Mormon presentation at BYU who perfectly articulated this present dilemma:

> We don't know where the Book of Mormon took place. We don't know where the Lamanites are. If we don't know who the Lamanites are, how can the Book of Mormon promise to bring them back? It's an identity crisis for many of us that [must] be understood[.][405]

However, if one looks to the past, clues exist about where the main Lamanite remnant is currently located. The Prophet Joseph Smith knew where to sent LDS missionaries to teach the Remnant: North America. As mentioned previously, Elder Parley P. Pratt belonged to the core group of missionaries sent by Joseph Smith to teach the Lamanites the gospel of Jesus Christ. At the conclusion of Elder Pratt's 1830-1831 mission to teach the remnant of Book of Mormon peoples, he listed off in his journal which North American tribes he and his mission companions visited.[406] Handsome Lake's tribe (the Seneca), the Huron, and the Delaware were all mentioned by Elder Pratt. The

Pottawatamie, Sauk & Fox, and Shawnee also received visits from Latter-day Saint missionaries. Each of these tribes were declared members of the Lamanite remnant by either Joseph Smith himself or other early LDS brethren. What follows is a description of these, and other, North American Indian tribes, their interactions with LDS missionaries, and evidence of their probable Lamanite heritage.

The Iroquois

Chief Sose Akwirranoron, an Onondaga of the Iroquois Confederacy (1905)

Accounts indicate that the Prophet Joseph Smith frequented the Mud Creek area of New York, a place where the Seneca and other Iroquois travelers would often gather and set up camp. Due to these interactions, Joseph Smith was no doubt familiar with the common physical traits of the Iroquois. Chief Sose Akwirranoron (featured above), an Onondaga Indian of the Iroquois Confederacy, was not alive when Joseph Smith interacted with members of the Iroquois, but he was born with common physical characteristics found among the Iroquois. Because of his appearance, he seems to have belonged to the Lamanite remnant. As could be expected, Chief Sose Akwirranoron possessed a high nasal bridge, a long and thin nose, and a long and narrow face. He did not possess the physical traits common to East Asians and Siberians.

When compared to other groups of indigenous Americans, members of the early Iroquois possessed atypical physical features. Their visages often matched those of prototypical Caucasian Middle-Easterners rather than East Asians and Siberians. Tribes from the Iroquois Confederacy possessed these traits long before European and Jewish Americans entered the New World. One of the earliest descriptions of the Iroquois was given by Sieur de Roberval, the first governor general of America's "New France" (Canada). Describing the Iroquois of the St. Lawrence (New York) region as they appeared in 1542, De Roberval noted:

> [I] declare unto you the state of the savages. They [the Iroquois] are a people of goodly stature and well made; they are very white, but they are all naked, and if they were appareled as the French are they would be as white and as fair, but they paint themselves for fear of heat and sunburning.[407]

Sieur de Roberval described the Iroquois as being "very white," not just "white." His account is suggestive of a description of the white Lamanites found in the Book of Mormon. Third Nephi reads, "And it came to pass that those Lamanites who had united with the Nephites were numbered among the Nephites . . . and their skin became white[cj] like unto the Nephites" (3 Nephi 2:14-15). The fact that the early Iroquois were "very white" increases the likelihood that they were members of the House of Israel and can be numbered among the members of the Lamanite remnant who descended from white Lamanites.

Historical records confirm the fact that members of the Iroquois Confederacy (formerly known as the Iroquois League) were known for their intelligence, oratory skills, powerfully built bodies, and prominent facial features—all traits possessed by many Book of Mormon peoples. For example, Teganissorens, a member of the Onondaga Iroquois who arrived in Quebec in May 1694, was like many other Iroquois diplomats of this time. He was so impressive that he dazzled his European contemporaries with his combination of powerful physical presence and impressive eloquence, even when he spoke in a tongue that they could not comprehend.

Cadwallader Colden, a writer and former Lieutenant Governor for the Province of New York, gave a high opinion of Teganissorens in the first part of his 1727 book, *The history of the Five Indian Nations depending on the Province of New-York in America*. Colden testified of his personal knowledge of Teganissorens' great talents. Concerning Teganissorens, he wrote:

> [Teganissorens] had for many years the greatest reputation among the Five Nations for speaking, and was generally employed as their Speaker, in their negotiations with both French and English. . . . He had a great fluency in speaking, and a graceful elocution, that would have pleased in any part of the world. His person was well made, and his features, to my thinking, resembled much the Bustos [busts] of Cicero.[408]

---

cj   The Lamanite's skin became white like the Nephite's skin probably due to genetic admixture.

Although pictures of Teganissorens may not exist, surviving busts of the Roman politician Cicero possess high nasal bridges, long and thin noses, and long, rectangular-shaped faces. Cicero's physical traits indicate he may have had Semitic ancestry, and thus, if Teganissorens resembled him, it is not unreasonable to infer that Teganissorens had Semitic ancestry as well.

An **Iroquois man** (probably Seneca), detail from an early photograph (1852)

The Iroquois man featured above possessed the Semitic-looking physical features that were common among the early Iroquois. Based on this information, it is very likely that he belonged to the Lamanite remnant. This Iroquois man's photo was taken in 1852, which is only shortly after the first photographs were invented, and not too long after Latter-day Saint missionaries first visited the Seneca in 1830. It is likely that, from the time the LDS missionaries visited the Iroquois to the date this photograph was taken, that the physical appearance of typical members of the Iroquois Confederacy did not change much.

Theyanoguin, a Mohawk leader of the Iroquois Confederacy (1691-1755)

Theyanoguin, also known as "Hendrick" (featured above), was an influential Mohawk leader of the Iroquois Confederacy, and one of the "Four Indian Kings" who visited Queen Anne in England in 1710.[409] As we can see from this picture above, Theyanoguin lacked the prototypical physical features of individuals from the Far East. He was known to have a prominent, thin and aquiline nose, a high nasal bridge, and a long and narrow face. These physical characteristics may indicate that Theyanoguin belonged to the Lamanite remnant. With his shorn head, Semitic features, and fierce appearance, he would certainly have had the typical appearance of a Lamanite of old as described in the Book of Mormon and depicted by prominent Latter-day Saint artists and filmmakers.

Like select members of his tribe, Theyanoguin was skilled in oratory. According to historical documents, the English were aware of the Iroquois' oratorical skills and were greatly impressed with them in eighteenth-century treaty councils.[410] In 1957, Wynn R. Reynolds, author of *Persuasive Speaking of the Iroquois Indians at Treaty Councils: 1678-1776*, examined 258 speeches by Iroquois at treaty councils between 1678 and 1776. He discovered that the Iroquois speakers resembled the ancient Greeks in their primary emphasis on ethical proof.[411] Based on how uncommon this practice was outside of the Old World, this connection would appear to tie them to similar ancient thinkers of the Mediterranean and Middle East.[ck]

The original framers of the US Constitution (most notably George Washington and Benjamin Franklin) greatly admired the concepts, principles, and governmental practices of the Iroquois.[412] Franklin once said of the North American Indians, "Savages we call them, because their manners differ from ours, which we think the perfection of civility; they think the same of theirs."[413] According to author Bruce Burton, the Iroquois had democracy before the United States of America. Other sources point to the civilized nature of many of the Iroquois. Former US President John F. Kennedy once wrote of the League of the Iroquois:

> The pioneers found that Indians in the Southeast had developed a high civilization with safeguards for ensuring peace. A northern

---

ck To learn more about ways in which the New World is connected to the Old World, refer to *Appendix 4: Old World Symbols in the New World*.

extension of that civilization, the League of the Iroquois, inspired Benjamin Franklin to copy it in planning the federation of States.[414]

Although currently unproven, a possible connection exists between the Iroquois language and Semitic languages. Handsome Lake—the man responsible for preparing the hearts of many of the Iroquois for Christianity—belonged to the hoya'ne' families in which the most honored Seneca title of "Skanyadariyoh" is vested.[415] Hoya'ne' means "noble" in Handsome Lake's native language, and the title "Skanyadariyoh" in his language is strikingly similar to "Sikandariyeh," an Arabic word for "Alexandria"—the ancient city in Iraq named after Alexander the Great. In Greek, "Alexander" (Αλέξανδρος) actually means "leader and protector of men," which in Arabic appears to be the same title ("Skanyadariyoh") that was used by the Iroquois to denote leadership and protection. Because of this possible connection, it is probable that the Iroquois title "Skanyadariyoh" originated in the Old World.

To find Arabic-like words in the Seneca lexicon is not unreasonable, especially since Arabic is a Semitic language and the Book of Mormon contains many words and names that stem from Semitic languages. According to Latter-day Saint scholars, Lehi's caravan most likely included Arabic-speakers.[416] If Ishmael, who traveled with Lehi and his family to the New World, was an Arab, then he and his family most likely knew Arabic and also Aramaic. This might explain why an Arabic word (and possibly other Arabic words) are found in the Seneca vocabulary. Semitic languages that originated in the Middle East such as Hebrew, Arabic, Akkadian, Amharic, Tigrinya, and Aramaic all have common origins.[cl]

---

cl   These languages (Hebrew, Arabic, Akkadian, Amharic, Tigrinya, and Aramai) have many

The Pottawatomie

Me-te-a, a Pottawatomie chief (1838)[417]

A number of Pottawatomie Indians appear to have belonged to the Lamanite remnant. Joseph Smith personally spoke with the Pottawatomie (as mentioned in *Chapter 1: Joseph Smith and Book of Mormon Peoples*) and informed them of a record of their forefathers, the Book of Mormon, and relayed to them that they had descended from Book of Mormon peoples. Featured above is "Me-te-a," a Pottawatomie chief (c. 1838). Because of the date of this painting, Me-te-a may have been among the Pottawatomie chiefs who visited Joseph Smith in Illinois on July 2, 1843. Me-te-a possessed a somewhat high nasal bridge, a long and pointy nose, a long and narrow face, and a long-headed (dolichocephalic) skull. Due to his appearance, Me-te-a seems to have had Middle Eastern ancestry and likely belonged to the Lamanite remnant.

---

cognates—words that have a common etymological origin.

The Sauk & Fox

Keeshewaa of the Sauk & Fox (1837)

Many Sauk & Fox Indians appear to have belonged to the Lamanite remnant. The Sauk & Fox, an Algonquian-speaking group comprised of two tribes, were visited at least twice by the Prophet Joseph Smith. Joseph Smith informed the Sauk & Fox of their forefathers, who he said had made the Book of Mormon.[418] Featured above is Keeshewaa of the Sauk & Fox, a man with Semitic physical features common to the tribe of Joseph of Egypt. Interestingly, Keeshewaa is wearing a keffiyeh-like head covering common to those found in the Middle East.[cm]

Charles C. Mann, an American journalist and author of *1491: New Revelations of the Americas Before Columbus,* said of the Algonquain language, which the Sauk & Fox spoke, "[T]he various Algonquain languages date back to a common ancestor that

---

cm A keffiyeh is a traditional Middle Eastern headdress fashioned from a square scarf, usually made of cotton, that is typically worn by Arabs, Kurds, and some Jews.

appeared in the Northeast a few centuries before Christ[.] . . . The ancestral language may derive from what is known as the Hopewell Culture."[419] This connection ties the Sauk & Fox to indigenous North Americans who lived during the time frame of the Book of Mormon.

Keokuk, "The Watchful Fox," a Sauk & Fox (1847)[cn]

Sometime before August 1841, Latter-day Saint missionaries met Chief Keokuk (featured above with the goatee) and provided him with a copy of the Book of Mormon. Early in August, Keokuk and a relatively large number of his people—the Sauk & Fox—

---

cn   Keokuk's photo was taken three years after Joseph Smith's second visit with the Sauk & Fox (1844). Keokuk may have joined his fellow Sauk & Fox to Nauvoo on the 22nd and 23rd of May, 1844. Joseph Smith wrote in his journal (on the 23rd) what he relayed to the Sauk & Fox people: "Great Spirit wants you to be united & live in peace. Found a book [presenting the Book of Mormon], which told me about your [fore]fathers & Great spirit told me[:] You must send to all the tribes you can, & tell them to live in peace, & when any of our people come to see you treat them as we treat you." Joseph Smith Journal, 23 May, 1844; cited in Joseph Smith, *An American Prophet's Record: The Diaries and Journals of Joseph Smith*, edited by Scott Faulring, Significant Mormon Diaries Series No. 1, (Salt Lake City, Utah: Signature Books in association with Smith Research Associates, 1989), p. 482. Aug. 12th, 1841, *History of the Church* vol. 4, pp. 401-402.

camped for a few days along the Mississippi River near Montrose, Iowa. In a grove, the Prophet Joseph Smith explicitly told Keokuk and the other Sauk & Fox that their forefathers had written the Book of Mormon.

Peatwy Tuck, a Sauk & Fox (1898)[420]

Peatwy Tuck (featured above), a Sauk & Fox Indian, appears to have the "look" of the Lamanite remnant as the other previously mentioned members of his Sauk & Fox tribe. He has the common physical traits of Semites from the House of Joseph of Egypt.

The Delaware

Chief Black Beaver, a Lenni Lenape (Delaware) Indian (1869)[421]

Groups of Delaware Indians appear to have belonged to the Lamanite remnant. William Penn (1644-1718), the founder of the Province of Pennsylvania, a philosopher, and an early Quaker, believed that the Delaware Indians (Lenni Lenape) were Semites.[co] Speaking of the Delaware, Penn claimed, "I'm ready to believe them [to be] of the Jewish race, I mean the stock of the Ten Tribes."[422] Because of his obvious Semitic appearance, Chief Black Beaver (featured above) is a good example of Penn's attestation.[cp]

---

co While this information does not prove William Penn's speculations correct, it certainly adds to the plausibility of his claim.

cp Semitic-looking Indians, such as Black Beaver, often transitioned the easiest into wearing European-style of clothing.

In 1693, Penn described the Delaware Indians and their distinct North American Indian language:

> The *Natives* . . . are generally tall, straight, well-built, and of singular Proportion; they tread strong and clever, and mostly walk with a lofty Chin; Of Complexion, Black, but by design, as the Gypsies in England. They grease themselves with Bear's fat clarified, and using no defense against Sun or Weather, their skins must needs be swarthy [(dark-skinned)]; . . . Their Language is lofty, yet narrow, but like the Hebrew; in Signification full, like Short-hand in writing; one word serveth in the place of three, and the rest are supplied by the Understanding of the Hearer: Imperfect in their [grammatical] Tenses, wanting in their Moods, participles, Adverbs, Conjunctions, Interjections. I have made it my business to understand it, that I might not want an Interpreter on any occasion. And I must say that I know not a Language spoken in Europe that hath words of more sweetness or greatness, in Accent and Emphasis, than theirs.[423]

Although no expert linguist, Penn was an educated man and well-versed in Hebrew. He made a solid connection between the Delaware language and ancient Hebrew.[cq]

Similarly, an early French missionary believed that Old World cultures would have admired the Delaware language and the Delaware's use of language.[cr] Pierre François Xavier de Charlevoix (1682-1761), a French Jesuit priest, traveler, and historian, remarked on the imaginative qualities, eloquent speech, and quick wit of the Delaware which he recognized during his time among them. He said:

---

cq Unfortunately, since no written records in the language of the Delaware have been discovered, a verification of Penn's comparison cannot be made to determine whether or not they were indeed similar.

cr To learn more about how the New World is connected to the Old World, refer to *Appendix 4: Old World Symbols in the New World*.

The beauty of their imagination equals its vivacity, which appears in all their discourses. They are very quick at repartee, and their harangues are full of shining passages which would have been applauded at Rome or Athens. Their eloquence has a strength, nature, and pathos which no art can give and which Greeks admired[.][424]

Others described the overall appearance, disposition, and intelligence of the Delaware. In 1634, Captain Thomas Young, one of the first explorers to interact with the Delaware, described the people as "very well proportioned, well featured, gentle, tractable, and docile."[425] Speaking of the Delaware "red men" specifically, Father La Jejune, one of the most devoted of the early French missionaries in North America, said that "in point of intellect," they "could be placed in a high rank."[426]

Like other North American Indian tribes, and certain Book of Mormon peoples, the early Delaware used vermilion face and body paint to adorn themselves. The Delaware were the first group of North American Indians in the late 17th century to be referred to as "red men" in recorded history.

In 1831, Joseph Smith sent Latter-day Saint missionaries to visit the Delaware "red men." It is quite possible that Chief Black Beaver (featured previously) was in attendance at the meeting.[cs] During their initial visit to the Delaware Indians, Latter-day Saint missionaries informed them of their Lamanitish heritage, and also explained to them that their forefathers had produced the Book of

---

cs   In 1831, Latter-day Saint missionaries Oliver Cowdery, Parley P. Pratt, and Frederick G. Williams visited the Delaware Indians on their reservation in Kansas. There they preached to them concerning the restored gospel of Jesus Christ.

Mormon. What these missionaries said to the Delaware strengthened William Penn's claim that they belonged to "the stock of the Ten Tribes." In Kansas, Latter-day Saint Elder Oliver Cowdery spoke to Kikthawenund,[427] a Delaware chief, and his accompanying council:

> Aged Chief and Venerable Council of the Delaware nation; we are glad of this opportunity to address you as our red brethren and friends. . . . This Book [the Book of Mormon] . . . was written in the language of the forefathers of the red men; therefore this young man [Joseph Smith], being a pale face, could not understand it; but the angel told him and showed him and gave him knowledge of the language and how to interpret the Book. So he interpreted it into the language of the pale faces, and wrote it on paper and caused it to be printed, and published thousands of copies of it among them, and then sent us unto the red men to bring some copies of it to them, and to tell them this news. So we have come from him, and here is a copy of the Book, which we now present unto our red friend, the Chief of the Delawares, which we hope he will cause to be read and known among his tribe; it will do them good.[428]

Elder Cowdery informed the Delaware that the Book of Mormon was written in the language of the forefathers of the Red Men. This statement is significant because it pinpoints North American Indians as descendants of Book of Mormon peoples. Indigenous North Americans have always been known as the only "red men." Strictly speaking, the description "red men" does not apply to other indigenous Americans.

The Huron

Kihue, a Huron (1930)[429]

A number of Huron Indians appear to have belonged to the Lamanite remnant. Some early Huron, including Kihue (featured above), possessed distinctively Semitic physical traits. Kihue had a long and aquiline nose, a high nasal bridge, a long and narrow face, and a long-headed (dolichocephalic) skull. Kihue, who claimed to be a full-blooded Huron, was born in 1837, which is shortly after Latter-day Saint missionaries first visited his people in Ohio. As is apparent from Kihue's physical features, the bloodline of Joseph of Egypt ran through the Huron. Thus it appears to be no coincidence that the first Latter-day Saint mission to the Lamanites included a visit to his tribe.

Elder Parley P. Pratt recorded in his journal the visit to the Huron (Wyandot) tribe:

We now pursued our journey for some days, and at length arrived in Sandusky, in the western part of Ohio. Here resided a tribe, or nation of Indians, called Wyandots [Huron], on whom we called, and with whom we spent several days. We were well received, and had an opportunity of laying before them the record of their forefathers, which we did. They rejoiced in the tidings, bid us God speed, and desired us to write to them in relation to our success among the tribes further west, who had already removed to the Indian territory, where these expected soon to go. Taking an affectionate leave of this people, we continued our journey to Cincinnati.⁴³⁰

According to this account, Latter-day Saint missionaries were treated well by the Huron. When informed of their Book of Mormon ancestry by the LDS Elders, the Huron were apparently pleased.

A **Huron** chief

Featured above is a Huron chief with Semitic physical features common to the tribe of Joseph of Egypt. He possessed a long and pronounced nose, a high nasal bridge, and a long and rectangular-shaped face. He had the "look" of individuals from Semitic and Israelite groups.

A **Huron** man (1880)

Another Huron man with common Semitic physical features is featured above. Once again, he possessed a long and thin nose, a high nasal bridge, and a long and rectangular-shaped face.

Like the language of the Delaware, the Huron language was created by an intelligent, complex people. Jesuit Father Sebastien Rasles, a missionary of the Society of Jesus in New France, wrote a letter to his brother about the Huron Indians and their language. In the missive, he said:

> The Huron language is the chief language of the Savages, and, when a person is master of that, he can in less than three months make himself understood by the five Iroquois tribes. It is the most majestic, and at the same time the most difficult, of all the Savage tongues.... [A Jesuit] [m]issionary is fortunate if he can ... express himself elegantly in that language after ten years of constant study.[431]

Pierre François Xavier de Charlevoix (1682-1761), a French Jesuit traveler often distinguished as the first historian of New France (Canada),[ct] was a man of learning and other considerable abilities. Although not an expert linguist, he appears to have paid much more attention to North American Indian languages than most travelers before him. In 1744, Father Charlevoix wrote in his *History of New France* validating the words of Jesuit Father Sebastien Rasles concerning the language of the Huron when he said:

> The Huron language has a copiousness, an energy, and a sublimity, perhaps not to be found in any of the finest languages we know of; and those whose native tongue it is . . . have such an elevation of soul, as agrees much better with the majesty of their language, than with the state to which they are reduced. Some have fancied they found a similarity with the Hebrew, others have thought it had the same origin with the Greek.[432]

The complex language of the Huron is likely to have ties to the Middle East and the Mediterranean, especially since Old World languages such as Hebrew, Arabic, and Aramaic are intricate and layered.[cu] In his statement, Father Charlevoix mentions that individuals before him had recognized the linguistic similarities between the Huron language and Hebrew, just as William Penn did. Both Charlevoix and Penn were intelligent and respectable men known for integrity. Although difficult to prove, the connection between Huron and ancient Hebrew is viable. Correlations can be found between the Huron and Hebrew languages and a number of learned individuals knew it. They found linguistic links that are difficult to dispute.

---

ct New France (modern-day Canada) then occupied much of North America known to Europeans.
cu To learn more about New World connections to the Old World, refer to *Appendix 4: Old World Symbols in the New World.*

The Shawnee

Lay-láw-she-kaw, **Goes Up the River**, an aged chief of the Shawnee (1830)

Certain Shawnee Indians appear to have belonged to the Lamanite remnant. Goes Up the River (featured above), a Shawnee chief, met with Latter-day Saint Elder Oliver Cowdery who visited his tribe in 1831. This encounter should make sense to Latter-day Saints because, as we can see from the photograph featured above, Goes Up the River lacked prototypical East Asian and Siberian physical characteristics, but did possess common physical features of Israelites. He wore a keffiyeh-like headpiece (featured above), which is easily recognizable as an accessory commonly adorning the heads of Middle-Easterners.

A highly intelligent and well-respected man named Elias Boudinot (1740-1821) recognized correlations between North

American Indian languages and the linguistics of Semites. Boudinot was an American Founding Father from New Jersey, and a highly-educated trustee of Princeton University. He became one of the Presidents of the Second Continental Congress, as well as an influential delegate from New Jersey in Congress. Boudinot was even a mentor of another highly-intelligent Founding Father, Alexander Hamilton, with whom he frequently corresponded (along with other noted Founders). Concerning the connection he recognized between some North American languages and the Hebrew language, Boudinot remarked,

> The [North American] Indian languages in general, are very copious and expressive, considering the narrow sphere in which they move; their ideas being few in comparison with civilized nations. They have neither cases nor declensions. . . . [which is] very similar to the Hebrew language. . . . [P]ublic speeches of the Indians, that the writer of these memoirs has heard or read, have been oratorical and adorned with strong metaphors in correct language, and greatly abound in allegory.[433]

As this quote suggests, Boudinot was acquainted with both Hebrew and North American Indian languages. His awareness of these two languages emerges in his linguistic commentary. He recognized that, just like Hebrew, the Shawnee language lacks cases and declensions. Boudinot also knew that select North American Indian languages and the Hebrew language were often both adorned with "strong metaphors," and "greatly abound[ed] in allegory." The Shawnee language and Hebrew definitely have layered aspects to them (i.e. literary devices such as metaphor and allegory). The writing styles of Hebrew prophets of old such as Isaiah (whose writings are found in the Bible and the Book of Mormon) often parallel the layered language styles and literary devices found in the recorded oratories of many early indigenous North American Indians who sat in councils with early European settlers and their interpreters. These language correlations, if legitimate, as they appear to be, are important connections between North American Indian languages and the languages of the descendants of Book of Mormon peoples.

Jewish Traditions Among the Shawnee

One early New World settler, James Adair, an 18th century American trader and writer, lived for almost 40 years among North American Indians, including the Chickasaw and Shawnee. Adair recognized peculiar similarities between the Chickasaw, Shawnee, and Semites. He wrote about these parallels in a history he compiled:

> I observed the Shawano [Shawnee] to be much fairer than the Chikkasah [Chickasaw]. . . . There is a record among the Chikkasah Indians, that tells us of a white child with flaxen hair, born in their country, long before any white people appeared in that part of the world. . . . Their eyes are small, sharp, and black; and their hair is lank, coarse, and darkish. I never saw any with curled hair[.] . . . Like the Jews, the Indians offer their first-fruits [Reishit Katzir, "the beginning of the harvest" in Hebrew culture]; they keep their new moons ["Rosh Chodesh" in Hebrew culture], and the feast of expiation [Yom Kippur, "Day of Atonement" in Hebrew culture] at the end of September or in the beginning of October; they divide the year into four seasons, corresponding with the Jewish festivals. According to Charlevoix and Long, the brother of a deceased husband receives his widow into his house as a guest, and after a suitable time considers her as a legitimate consort. There is also much analogy between the Hebrews and Indians in that which concerns various rites and customs, such as the ceremonies of

purification, the use of the bath, the ointment of bear's grease, fasting, and the manner of prayer. The Indians likewise abstain from the blood of animals, as also from fish without scales; they consider divers quadrupeds unclean, as also certain birds and reptiles; and they are accustomed to offer as a holocaust the firstlings of the flock. Acosta and Emanuel de Moraes relate that various nations allow matrimony with those only of their own tribe or lineage, this being, in their view, a striking characteristic, very remarkable and of much weight. But that which most tends to fortify the opinion as to the Hebrew origin of the American tribes is a species of ark, seemingly like that of the Old Testament. This the Indians take with them to war. It is never permitted to touch the ground, but rests upon stones or pieces of wood, it being deemed sacrilegious and unlawful to open it or look into it. The priests scrupulously guard their sanctuary, and the high priest carries on his breast a white shell adorned with precious stones, which recalls the urim of the Jewish high priest, of whom we are also reminded by a band of white plumes on his forehead.[cv]

Adair further expounded upon some of the language equivalents and cultural practices among Hebrews (which includes Semitic groups) and select indigenous North Americans:

> [They] celebrate the first fruits with religious dances, singing in chorus these mystic words: YO MESCHICA, HE MESCHICA, VA MESCHICA, forming thus, with the first three syllables, the name of Je-ho-vah, and the name of Messiah thrice and pronounced, following each initial. The use of Hebrew words was not uncommon in the religious performances of the North American Indians, and Adair assures us that they called an accused or guilty

---

[cv] "The hotter, or colder the climate is, where the Indians have long resided, the greater proportion have they either of the red, or white, colour. . . . I took particular notice of the Shawano [Shawnee] Indians, as they were passing from the northward, within fifty miles of the Chikkasah [Chickasaw] country, to that of the Creeks [Ocmulgee Creek in Georgia]; and, by comparing them with the Indians which I accompanied to their camp . . . I never saw any with curled hair, but one in the Choktah country, where was also another with red hair; probably, they were a mixture of the French and Indians." *A History of the North American Indians, Their Customs,* James Adair (1775), pp. 4, 307-308.

person haksit canaha, 'a sinner of Canaan'; and to him who was inattentive to religious worship, they said, Tschi haksit canaha, 'you resemble a sinner of Canaan'.[434]

The Cherokee

Ayunini, a Cherokee (1888)[435]

Many Cherokee appear to have belonged to the Lamanite remnant. Featured above is Ayunini, a Cherokee man, who possessed multiple physical characteristics common among Semites. It was common for him to wear a keffiyeh-like headpiece (featured above), which is easily recognizable as an accessory commonly adorning the heads of Middle-Easterners.

Before the Cherokee walked the "Trail of Tears," they were a powerful and well-represented tribe. They were described in this manner in 1792:

> The Cherokees are yet taller and more robust than the Muscogulges [Muskogee], and by far the largest race I ever saw. They are as comely as any, and their complexions are very bright, . . . [T]he Cherokees, have fine features, and are every way handsome men. Their noses are very often aquiline; they are well limbed, countenances upright, and their eyes brisk and fiery. . . . I have seen Indian infants of a few weeks old; their color was like that of a healthy, male, European countryman or laborer of middle age,

though inclining a little more to the red or copper tinge; but they soon become of the Indian copper. I believe this change comes naturally, as I never, from constant inquiry, could learn that the Indians had any artificial means of changing their color.[436]

The size and strength of the Cherokee fit the descriptions Nephi and other Book of Mormon prophets gave of themselves. Before 1923, author William Harlen Gilbert, Jr. also described the pre-"Trail of Tears" Cherokee in his book, *The Eastern Cherokees*:

> Most authorities concur in the opinion that both physically and temperamentally the older Cherokees made a most favorable impression. They delighted in athletics and excelled in endurance of intense cold. Well featured and of erect carriage, of moderately robust build, they were possessed of a superior and independent bearing. Although grave and steady in manner and disposition to the point of melancholy and slow and reserved in speech they were withal frank, cheerful, and humane, as well as honest and liberal. ... The men seem to be lighter ... in the face and more approaching the white type of feature. ... The long black hair of the women is in many cases rather attractive. The skin color is a variable brown tending toward lighter shading. ... the Mongolian [epicanthic] fold appears in the eyes of the females occasionally. ... Lighter eye coloring than is usual with dark races appears now and then.[437]

Gilbert mentions in his quote that the skin color of the Cherokee was "more approaching the white type." Although the Cherokee were one of the North American tribes that intermarried with whites more quickly than other tribes, they were of a lighter shade even before genetic admixture transpired. Gilbert also mentions that East Asian and Siberian physical traits (i.e. epicanthic eye folds) only occurred occasionally among the Cherokee women. This distinction is worthy of notice because of how common hooded eyes are currently among many Indian tribes all across the Americas. Because of the Cherokee's collective appearance, many of them appear to belong to the Lamanite remnant.

The Mandan

Spotted Bull, a Mandan (1908)

In 1830, with a strong desire to paint indigenous Americans, George Catlin, a pictorial historian and a frontiersman, began a journey into North American Indian territory.[438] Catlin ascended the Missouri River to Fort Union in 1832, where he spent several weeks among indigenous groups, which included the Mandan—a tribe relatively untouched by European civilization.[cw] Featured above is Spotted Bull, a Mandan. By appearance alone, Spotted Bull could easily be mistaken for a Semite with his long aquiline nose and high nasal bridge. He appears to have possessed the large stature of a Nephi, Moroni, or Mormon of the Book of Mormon. Many of Spotted Bull's people appear to have descended from Book of Mormon peoples.

---

cw   George Catlin visited eighteen tribes, including the Pawnee, Omaha, and Ponca in the south and the Mandan, Hidatsa, Cheyenne, Crow, Assiniboine, and Blackfeet to the north.

Photographs and descriptions of the Mandan of North America indicate they may have had Israelite blood. Records show that the Mandan were atypical American Indians who looked and behaved like civilized Semites. As a people, they were reminiscent of the Lamanites who became more righteous than the Nephites of old.[439]

For several months in 1832, George Catlin lived among the Mandan near present-day Bismarck, North Dakota.[440] What he found among the tribe surprised him. The "Mandans," "Mi-ah'-ta-nees," or "people on the bank,"[441] differed heavily from the more primitive tribes he had previously associated with in North America. "I am amongst a strange and peculiar people",[442] said Catlin, referring to the Mandan. Catlin was bemused by Mandan complexions and hair colors that differed greatly from the typical locks and skins of North American Indian tribes.[443] He was accustomed to Indians that were "dark copper-colored, with jet black hair[.]"[444] In 1833, Catlin commented on the uncommon physical attributes of the Mandan:

> [They are a] strange yet kind and hospitable people. . . . A stranger in the Mandan village is first struck with the different shades of complexion and various colors of hair which he sees in a crowd about him, and is at once almost disposed to explain that 'these are not Indians. . . . I am fully convinced that they have sprung from some other origin than that of the North American tribes, or that they are an amalgam of natives with some civilized race.'[445]

Catlin could tell they were not prototypical Indians of North America. He had been exposed to many North American Indian tribes on his travels and was, as he said, "fully convinced" that the Mandan stemmed from "some other origin than that of the North American tribes."[446] Catlin also provided the option that the Mandan were "an amalgam of natives with some civilized race."[447] One-fifth to one-sixth of the Mandan were "nearly white" in skin color with light blue eyes, according to Catlin, and yet lacked a known history of genetic admixture with European settlers.

Other individuals, besides Catlin, wrote of the impressive Mandan people. Brigadier General W. H. Jackson substantiated the words of Catlin in 1877 when he wrote about his experience with these atypical American Indians:

> In their personal appearance, prior to the ravages of the small-pox, they were not surpassed by any nation in the Northwest. The men were tall and well made, with regular features and a mild expression of countenance not usually seen among Indians. The complexion, also, was a shade lighter than that of other tribes, often approaching very near to some European nations [.]. . . Another peculiarity was that some of them had fair hair, and some gray or blue eyes, which are very rarely met with among other tribes. A majority of the women, particularly the young, were quite handsome, with fair complexions, and modest in their deportment.[448]

Both Catlin and Jackson were known to their contemporaries as upright and honest gentlemen. They each valued honesty, kindness, and hospitality. It was obvious to these good men that the Mandan were also a people of good character. Concerning the character traits of the Mandan, Catlin recalled:

> [A] better, more honest, hospitable and kind people, as a community, are not to be found in the world. No set of men that

ever I associated with have better hearts than the Mandans, and none are quicker to embrace and welcome a white man than they are—none will press him closer to his bosom, that the pulsation of his heart may be felt, than a Mandan; and no man in any country will keep his word and guard his honour more closely."[449]

The Mandan cared so deeply about decency and propriety,[450] according to Catlin, that the location in the river where their women bathed was protected by "sentinels, with their bows and arrows in hand to guard and protect this sacred ground from the approach of boys or men from any direction."[451] In 1877, Jackson corroborated Catlin's account concerning the character traits of the Mandan:

> They were noted for their virtue. This was regarded as an honorable and most valuable quality among the young women, and each year a ceremony was performed, in the presence of the whole village, at which time all the females who had preserved their virginity came forward, struck a post, and challenged the world to say aught derogatory of their character. In these palmy days of their prosperity much time and attention was given to dress, upon which they lavished much of their wealth. They were very fond of dances, games, races, and other manly athletic exercises. They were also a very devotional people, having many rites and ceremonies for propitiating the Great Spirit.[452]

While interacting with the Mandan, Catlin also noted that they had "advanced farther in the arts of manufacture" than any other Indian nation he had ever before seen.[453] The Mandan were highly civilized. They built lodges equipped with "more comforts and luxuries of life" than any other American Indian lodgings of the day.[454] This lifestyle appears to mirror the affluent lifestyles found throughout the Book of Mormon during times of prosperity.

The Sioux

Red Fly, a chief of the Oglala Sioux (1900)[455]

Featured above is Chief Red Fly of the Oglala Sioux. Chief Red Fly possessed a long and protracted nose, a high nasal bridge, and a rectangular-shaped face. Because of his Semitic appearance, it is possible that he belonged to the Lamanite remnant. He looked nothing like prototypical East Asians and Siberians. If one were to take away Red Fly's headdress and Native American clothing and replace them with Middle Eastern garb, he would fit in just as easily among those of the Middle East.

Standing Bear, a Lakota Sioux (1904)[456]

Standing Bear (featured above) was a prominent Lakota Sioux who possessed many physical traits common to Semites, especially members of the bloodline of Joseph of Egypt. As we can see from this photograph above, Standing Bear possessed a long and aquiline nose, a high nasal bridge, a narrow and rectangular-shaped face, and a long-headed (dolichocephalic) skull—all traits that indicate he may have belonged to the Lamanite remnant.

Known to the white man as Luther Standing Bear (1868 - 1939), he was a highly intelligent Oglala Lakota chief. Due to Standing Bear's level of intelligence, good character, assiduousness, high conscientiousness, agreeableness, and hard work, he accomplished much good in his lifetime. The contributions Standing Bear (and other great North American Indians) made to the world account for a partial fulfillment of the LDS doctrine contained in the *Doctrine and Covenants:* "Before the great day of the Lord shall come, ... the Lamanites shall blossom as the rose" (D&C 49:24).

Standing Bear was known as a Native American author, educator, philosopher, and actor of the 20th century. Standing Bear was educated at the Carlisle Indian Industrial School, an Indian boarding school in Pennsylvania. At Carlisle, he learned English and became adept at writing in the language. He authored a few notable books: *My People the Sioux* (1928), *Land of the Spotted Eagle* (1933), and *Stories of the Sioux* (1934).

Standing Bear became an actor and starred in a few western films as an old man. He was featured in *The Sante Fe Trail* (1930), *The Circle of Death* (1935), and *Texas Pioneers* (1932). His name even shows up online among members of the Actor's Guild of Hollywood. Standing Bear also became a dancer and horseback rider for Buffalo Bill's Wild West Show.[457] He was definitely a well-rounded individual with many talents.

Spotted Tail, a Sioux chief (1833-1881)[458]

Born in 1833, near Fort Laramie, Wyoming, Spotted Tail (featured above) was a Sioux warrior and a chief. Spotted Tail was a fierce fighter, but he strongly supported compromise to avoid conflict, including clashes with the white man. He skillfully employed diplomacy several times in his life, allowing his tribe to steer clear of warfare. Spotted Tail may have belonged to the Lamanite remnant because he possessed the common physical features of the House of Joseph of Egypt: a long and aquiline nose, a high nasal bridge, a rectangular-shaped face, and a long-headed (dolichocephalic) skull. He had the prototypical "look" of men from the literal House of Israel.

Ta-oyate-duta, aka "His People Are Red," a Mdewakanton Dakota Sioux (1858)

Featured above is "Ta-oyate-duta," which means "His People Are Red" in the Siouan language. His name references the term "red men"—a term used much more commonly in the 1800s than today. Like many other Sioux, Ta-oyate-duta possessed Semitic physical features. Because of his appearance and large physical stature, it is possible that Ta-oyate-duta was related to Book of Mormon peoples and certainly had the "look" of many Semites and Israelites.

Short Bull, an Oglala Sioux (1909)[459]

Short Bull, an Oglala Sioux, is featured above. He also possessed the common physical traits of Semites and the Lamanite remnant: a long and prominent nose, a high nasal bridge, and a rectangular-shaped face.

From the example of Short Bull and the many other examples in this chapter of Semitic-looking North American Indians, we can overtly recognize that many indigenous peoples from North America share many common physical and cultural traits with people who originated in the Middle East. The next chapter, *Chapter 9: Appearance and Genetics*, delves further into Semitic appearance of indigenous Americans by discussing how their Semitic-looking appearance connects with Middle Eastern genetic markers.

9

Appearance and Genetics

DNA links all life through the code and the more closely related the two species are physically, the more similar their codes. —**Richard Dawkins, British ethologist and evolutionary biologist (2009)**[460]

When several traits are analyzed at the same time, forensic anthropologists can classify a person's race with an accuracy of close to 100% based on only skeletal remains. —**Neven Sesardić, writer known for his work on heritability and race (2010)**[461]

[W]e testify that Joseph Smith was inspired when . . . he translated the Book of Mormon . . . and it is our firm belief that scientific investigation and discovery will confirm our testimony, rather than weaken or repudiate it. —**Joseph F. Smith, Latter-day Saint Prophet (1913)**[462]

Of course our genes, circumstances, and environments matter very much, and they shape us significantly. —**Elder Neal A. Maxwell, Latter-day Saint Apostle (1996)**[463]

Latter-day Saint leader and songwriter W. W. Phelps penned the words to an old LDS hymn entitled "O Stop and Tell Me, Red Man."[464] The first line of this hymn asks an important question: "O stop and tell me, Red Man, who are ye?" Who are the red men? From which ancient people did they originate?

Current scientists merge natives from North, Central, and South America into one group they call "Amerindians," and claim that everyone in this collective originated in East Asia and Siberia.[465] This aggregation of natives makes Semitic-looking indigenous groups of North America (especially tribes known as "red men") appear genetically similar to all the other Amerindians because their statistics are subsumed in the total data gathered across the continent. Semitic-looking North American Indians should not be clustered with most Amerindians because they often possess DNA that differs from the majority. This chapter discusses the genetic connection between Semitic-looking indigenous North Americans and Semitic groups.

Physical Anthropology

Physical anthropology is concerned with the comparative study of human variation and classification, especially through measurement and observation. For example, if something looks like a duck, swims like a duck, and quacks like a duck, then it is most likely a duck. The duck test is a humorous way to implement inductive reasoning in order to understand how physical anthropological methods can be used to track the appearance of indigenous North Americans over time. These methods flesh out how appearance and genetics are highly correlated. Richard Dawkins, British ethologist and evolutionary biologist, said in 2009 concerning the correlation, "DNA links all life through the code and the more closely related the two species are physically, the more similar their codes."[466] This means that once a person knows the physical traits that are associated with the DNA of a certain race, they can make an accurate determination of persons who belong to it simply by observing physical appearance alone.

One might consider the following example: If an average person were in a room full of people and asked to point out the only individual with Down Syndrome, ten times out of ten that person could do it. Why? These individuals are highly recognizable. When a person spots someone with Down Syndrome they are recognizing verifiable aspects of their DNA. It is like spying on that person's genetics. When one recognizes the physical appearance of Down Syndrome, an extra copy of chromosome 21 is identified.[467] And as one consistently spots individuals with Down Syndrome, the cogency of the strong link between physical appearance and DNA is corroborated.[cx]

Comparing identical twins is another way to demonstrate the strong link between genetics and appearance. When someone encounters identical twins, it is uncommon for that person to say to themselves, "they must have different genetics." It is most common for individuals to think the opposite; when one spots individuals who appear alike, relatedness is often automatically supposed. The whole animal kingdom has been organized and categorized in this same way by taxonomists. Taxonomists recognize similar appearances and

---

cx   Like other species, humans possess various physical traits that are highly correlated with specific dispositional traits. For example, in the Netherlands, Dutch researcher David Terburg and his team at Utrecht University found that higher levels of testosterone lead to social dominance, feelings of superiority, indefatigability, strength, and anger. Testosterone Affects Gaze Aversion From Angry Faces Outside of Conscious Awareness, David Terburg et. Al (2011). This is why individuals who are genetically predisposed for strong and bulky musculature tend to be more aggressive than thin, less muscular, individuals. People who are visibly muscular typically possess higher levels of testosterone—a hormone that promotes aggression automatically. Research suggests that testosterone can make people more poised for aggression, even if they're not feeling feisty. Testosterone promotes aggression automatically, Christie Nicholson, June 9, 2012. When an individual spots someone with a bulky musculature, it is easy for the observer to know that the person with big muscles possesses more testosterone than a skinny individual.

shared characteristics in organisms, humans included of course, and simply name groupings that already exist. By doing so, scientists help the world to understand how the various species are genetically similar or dissimilar.

Physical and genetic connections similar to these examples can be used to link select North American Indian tribes to their Semitic ancestors. In fact, these similarities can even be determined at the skeletal level. As Neven Sesardić, a writer known for his work on heritability and race, has explained, "When several traits are analyzed at the same time, forensic anthropologists can classify a person's race with an accuracy of close to 100% based on only skeletal remains."[468] Along these same lines, we will begin by observing similarities that exist between the cranial morphologies (shapes and sizes of skulls) of Semitic skulls and the skulls of ancient indigenous peoples of North America dating back to the time frame of the Book of Mormon.

Kihue, a full-blooded Huron & **Yuya**, vizier of Egypt (1390 BC)

Although we do not know his genetic profile, Kihue, a Huron, could easily have shared similar genetic markers with Yuya of Egypt (both featured above). Given how similar he looks to Yuya's mummy, this would make sense. As is apparent above, when pictures of these two men are placed side-by-side, a family resemblance emerges.[469]

Because Joseph of Egypt is a descendant of Shem, and his lineage is *the* main genealogy of Book of Mormon peoples, it should come as no surprise to Latter-day Saints that his bloodline has a distinct Semitic phenotype. Kihue's physical traits give us the impression that he could have belonged to the bloodline of the Lamanite remnant through the genetic lineage of Joseph of Egypt. Both he and Yuya had the "look" of Semites and Israelites. Yuya looked much more like a Semite than an Egyptian, and Kihue was much more Semitic than the prototypical American Indian of today. As we have discussed before, it is interesting to note that no known Mayans, Aztecs, or Olmecs look anything like Yuya, which most likely indicates that select indigenous North American Indians like Kihue are genetically unique and different from indigenous Mesoamericans and their East Asian ancestors.

Craniology and Genetics

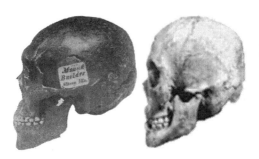

An Illinois **Hopewell skull** (1 AD-400 AD)[470]
& a prototypical **East Asian skull**

Cranial morphology differs among the various ethnicities in the world. Wide and flat-faced skulls with low nasal bridges are commonly found among East Asians and Siberians. An example of this type of skull is featured in this section on the right. Semitic skulls, on the other hand, are prototypically long-faced with narrow face shapes, and high nasal bridges. The Hopewell skull featured in this section on the left, which dates back to the time frame of the Book of Mormon, closely resembles a prototypical Semitic skull. It possesses every physical feature common to skulls of prototypical Semites. Individuals of East Asian and Siberian ancestry almost never possess craniums like this Hopewell skull.

The Hopewell people belonged to an ancient North American Indian culture that flourished along rivers in the northeastern and midwestern United States from 200 BC-500 AD (the Middle Woodland Period). The name for these people, "Hopewell," did not originate from the ancient culture, but comes from Mordecai Cloud Hopewell, the farmer whose land archaeologists were excavating in Ross County, Ohio.

Many modern-day Native Americans possess craniums like the flat-faced skull featured in this section (a skull type common among East Asian and Siberian peoples). The fact that a number of modern Native American skulls are so dissimilar to ancient ones of the Hopewell Culture indicate that many Native Americans do not belong to the Lamanite remnant. As Walter Neves, a Brazilian anthropologist, archaeologist and biologist from the University of São Paulo, Brazil, has attested, prototypical modern-day "Native Americans tend to exhibit a cranial morphology similar to late and modern Northern Asians (short and wide neurocrania; high, orthognatic and broad faces; and relatively high and narrow orbits and noses)."[471]

Skulls belonging to ancient and modern-day indigenous peoples from the Great Lakes region and the Great Plains area, however, tend to have skulls that are not short and wide. These groups of North American natives often possess cranial morphologies most similar to Caucasian Semites. Ancient North American skulls also most often match the skulls that appear like that of Yuya of Egypt's.

Cranial morphology categories work in the same way that taxonomy categories do (i.e. similar-looking skulls are categorized together just as similar-looking organisms are also put into groups). Similar-looking skulls share ethnic similarities as well as genetic similarities. In 1945, archaeological data on skulls was collected by archaeologists William S. Webb and Charles E. Snow. They studied ancient Ohio Hopewell skulls (200 BC-400 AD) dating back to the time of the Book of Mormon. Webb and Snow found that a great majority of them possessed long-headed (dolichocephalic) skulls, which is a skull type that is virtually non-existent among ancient and modern East Asians and Siberians.[472] Of the measured Ohio Hopewell skulls, 80% were dolichocephalic (long-headed). Only 10-15% were brachycephalic (short-headed) skulls, and the rest were unidentifiable.[473] By what appears to be no coincidence, the skull type of 8 out of every 10 Ohio Hopewell skulls that date back to the time of the Book of Mormon match the skull type of Yuya of Egypt.

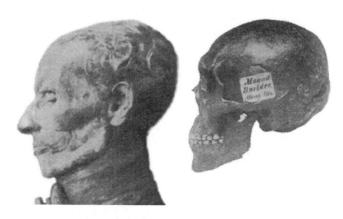

Yuya of Egypt (1390 BC)
& an Illinois **Hopewell skull** (1 AD-400 AD)[474]

Long-headed (dolichocephalic) skulls typically possess protruding occiputs (backs of the head), which are very similar to Yuya of Egypt's skull and also the Illinois Hopewell skull (both featured above). These skull types, which are extremely common among the ancient Hopewell, and certain American Indian tribes living near the Great Lakes region and the Great Plains, are virtually non-existent in Mesoamerica. Most skulls of ancient and modern indigenous peoples of Central America and South America are brachycephalic (short-headed), just like other groups in the world who visually appear East Asian and Siberian.

This similarity in cephalic index and skull type between Yuya and an ancient skull from an indigenous group of North America dating back to the time of the Book of Mormon further corroborates the position that certain indigenous North Americans belong to the Lamanite remnant.

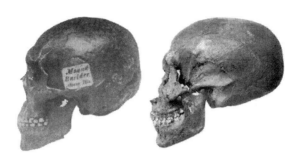

An Illinois **Hopewell skull** (1 AD-400 AD)⁴⁷⁵
& **Pharaoh Akhenaten**, an ancient Egyptian (d. perhaps in 1336 BC)

This Illinois Hopewell skull (featured above) that dates back to the time of the Book of Mormon is not only the same skull type as Yuya of Egypt, but it almost exactly matches the skull type of Yuya's descendant, Pharaoh Akhenaten. Pharaoh Akhenaten of Amarna was the grandson of Yuya and a unique pharaoh of ancient Egypt of the 18th Dynasty. It is uncanny how Akhenaten's skull, that dates back to approximately 1550-1292 BC, closely resembles the Illinois Hopewell skull. Both of these skulls are dolichocephalic (long-headed) and possess other physical similarities.

In the 1800s, archeologist Daniel G. Brinton gathered data on skulls from across the Americas. He measured 77 skulls of Algonquin Indians of North America and found that approximately 70% of them were dolichocephalic (long-headed).⁴⁷⁶ Among the Iroquois and Cherokees, he discovered that dolicocephalism heavily prevailed as well.⁴⁷⁷ The size of these North American Indian skulls is consistent with the size of individuals from the

Book of Mormon and the Ohio Hopewell, and the size of North American Indians who possess the genetic markers and the "look" of Israelites.[cy] Conversely, Brinton discovered that the skulls of the Maya of the Yucatan were almost all brachycephalic (relatively broad and short).[478] This finding is consistent with the measurements of the great majority of skulls from East Asia and Siberia.

---

cy Most Hopewell skulls from the time of the Book of Mormon match the size and look of many modern North American Indian tribe skulls. Both groups of skulls tended to be long-headed (dolichocephalic) and matched the size of Yuya of Egypt's skull and his descendant Akhenaten's skull. On the other hand, East Asian skulls tended to be brachycephalic (relatively broad and short). These types of skulls are predominantly found among the majority of indigenous Mesoamericans.

10

Genetic Markers of Semites, East Asians, and Siberians

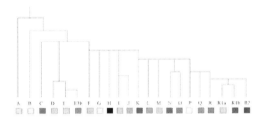

He [angel Moroni] told me of a sacred record which was written on plates of gold[.] . . . [H]e [also] said the Indians were the literal descendants of Abraham. —**Joseph Smith, Jr., Latter-day Saint Prophet (1835)**[479]

Native Americans may have a more complicated heritage than previously believed. . . . Nearly one-third of Native American genes come from west Eurasian people linked to the Middle East and Europe, rather than entirely from East Asians as previously thought, according to a newly sequenced genome. —**Brian Handwerk, writer, *National Geographic* (2013)**

When haplogroups and autosomal DNA markers of indigenous peoples of the Americas are used as guides, it is less of a struggle for Latter-day Saints to locate the Lamanite remnant. Haplogroups are sets of similar combinations of closely linked DNA sequences on one chromosome that are often inherited together, and autosomal DNA includes markers on autosomes (chromosomes that are not sex chromosomes) that indicate the percentage of an individual's autosomal DNA inherited from different parts of the world. Haplogroups and autosomal DNA markers can be analyzed with a saliva, blood, or a bone marrow sample taken from a living or deceased person. However, sometimes it is difficult to procure DNA data from very old bones since DNA often breaks down over time.

Autosomal DNA

Autosomal tests utilize DNA from the 22 pairs of autosomal chromosomes and therefore can be more thorough than mere mitochondrial DNA (mtDNA) and Y-chromosomal (Y-DNA) tests in determining the origin of American Indian DNA. Concerning the subject, *AncestryDNA* has attested:

> Unlike the Y-chromosome or mitochondrial DNA tests, the new AncestryDNA autosomal test looks at a much broader range of your DNA, which helps identify matches throughout your entire family tree—along both your paternal and maternal sides.[480]

That is to say, when mtDNA or Y-DNA tests (maternal and paternal DNA tests) are used to discover specific haplogroups of individuals, many ancestors are completely left out of a family tree. For this reason, autosomal tests are more comprehensive than mitochondrial and chromosomal DNA tests.

Using autosomal tests, we can determine genetic similarities. For example, with these tests we learn that Europeans, Middle Easterners, and Russians all share common Middle Eastern and Semitic ancestry. Tests confirm that the group with the shortest autosomal genetic distances from Ashkenazi Jews (Germans with Middle Eastern DNA) are the Italians, followed by the Greeks, Spaniards, Germans, Israeli Druze, Palestinians, Irish, and Russians—all groups that often possess Semitic genetic lineages.[481]

Autosomal tests can provide key information not found in other kinds of DNA tests. In 2014, Maanasa Raghavan, a researcher in paleogenomics, learned from autosomal tests that "[T]he Y chromosome of MA-1 [found in an ancient indigenous boy in North America] is basal to modern-day western Eurasians [which include Middle Easterners and Near Easterners] and near the root of most Native American lineages."[482] This autosomal evidence also indicates that MA-1 (Y-DNA) has no close affinity to east Asians.[483]

Although this data was collected from a bone that pre-dated the Book of Mormon by thousands of years, it is significant that geneticists discovered East Asia and Siberia were not the origin of all American Indians. This fact opens the door to the possibility of transatlantic voyages from the Old World to the New World. It also increases the plausibility that the stories contained in the Book of Mormon are true.

It just so happens that the largest concentrations of Semitic-looking individuals on the American continent with Middle Eastern autosomal DNA markers and haplogroups are found among indigenous North Americans. The most dense populations of these Semitic-looking groups center around the Great Lakes region and the Great Plains of North America.[cz]

Joseph, not necessarily Judah

Mainstream scholars and geneticists have made the assumption that the Book of Mormon is a book about Jews who sailed to the New World; a claim that has led to the unreasonable assertion that Book of Mormon peoples must possess prototypical Jewish DNA (i.e. haplogroups common among the tribe of Judah) for the book to receive consideration. By their reasoning, since prototypical Jewish DNA was not found in statistically significant amounts among native populations in the Americas, the Native Americans are not related to ancient Jews. The problem with this reasoning is that they are looking for descendants of Judah rather than descendants of Joseph of Egypt.[484]

By his own hand,[da] the Prophet Joseph Smith wrote in his journal that Angel Moroni had told him the American "[I]ndians were *literal* descendants of Abraham."[485] Chief Captain Moroni of the Book of Mormon claimed, "Behold, we are a remnant of . . . the seed of Joseph, whose coat was rent by his brethren into many pieces" (Alma 46:23).[486] Book of Mormon peoples predominantly belonged to the lineage of Joseph of Egypt (the bloodline of Abraham's great-grandson), which means they were mostly

---

cz If these indigenous North Americans did not arrive in the north from Mesoamerica or South America after the Book of Mormon ended, then the chances of North America being the primary setting of the Book of Mormon increase exponentially.

da Rather than use a scribe to write in his journal for him (which was very common for him to do), Joseph Smith actually wrote this sentence with his own hand in his journal.

Josephites, not Jews (from the bloodline of Judah).[db]

Although ancient Jewish DNA is probably included in bloodlines of Book of Mormon peoples, the Book of Mormon is mostly a narrative about the lives of Joseph of Egypt's descendants. According to the Book of Mormon, the Mulekites hailed from the Holy Land via the House of Judah before they created settlements such as Zarahemla in the New World. They could have had common Jewish DNA. However, the majority of the Book of Mormon is not about the Mulekites; it is overwhelmingly about the Nephites and the Lamanites, who were mostly from Joseph of Egypt's lineage. Lehi and his caravan—the group from which the Nephites and Lamanites originated—were "Jews," but only if the term "Jew" is used generically to refer to anyone belonging to the Hebrew religion of Abraham, Isaac, and Jacob. Most Nephites and Lamanites claimed ancestry from Joseph of Egypt's bloodline, not Judah's.

Mormon, the prophet who abridged the Book of Mormon record, differentiated the Jews from the Lamanites when he stated in the Book of Mormon: "The record of the people of Nephi, and also of the Lamanites—written to the *Lamanites,* who are a remnant of the house of Israel; and *also* to Jew and Gentile."[487] The Lamanites, Jews, and Gentiles are categorically separated by Mormon in the Introduction into three distinct groups. This is one of the reasons why the Lamanites and Jews are not necessarily

---

db   Latter-day Saint scholar Hugh Nibley implied in his works that Lehi's family was not necessarily genetically Jewish. Nibley said, "Nephi always speaks of 'the Jews who were at Jerusalem' (1 Nephi 2:13) with a curious detachment, and no one in 1 Nephi ever refers to them as 'the people' or 'our people' but always quite impersonally as 'the Jews.'" Hugh W. Nibley, "Lehi and the Arabs."

genetically matched. The Lamanite remnant mostly belong to the tribe of Joseph of Egypt and the Jews are largely from the tribe of Judah. Hence the Book of Mormon is not predominantly a record of ancient Jews, but of ancient Josephites who traveled to the New World in search of a promised land.[dc] This is an important distinction. Joseph Smith claimed on a few occasions that the Lamanite remnant were predominantly related to Judah's brother: Joseph of Egypt.[dd] In 1833, for example, Joseph Smith stated that "The Book of Mormon is a record[.] . . . By it, we learn that our . . . Indians, are descendants from that Joseph that was sold into Egypt."[488]

Despite the fact that the Book of Mormon, the Prophet Joseph Smith, and other prominent Latter-day Saints have explicitly claimed the Book of Mormon is predominantly about Joseph of Egypt's descendants, scholars and geneticists have compared common Jewish haplogroups (distinct groups of genetic markers) with common haplogroups belonging to indigenous peoples of the Americas. After comparing prototypical Jewish genetic lineages with those of American Indians, geneticists discovered that the majority of indigenous groups of the Americas lacked common Jewish haplogroups. These results should not necessarily be

---

dc "*To the Red Men of America* . . . You are a branch of the house of Israel. You are descended from the Jews, or, rather, more generally, from the tribe of *Joseph*, which Joseph was a great prophet and ruler in Egypt." Parley P. Pratt, Latter-day Saint Apostle (1851).

dd Early in the Book of Mormon, the prophet Lehi, attested, "I am a descendant of Joseph who was carried captive into Egypt. . . . [G]reat were the covenants of the Lord which he made unto Joseph." (2 Nephi 3:4). Nephi, son of Lehi, informs us that his father found out their genealogy from an ancient record they retrieved while in Jerusalem. Nephi states, "And it came to pass that my father, Lehi . . . found upon the plates of brass a genealogy of his fathers; wherefore he knew that he was a descendant of Joseph; yea, even that Joseph who was the son of Jacob [Israel], who was sold into Egypt, and who was preserved by the hand of the Lord, that he might preserve his father, Jacob, and all his household from perishing with famine" (1 Nephi 5:14).

surprising since they compared indigenous American DNA to the DNA of Judah's tribe (the Jews).

Latter-day Saint geneticists and academics should consider clamoring for Egypt to release Yuya of Egypt's official autosomal DNA and his paternal and maternal haplogroups. Once this information is released, Book of Mormon critics will have less empirical ammunition to use against Latter-day Saints. If Yuya's genetic lineage is out in the open, the plausibility of Joseph Smith's claim about the Book of Mormon could increase substantially.

Joseph and Judah may have had completely different haplogroups. Although Judah and Joseph were both related to Shem and share the same father, Israel, they were born to different mothers. Joseph of Egypt was born to Rachel, whereas Judah's mother was Leah. This could have resulted in their descendants possessing different haplogroups. Also, because of the normal rate of mutation in genetics, the two brother's paternal lineages (Y-DNA haplogroups) may differ among their descendants, thus making it appear as if their descendants are less related than they really are. DNA polymerase enzymes are known to make mistakes at a rate of approximately 1 per every 100,000 nucleotides. This rate may seem insignificant until you consider how much DNA a cell contains. "In humans, with our 6 billion base pairs in each diploid cell, that would amount to about 120,000 mistakes every time a cell divides!"[489] Thus, the tribes of Judah and Joseph have likely drifted apart genetically due to mutations.

That is why research that supposedly proves the Book of Mormon false is debatable. Rather than using common haplogroups of Judah's descendants, geneticists should consider comparing common Semitic autosomal DNA and haplogroups (which are not limited to common Jewish haplogroups) with indigenous peoples of select parts of North America. DNA samples from *every* indigenous group from the entire American continent does not need to be included in the comparison. A new picture of the genetics of North American Indians will emerge if geneticists compare the *right* autosomal DNA and haplogroups of Semites with that of individuals within the correct region of North America where the actual Lamanite remnant live(d).

Another explanation for this possible drift is that admixture between Israelite paternal lineages and Gentile nations could have diluted the DNA of Judah's and Joseph's descendants so that some of them have different paternal lineages. Although the two groups are technically less "related" if they belong to different haplogroups, this does not mean that both groups cannot be Israelites. The two paternal lineages of Haplogroup I and Haplogroup J derive from one single Y-DNA haplogroup: Haplogroup IJ.[de] This haplogroup split into two separate haplogroups, which means that mutations occurred after Haplogroup IJ entered the genetic landscape. Haplogroup J is currently recognized as the main "Jewish" paternal lineage, even though haplogroups I and J both originate from the Middle Eastern Haplogroup IJ.[490] Both haplogroups I and J,[df] which can be found in

---

[de] Haplogroups: sets of similar combinations of closely linked DNA sequences on one chromosome that are often inherited together.

[df] Even though Haplogroups I and J come from the same paternal lineage, haplogroup IJ (a Y-DNA lineage thought by scholars to originate in the Middle East), it is claimed that haplogroup I (Y-DNA) is a European paternal lineage. What they fail to mention is that it has become a European paternal lineage, when in fact it originated in the Middle East. The mutation may have occurred in Europe, but the mutation's location does not make the haplogroup less Semitic in origin. Haplogroup I (Y-DNA) and haplogroup J (Y-DNA) are equally Semitic.

statistically significant amounts in certain parts of western Eurasia, are Middle Eastern in origin. This fact is rarely, if ever, discussed by scientists in scientific journals. A number of scientists state that Haplogroup J is a Semitic paternal lineage, and Haplogroup I is a European paternal lineage, even though Haplogroup I is actually Middle Eastern in origin.[491]

Haplogroup J, which is considered by many scholars and academics to be the prototypical "Jewish" male haplogroup, was compared with Native American DNA by geneticists and other scientists.[492] When theses two groups failed to indicate statistically significant genetic similarities according to their misguided comparison, the media took the results and claimed that the research supposedly "disproved" the Book of Mormon.

One of the main reasons why this comparison is faulty is because the highest percentages of 'J' are not found among Jews. The highest percentages of 'J' are actually found among the Ingush of the Caucasus who mostly inhabit the Russian republic of Ingushetia. Of the 143 Ingush Caucasians who had their DNA tested, 91.6% belonged to paternal lineage Haplogroup J.[493] The average Jew does not necessarily belong to 'J,' but members of the Jewish priestly class (the Ashkenazi Cohanim) do have extremely high percentages. Of the Ashkenazi Cohanim who were tested, 87% were Haplogroup J.[494] Of the 62 Yemenis of the Arabian peninsula, 82.1% belonged to the paternal lineage Haplogroup J.[495] Hence Haplogroup J should not be the determining factor when studying Native American DNA and determining if the Book of Mormon is true.

Haplogroup I,[dg] considered by many scientists to be a typical European paternal lineage is predominantly found among Bosnian Croats (73%),[496] Bosniaks (58.5%),[497] Swedes (42%),[498] and Norwegians (40%)—all European groups that often have Semitic ancestry.[499] Because of their genetic origins, many of these European groups physically appear very similar to the Ingush Caucasians, Ashkenazi Cohanim, and Yemenis of the Arabian peninsula. This makes sense because everyone from Haplogroup I came from Haplogroup IJ, which originated in the Middle East.

Some of Joseph of Egypt's descendants may belong to paternal lineage Haplogroup J, but more of them probably belong to Haplogroup I—a predominantly Scandinavian paternal lineage—and/or Haplogroup R1 (which is a paternal lineage commonly found among Ashkenazi Jews, other Europeans, and also American Indians). Haplogroups I and R1 are no less Semitic than Haplogroup J, and yet current scholars claim that these paternal lineages are merely "European." A better way to label them would be that they are Semitic European paternal lineages. Better still, they could be considered unrecognized Semitic haplogroups, possessed by people who were scattered after leaving the Holy Land and migrating to Europe.

American Indian Paternal DNA: R1 & Q

Certain DNA markers have been discovered in high percentages among select indigenous North American Indian tribes that are also common among Semites. For example, several of the

---

dg Haplogroup I (Y-DNA) remains unrecognized by scholars as a common Semitic haplogroup, even though it is equally Semitic as haplogroup J (Y-DNA).

Ojibwe from North America had their DNA tested and nearly 80% belonged to Haplogroup R1—a paternal lineage that commonly belongs to Semites.[500] It is data like this that has caused scholars to rethink what has previously been a virtually impregnable stance about which ancient groups could have populated the Americas. Until recently, scholars have only accepted the idea that the Americas were originally populated by East Asians and Siberians who had crossed the Bering land bridge into the New World. Now consideration is starting to be given for other explanations.

In 2008, anthropologist Ripan Singh Malhi stated that any non-Siberian and East Asian DNA presence among American Indian populations was a result of recent European admixture. Particularly, Malhi claimed that American Indians had mated with early European explorers and settlers, and the children that came from the admixture inherited common "European" haplogroups. Recent studies indicate that Malhi and his team were incorrect in their assertion.

New evidence suggests a more complex genome for many North American Indians than was once supposed. What mainstream geneticists once thought were recent "European" DNA markers among indigenous North American Indians are actually much older than they ever thought possible. Much of this collective genome came from a region that includes the Near East and the Middle East.

In a 2014 study, ancient DNA expert Morten Rasmussen and his team discovered that a third of the collective genome of North American Indians did not originate from Siberia or East Asia.[501]

Furthermore, the study noted that this portion of the collective North American Indian genome did not derive from admixture with Europeans who had arrived in the New World beginning with John Cabot—an Italian-born English explorer—who first set foot in North America in 1497. This evidence supports a main point of this book: statistically significant amounts of North American Indians are actually related to ancient Semitic peoples from the Middle East and Europe.[502]

The paternal lineage Haplogroup R1 has been found among indigenous Mesoamericans, but only in insignificant amounts. Almost all indigenous Mesoamerican males possess paternal lineage Haplogroup Q, which stems from Siberia and East Asia.[dh] Of the 71 modern-day Maya of Mesoamerica whose DNA was tested, only 12.7% belonged to Haplogroup R1.[503] Among the Zapotec of Mexico whose DNA was tested, 6% were Haplogroup R1.[504] Of the Uto-Aztecan-speaking Nahua of Mexico whose DNA was tested, 0% were Haplogroup R1.[505]

Because these figures fail to demonstrate a significant presence of 'R1' or any other common Semitic haplogroup among indigenous Mesoamericans, it appears that this group of people has only a small connection to what could conceivably be Book of Mormon lineages. Conversely, high percentages of Haplogroup R1 (Y-DNA) found among the Ojibwe and other North American tribes tie these groups to bloodlines common among Israelites.

---

dh Common paternal lineages (Y-chromosome DNA) among Siberians of North Asia include Haplogroups C, N, O3, P, Q, R1a, R1b, and R1. For more information on common maternal (mtDNA) lineages among East Asians and Siberians, refer to *Appendix 1: Genetic Markers Expansion* of this book.

Haplogroup R1 is the only significant paternal lineage found in the Americas that is commonly possessed by Israelites around the world. It is the only set of paternal genetic markers that appears to increase the plausibility that the Book of Mormon is what the Prophet Joseph Smith claimed it was—an ancient record of Israelites who inhabited the New World. Because indigenous North Americans have statistically significant amounts of Haplogroup R1, and those who belong to it often have common Semitic physical features, 'R1' is a noteworthy haplogroup.

Despite the fact that negligible amounts of Semitic DNA exist among indigenous Mesoamericans, and most Mesoamericans look East Asian and Siberian, certain prominent Latter-day Saint scholars continue to purport that many indigenous peoples currently living in Southern Mexico and Guatemala are related to Nephites, Lamanites, Mulekites, and Jaredites. Even some Latter-day Saint patriarchs in Mesoamerica have convinced themselves that Lehi's literal seed, through Joseph of Egypt, is prevalent in Mexico and Guatemala,[di] despite the fact that almost all indigenous Mesoamericans appear East Asian, and possess DNA common to East Asia and Siberia.

Apostle Mark E. Petersen stated in his book *Joseph of Egypt*, "In Mexico, patriarchs report that 75 percent of the blessings indicate the lineage of Manasseh and 25 percent, of Ephraim. Only two or three blessings out of hundreds given mention any other tribe."[506] It is unclear why many of these blessings were given to

---

di Some Latter-day Saint patriarchs in Mesoamerica believe Lehi's literal seed through Joseph of Egypt is prevalent in Mexico and Guatemala (unless these patriarchs weren't declaring the literal lineage of Joseph of Egypt but were instead declaring that individuals were adopted into the tribe of Joseph of Egypt).

indigenous Mexicans living in regions that Mesoamerican enthusiasts have claimed are actual Book of Mormon lands. And yet, even if these blessings were only given to indigenous Mesoamericans, 3 out of 4 Mexicans were assigned to Manasseh, which means that the patriarchs of the area believed Mexicans were mostly from the Lamanite remnant of the Book of Mormon.

Of all the human depictions in ancient Mesoamerican artifacts dating back to the time of the Book of Mormon, only a trifling amount even remotely resemble Semites of Joseph of Egypt's lineage. Furthermore, the ancient remains of indigenous Mesoamericans dating back to Book of Mormon times mostly possess East Asian or Siberian genetics. Not many logically consistent reasons exist for how indigenous Mexicans and Guatemalans could literally belong to Joseph of Egypt's lineage. Of course, indigenous Mesoamericans can always be adopted into the House of Israel, but when it comes to literal bloodlines, these groups infrequently appear to have descended from Joseph of Egypt.

Approximately 1,600 years have passed since the Book of Mormon ended. During that time it is possible that some of the main Lamanite remnant migrated southward and admixed with indigenous Mesoamericans and even indigenous South Americans, which would make them a part of the genetic Lamanite remnant. That being said, DNA evidence suggests that if admixture did occur it was rare, and in some cases almost non-existent, like in the example of the ancient Maya of Xcaret in Mexico whose mitochondrial DNA showed no (0%) Semitic genetic markers.[507] If large percentages of indigenous Mesoamericans actually did belong to the literal Lamanite remnant, they would look more like other members of Joseph of Egypt's lineage and also possess more Semitic DNA than they currently do.

Examples of Paternal Lineage 'R1'

Chief Standing Bear, a Sioux[508]
& a **Bedouin** Arab from the Dead Sea area (Jordan)[509]

Featured above is Chief Standing Bear, an Oglala Sioux, and a Bedouin Arab from the Dead Sea region of Jordan. Both men possessed thin and long noses with high nasal bridges, and long and rectangular-shaped faces—all physical traits common to Semitic males. It is probable that Chief Standing Bear and this Bedouin Arab belonged to paternal lineage Haplogroup R1 because of how common the haplogroup is among the Sioux and Arabs from the Dead Sea region of Jordan.

Half of the Sioux whose DNA was tested belonged to paternal lineage Haplogroup R1.[510] Of the 45 Arab men living in the Dead Sea area of the Near East who had their DNA tested, 44.44% belonged to paternal lineage Haplogroup R1.[511]

Hole in the Sky, an Ojibwe (1860)[512]

Featured above is Hole in the Sky, an Ojibwe chief of North America.[513] It is highly likely that Hole in the Sky belonged to Haplogroup R1 (Y-DNA) because of his appearance and the high prevalence of paternal lineage 'R1' among the Ojibwe. Hole in the Sky possessed a thin and long nose with a high nasal bridge, and a long and rectangular-shaped face—all facial features that indicate he was more Middle-Eastern than East Asian and Siberian.

Of the Ojibwe whose DNA was tested, nearly 8 out of 10 belonged to Haplogroup R1 (Y-DNA).[514] To date, this percentage of Haplogroup R1 (Y-DNA) is the highest in the world. Interestingly, the highest percentage of this haplogroup located outside of North America is found in the Middle East.

Someone Traveling, an Ojibwe (1901)[515]

"Someone Traveling," an Ojibwe man, is featured above. His long nose with a high nasal bridge and long and rectangular-shaped face indicate that he most likely belonged to Haplogroup R1 (Y-DNA): the most common haplogroup among the Lamanite remnant. Since paternal lineage 'R1' was found in nearly 8 out of 10 Ojibwe who had their DNA tested, it is highly likely that "Someone Traveling" also belonged to 'R1.'[516] Someone Traveling's Semitic-looking physical features also denote that he may have been related to Joseph of Egypt. He certainly had the prototypical "look" of those who belong to the genetic lineage of Joseph of Egypt.

Mico Chlucco the Long Warrior, or "King of the Seminoles" (1792)[517]

Featured above is Mico Chlucco, a Seminole king of the Creek Confederacy. The Creek were a Muskogean-speaking group of Florida.[518] Of the Seminole men who had their DNA tested, 50% belonged to paternal lineage Haplogroup R1.[519] If this percentage held true in Mico Chlucco's day, it is possible that he belonged to paternal lineage 'R1.' Mico Chlucco possessed a high nasal bridge, a long and aquiline nose, a rectangular-shaped face, and a long-headed (dolichocephalic) skull: all common traits of those who belong to 'R1' and Semites of the House of Israel via Joseph of Egypt's lineage. Because of his Semitic-looking appearance, Mico Chlucco may have belonged to the Lamanite remnant.

A **Passamaquoddy** man, Algonquin peoples (1875)

A Passamaquoddy man of North America who lived in the 1800s is featured above. The Passamaquoddy are an indigenous group that belong to the Algonquin peoples. This man featured above possessed an aquiline nose, a long and rectangular-shaped face, and a high nasal bridge—all traits common to Semitic peoples of the House of Israel through the lineage of Joseph of Egypt.

Of the 155 Algonquin Indians of Northeast North America who had their DNA tested, 38.1% belonged to paternal lineage Haplogroup R1.[520] It is possible that this Passamaquoddy man belonged to Haplogroup R1 since nearly 4 out of 10 Algonquin men whose DNA was tested belonged to the haplogroup. This Passamaquoddy man's common Semitic physical features also signify that he may have belonged to the Lamanite remnant.

Two North American Indians from the **Dogrib** (Tłı̨chǫ) tribe

Two Dogrib Indians (featured above) may have belonged to paternal lineage Haplogroup R1, since 40% of the Dogrib males whose DNA was tested belonged to this particular haplogroup.[521] The common Semitic physical features of these Dogrib Indians suggest they may have belonged to 'R1' and also the Lamanite remnant.

A **Narragansett-Mohegan** man, Algonquin peoples[522]

Featured above is a bearded Narragansett-Mohegan man of the Algonquin peoples—an indigenous group of North America. The visage of this man is Semitic-looking, which is a common phenotypic pattern among Algonquin tribes of North America. This man possessed a long aquiline nose with a high nasal bridge and a long and rectangular-shaped face. Of the 155 Algonquin Indians of Northeast North America whose DNA was tested, 38.1% belonged to Haplogroup R1.[523] Since almost 40% of the Narragansett-Mohegan possessed these genetic markers,[524] and because this man had the common look of Semites, it is possible he may have belonged to Haplogroup R1 and also the main Lamanite remnant.

A Cree man, **Algonquin** peoples (1875)[525]

An Algonquin man of the Cree tribe is featured above. He possessed common Semitic physical characteristics including a long nose, a high nasal bridge, and a long and rectangular-shaped face. Since so many Algonquin men who had their DNA tested were paternal lineage Haplogroup R1, and because the Cree man featured above looks like others who possess 'R1,' it is possible that he belonged to Haplogroup R1. Also, by appearance alone, this Cree man looks like he may have belonged to the main Lamanite remnant.

Obtossaway, an Ojibwe (1903)

Obtossaway (featured above), another Ojibwe man, most likely belonged to paternal lineage Haplogroup R1 due to his Semitic appearance and the high prevalence of 'R1' among Ojibwe males. From Obtossaway's photograph, we can see his long and aquiline nose, long and rectangular-shaped face, and high nasal bridge—all traits common among Semites and the Lamanite remnant. Obtossaway's appearance, and possible paternal genetic lineage, make him a likely candidate for the Lamanite remnant: he looks like an individual who belonged to the House of Israel through Joseph of Egypt's lineage.

Wizi, a Yanktonai Sioux (1884)[526]

Wizi (featured above) was a Yanktonai Sioux of North America.[527] He lived in the 1800s and possessed unmistakable physical features uncommon to the majority of East Asians and Siberians. As we can see from this photograph, he possessed a long and aquiline nose, a high nasal bridge, a long and rectangular-shaped face, and a long-headed (dolichocephalic) skull. He appears to have been a literal descendant of Joseph who was sold into Egypt because he looks so much like others from the seed of Joseph of Egypt. Because of his Semitic-looking appearance, it is likely that Wizi possessed Middle-Eastern DNA rather than genetic markers common to other ethnicities. Fifty percent of Sioux men whose DNA was tested belonged to paternal lineage Haplogroup R1. Because Wizi possessed common physical characteristics of Semites and Israelites, it is likely he belonged to Haplogroup R1.[528]

Examples of Paternal Lineage 'Q'

A **Siberian Ket** man & a **Navajo** medicine man

Members of Haplogroup Q are often easy to differentiate between other paternal lineages, such as Haplogroup R1, because members of Haplogroup Q tend to appear East Asian and Siberian. In contrast to Haplogroup Q, it is common for members of paternal lineage Haplogroup R1 to look Semitic. Featured above is a prototypical-looking Siberian Ket and also a North American Navajo man. Both men follow the common physical patterns of Haplogroup Q: jet black hair, high cheekbones, flat face, flat and wide nose, and hooded eyes (eyes with epicanthic eye folds).[529]

Because of how common it is for members of paternal lineage Haplogroup Q to appear prototypically East Asian, it should come as no surprise that the specific subgroup Haplogroup Q1a.[530] is mostly found among Mongolians, Siberians, and the American Indians who all predominantly appear East Asian.[531]

Of all those who belong to Haplogroup Q, the highest percentage is found among the Siberian Ket of Northern Asia. Of the 48 Siberian Ket men whose DNA was tested, 93.7% were Haplogroup Q.[532] Since virtually all Kets belong to the same lineage, it is safe to say that the Ket man featured in this section belongs to Haplogroup Q. He possessed the common physical patterns of Haplogroup Q: jet black hair, high cheekbones, flat face, flat and wide nose, and hooded eyes (epicanthic eye folds).[533]

Some indigenous North Americans, such as the Navajo, follow similar DNA patterns of the Siberian Ket. Of the 78 Navajo men of the Southwest United States who had their DNA tested, 92.3% belonged to Haplogroup Q.[534] Not surprisingly, the Navajo man featured in this section possessed jet black hair, high cheekbones, a flat face, a flat and wide nose, and hooded eyes (epicanthic eye folds).[535]

Inuit people of Alaska (Eskimos)

From this photograph above, it is not difficult to see how most, if not all, of these prototypical Inuits of Alaska possessed common East Asian and Siberian phenotypes: jet black hair, flat faces, flat and wide noses, and hooded eyes (eyes with epicanthic eye folds). This appearance is most common among indigenous Americans who belong to paternal lineage Haplogroup Q. Statistics corroborate this claim. Of the 60 Inuit (Eskimos) who had their DNA tested, 80% belonged to Haplogroup Q.[536] The percentage of Inuit men who tested as Haplogroup Q closely resembles the Siberian Ket and the North American Navajo, which indicates common paternal origin. Hence, the Inuit belong to East Asia and Siberia, not the Middle East, and have virtually no chance of being a part of the literal bloodline of Lehi and his traveling group who came from Jerusalem.

A **Lacandon Maya** leader, Mexico

Featured above is a modern-day Lacandon Maya leader from Mexico. He possesses common physical characteristics of the Maya: jet black hair, flat face, flat and often wide nose, and hooded eyes. His visage matches many of the phenotypes of the Siberian Ket, North American Navajo, and Alaskan Inuit. By no coincidence, these groups, especially the Inuit and Maya, share very similar paternal lineages. Of the 71 modern-day Maya of Mesoamerica whose paternal lineage (Y-chromosome DNA) was tested, 87.2% belonged to Q.[537] It is not mere happenstance that Inuits and Maya both look similar and share almost identical paternal lineages. These statistics clearly reflect the obvious fact that these groups bear virtually no visible resemblance to the previously discussed prototypical Semitic peoples who belong to Joseph of Egypt's lineage.

An **Apache** man & **Ndee Sangochonh**, also an **Apache** man

Featured above are two Apache men with common East Asian and Siberian physical traits. In a genetic study of 96 Apache Indian men of North America, 78.1% of the men who were tested belonged to the paternal lineage Haplogroup Q.[538] Not surprisingly, Apaches follow the common phenotypic pattern of the Siberian Ket, North American Navajo, Alaskan Inuit, and Mesoamerican Maya: jet black hair, high cheekbones, flat faces, flat and wide noses, and eyes with epicanthic eye folds. Featured above are two Apache men who possess these common East Asian and Siberian physical traits. Since 8 out of 10 Apache men who were tested belong to Haplogroup Q, and due to the fact that they appear East Asian and Siberian, it is highly probable that these Apache men featured above belong to Haplogroup Q. Their visages appear nothing like the majority of Semitic peoples, especially ones who belong to the tribe of Joseph of Egypt, which indicates that they are most likely not members of the Lamanite remnant.

Siberian Selkup people

The photograph above features a group of Siberian Selkup people next to a common North Asian dwelling. One cannot help but think of a Native American tipi when viewing this picture. The Selkups share the same familiar look of most indigenous Americans: jet black hair, high cheekbones, flat faces, flat and wide noses, and epicanthic eye folds. It is not too difficult to imagine how they would be related to Native Americans, and in fact, the DNA of these two groups match. Of the 131 Selkup men whose DNA was tested, nearly 70% of them belonged to Haplogroup Q.[539]

From the many photographs and ample DNA evidence given here, one can see that the majority of male Ket, Navajo, Inuit, Maya, Apache, and Selkup all share similarities both physically and genetically. These similarities demonstrate the power of the connection between phenotype and genotype; if two groups look alike, the likelihood that they share similar DNA and haplogroups increases dramatically.

Semites, Mesoamericans, and North Americans

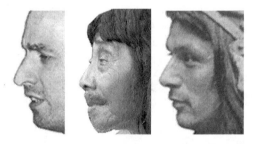

A **Jew**,[540] a **Mayan**, Mexico[541] & an **Iroquois**, Great Lakes region of North America[542]

The common physical traits of the main Lamanite remnant become fairly evident when a known Semite, a prototypical

Mayan, and a characteristic North American Indian from the Great Lakes region are compared with one another. Featured above are a Jewish man with typical Jewish genetics, a Mayan man from Mexico, and an Iroquois man. When physical comparisons are made, the resemblance between the genetically related groups becomes unequivocal. The Mayan man featured above does not resemble the Jewish man in any way. This indigenous Mesoamerican looks more like an East Asian or a Siberian than a Semite.[dj] On the other hand, the Iroquois man's profile is very similar to the Jewish man's profile, which implies likely genetic similarities.

This Mayan man possesses facial traits which are those most typically representative of the indigenous groups across the American continent. Although very uncommon among the very specific groups of indigenous North Americans mentioned previously in this chapter, these East Asian and Siberian-looking facial traits are so prevalent in the Americas that it is no wonder geneticists who are cognizant of LDS beliefs are puzzled by the assertion made by Joseph Smith that the Book of Mormon was a book about ancient Israelites in the New World. The prevalence of this non-Semitic phenotype in the Americas is difficult to ignore.

However, as demonstrated, a number of North American tribes from the Great Lakes and Great Plains regions do not fit in with the majority of indigenous Americans, and DNA tests confirm this assertion. This uniqueness of certain groups of

---

dj   To learn more about this strong connection between the common appearance of indigenous Mesoamericans and the appearance of East Asians, refer to *Appendix 1: Genetic Markers Expansion*.

indigenous North Americans points to a Middle Eastern origin and possibly Joseph of Egypt's genetic lineage, leading one to conclude that they may very well be the main remnant of the Lamanites of old.

11

Lehi and Arabs

Lehi possesses in a high degree the traits and characteristics of the model [Arab] *sheikh* of the desert. He is generous, noble . . . devout, and visionary, and he possesses a wonderful capacity for eloquence and dreams. As to the dreams, when the Arabs wander, they feel they must be guided by dreams, and their *sheikhs* are often gifted dreamers. —**Hugh Nibley, Latter-day Saint scholar (1988)**[543]

[T]he Hebrew origin of the native American races is fundamental as testimony to the truth of the Book of Mormon. . . . I next call attention to the evidences of the Hebrew origin of the native Americans . . . The opinion that the [American Indians] are of Hebrew origin is supported by similarities in character, dress, religion, physical peculiarities, condition, and customs.
—**BH Roberts, Latter-day Saint historian and General Authority (1895)**[544]

Lehi fled the great city of Jerusalem in search of a promised land after he had received a prophetic vision of its destruction.[545] To make his escape, it is likely that he traveled along the Arabian Peninsula with his caravan, which was comprised of his own family, Ishmael's family, and Zoram—a former servant of Laban.[546] The Book of Mormon states that before Lehi departed

Jerusalem, he learned of his genealogy from the Brass Plates—an important record that his sons had commandeered from Laban. Lehi discovered that he belonged to the seed of Joseph of Egypt through the tribe of Manasseh—a tribe with ancient ties to Arabians. Concerning Lehi's genetic bloodline, Latter-day Saint scholar Hugh Nibley once remarked:

> [O]f all the tribes of Israel, Manasseh was the one which lived farthest out in the desert, came into the most frequent contact with the Arabs, intermarried with them most frequently, and at the same time had the closest traditional bonds with Egypt.[547]

Furthermore, when Hugh Nibley taught at Brigham Young University, he asserted in one of his courses that "'Ishmael' [of Lehi's caravan] you can be sure is [an] Arab."[548] Ishmael is an Arabic name and the namesake of Abraham's second son, which could give it strong Arabian ties.

Of all the indigenous peoples of the Americas, indigenous North Americans appear to most closely resemble Arabs from the genetic lineage of Joseph of Egypt in physical characteristics and temperament traits. In 1774, Thomas Harmer, a minister and author, recognized this physical likeness between these two seemingly disparate groups. Harmer remarked concerning the similarities, "[T]here is some resemblance between the Arabs and the Indians of North America."[549]

What follows is a selection of some possible North American descendants of Lehi compared with a number of Arabian men and women. The illustrations in this chapter demonstrate the obvious resemblance between the two groups and further support the claim that these groups of indigenous North Americans are likely part of the main Lamanite remnant.

Does Everything, a Crow[550] & **Edward Wadie Saïd**, a Palestinian Arab[551]

Does Everything, an Apsáalooke (Crow) Indian, and Edward Wadie Saïd, a Palestinian Arab born in the city of Jerusalem in Mandatory Palestine, are featured above. As we can see from their photographs, these men both possessed prototypical physical features common to the Middle East and surrounding areas. Their noses were prominent, long and thin; their nasal bridges were high; and their faces were long and rectangular-shaped. Doubtless these are the prototypical faces of Semites.

Many Palestinians like Saïd share the same groups of genetic markers as Semites, especially Jews, because they come from similar family trees.[552] Although Crow Indians do not often possess common Semitic genetic lineages, a number of them do. Because of his Semitic-looking physical characteristics, Does Everything appears to have belonged to the main Lamanite remnant.

A **Sabaean woman** of Southern Arabia⁵⁵³ & **Does Everything**, a Crow⁵⁵⁴

Above, Does Everything is featured next to a Yemenite Sabaean of Southern Arabia. Sabaeans of Southern Arabia possess common physical similarities with some Crow Indians. Arabs of Yemen, especially Sabaeans, are almost all Semitic in a physical and genetic sense.⁵⁵⁵ The South Arabian woman featured above and Does Everything share Semitic facial features including similar prominent long and thin noses, high nasal bridges, and long and narrow faces. Even the necklace styles of these two are uncannily similar.

Sabaeans often possess genetic markers[dk] and physical traits of Semites, especially ones common to Jews. Does Everything and this Sabaean woman possessed physical traits that are virtually non-existent among East Asians and Siberians.

---
dk   Haplogroups are groups of genetic markers.

An ancient Southern Arabian **Sabaean** coin (250 BC-201 BC)[556]

Featured above is a depiction of a Semitic man on an ancient Sabaean coin that dates back to the late third century BC. It is clear from the facial features on this artifact that this man had a long nose with a high nasal bridge: traits common among individuals from the Middle East. If one did not know that the coin depicted an ancient Sabaean, the face on the coin could easily pass for an indigenous North American from the Great Lakes or Great Plains regions—all it needs is a feathered headdress!

Ancient Sabaeans from Southern Arabia often belong to Semitic bloodlines. Because of their family trees, many of them possessed similar physical traits common to Semites (like the Semite featured above on the ancient Sabaean coin). When a comparison is made between ancient and modern Sabaeans, we can observe that the Semitic facial features of the ancients were passed on to their modern-day descendants.

Nabataean Kingdom coin, **Aretas IV** (9 BC-40 AD)[557]
& a **US Indian** $5 coin

Nabataeans physically resemble Sabaeans. The Nabataeans were a group of Semitic people who had established themselves in the region around Petra (or Reqem, as it was known to them), but they still maintained a largely nomadic lifestyle, traveling seasonally across the desert with their tents and herds to find water and fresh pasture.[558]

Select groups of indigenous North Americans look like Sabaeans and Nabataeans. This should come as no surprise since all three groups are Semitic in origin. Featured above is a Nabataean coin displaying the profile of King Aretas IV (9 AD-40 AD) and a North American Indian on a $5 gold US minted coin. Both King Aretas IV and the US American Indian had high nasal bridges, long and thin noses, and—what appear to be—long and rectangular-shaped faces.

An **Arab woman** of North Africa (early 1900s)
& a **Salish** woman (1900)

A number of Salish Indians of North America also resemble Arabs.[dl] An Arab woman of North Africa and a Salish woman are featured above.[559] Both women possess characteristically Semitic facial features: high nasal bridges, and long and thin noses. The clothing of these women also denotes an Old World style, despite the fact that current scholars do not tie indigenous North Americans to the Old World.[dm] Clearly, these women possess a look that is neither from East Asia, Siberia, or Mesoamerica.

---

dl   The Salish Indian tribe were incorrectly called the Flathead Indians by the first Europeans who came in contact with them.

dm   To learn more about New World connections to the Old World, refer to *Appendix 4: Old World Symbols in the New World.*

Nedshib Huri, a Lebanese Arab (1911) & **American Horse**, a Sioux (1840-1908)

Nedshib Huri, an Arab from Shuafat, Lebanon, and Wašíčuŋ Tȟašúŋke (AKA American Horse),[dn] a famous Oglala Lakota Sioux chief, are featured above. These two men had many phenotypic similarities common among Middle Easterners, especially Semites. Both men possessed high nasal bridges, large noses, and long rectangular-shaped faces. Their cheekbones are also similar, and their eyes are almost exact replicas of one another.

The Sioux have one of the largest percentages of Semitic-looking men and women when compared to other Indian tribes of North America. It is also no coincidence that the Sioux have a large percentage of ancient Near Eastern and Middle Eastern (ancient Western Eurasian) blood within their ranks.

---

dn Wašíčuŋ Tȟašúŋke, commonly known as "American Horse" actually means, "He-Has-A-White-Man's-Horse." He was a chief, statesman, educator, and historian.

Red Fly, an Oglala Sioux Chief (1900)[560]
& an **Arab man** from Yemen[561]

Featured above are two men who appear to be blood relatives: Red Fly, an Oglala Sioux chief, and an Arab man from Yemen.[do] This Sioux chief and Arab man share many similar Semitic facial features including prominent long and thin noses, high nasal bridges, and long and rectangular-shaped faces. They both share similar cheekbones as well. From these photographs, one can see that they share many of the same physical characteristics common to the Middle East. Not only do the Sioux and Arabs share similarities in appearance, but they also seem to share similar paternal lineages. Because of the Middle Eastern look of these men, and due to how common paternal lineage 'R1' is among both the Arabs and the Sioux, they likely belong to a similar haplogroup.

---

do   Yemen is an Arab country in Southwest Asia, occupying the southwestern to southern end of the Arabian Peninsula.

Little Wound, a Sioux[562]
& an **Arab** man

Featured above is Little Wound, a Sioux Indian, and an Arab man from the Middle East. Little Wound and this Arab man, although not doppelgängers, share a number of facial features common among Middle Easterners. They both possess similar prominent long and narrow noses, high nasal bridges, and long and rectangular-shaped faces. Little Wound does not possess common physical features of prototypical East Asians and Siberians. His Semitic physical features indicate that he belongs to a Semitic group, most likely of the House of Israel. It is highly likely that he belonged to a haplogroup that is common to Manassehites (the main genetic lineage of the Lamanite remnant).

An **Arab** with a keffiyeh
& **Great King Steeh-tcha-kó-me-co**, a Creek Indian (1834)⁵⁶³

A number of Creek Indians resembled Semitic-looking Arabs and other Old World peoples.ᵈᵖ When American painter John Trumbull sketched Creek chiefs visiting New York in 1790, contemporaries were astounded by what their eyes beheld. These North Americans were said to resemble Roman senators "for their dignity of manner, form, countenance and expression."⁵⁶⁴ Europeans in America were also surprised to discover that Creek Indians, and other tribes such as the Delaware (Lenni Lenape), Choctaw, and Shawnee wore Arabian-like keffiyehs.ᵈᵠ A Wadi El-Rayan Arab man and Great King Steeh-tcha-kó-me-co, a Creek Indian (both featured above) are both wearing Middle Eastern headdresses.⁵⁶⁵

---

dp   To learn more about ways in which the New World is connected to the Old World, refer to *Appendix 4: Old World Symbols in the New World*.
dq   Keffiyehs are Middle Eastern headdresses consisting of a square of cloth folded to form a triangle, typically worn by both Arab men and some Kurdish men.

Hole in the Sky, an Ojibwe (1825-1868)[566]
& a Bedouin woman (1898)[567]

The Ojibwe are a North American Indian tribe with a number of Semitic-looking men and women among them. Featured above is Hole in the Sky, an Ojibwe chief, and also a Semitic-looking Bedouin woman[568] who most likely possessed Semitic blood. Both Hole in the Sky and the Bedouin woman possess many physical traits of individuals from the Middle East.

Concerning the Bedouin,[569] LDS scholar Hugh Nibley has claimed that "Ishmael of the Book of Mormon has descendants in the New World who closely resemble the Ishmaelites of the Old World,[dr] which includes many Bedouin."[570]

As apparent from the photo of Hole in the Sky,[571] he was a handsome man and symmetrically formed. He possessed much grace of manner and natural refinement. Hole in the Sky was also an astute student of diplomacy.[572] His name, which could be synonymous with "opening to the heavens" or "holy oracle," was most likely given for his natural intuitive abilities and natural spirituality.

Hole in the Sky possessed many admirable qualities. Reverend Claude H. Beaulieu, who met Hole in the Sky, described him thusly:

> Hole-in-the-Day [Sky] was a man of distinguished appearance and native courtliness of manner. His voice was musical and magnetic, and with these qualities he had a subtle brain, a logical mind, and quite a remarkable gift of oratory. In speech he was not impassioned, but clear and convincing, and held fast the attention of his hearers. . . . He dressed well in native style with a touch of civilized elegance, wearing a coat and leggings of fine broadcloth, linen shirt with collar, and, topping all, a handsome black or blue blanket.[573]

Hole in the Sky was civilized; he mingled easily in white society. During his many trips to Washington DC, and St. Paul, Minnesota, he was popular with both men and women, but was

---

dr   To learn more about how the New World is connected to the Old World, refer to *Appendix 4: Old World Symbols in the New World.*

especially popular with women.⁵⁷⁴ After his visits, white men in both the nation's capital and the territorial capital began to look upon him as a spokesman for all his people.⁵⁷⁵

Elk Skin, a Yakima Indian & a **Bedouin** woman

Elk Skin, a Yakima North American Indian, and a Bedouin woman are featured above. Both of these women share similar physical characteristics and somewhat similar clothing styles. Even their hair styles resemble one another. Although uncommon in the modern age due to Muslim influence, braided hair was common among the Bedouin in years past. Braids are not exclusive to the Bedouin and North American Indians, but the shared hairstyle adds to the commonalities between the two women.

As the photographs in this chapter have demonstrated, an Arab presence exists among select tribes of North American Indians. Many Amerindians of the north have the coloring, high nasal bridges, long and thin noses, long and rectangular-shaped faces, eyes, and other traits of Arabs of the Middle East. These similarities in appearance (phenotype) and genetics (genotype) help one to see just how connected certain groups of North American Indians are to Middle-Eastern peoples, especially Semites.

12

North American Indians and Jews

And then shall the remnant of our seed know concerning us, how that we came out from Jerusalem, and that they are descendants of the Jews. —**Nephi, Book of Mormon Prophet (559 BC-545 BC)**[576]

These natives belong to the house of Israel[.] . . . [T]hey are descendants of the Jews. —**Brigham Young, LDS President (1864)**[577]

Individuals with genetic markers common to the House of Israel typically share a family resemblance. The descendants of Joseph of Egypt and Judah frequently possess physical commonalities for this very reason.[ds] Because the Lamanite remnant is Semitic in origin, and predominantly of the tribe of Joseph of Egypt, it makes sense that members of this group would

---

ds Joseph of Egypt and Judah were half-brothers—they shared the same father but had different mothers.

look like other Semites, including Jews. For whatever reason, members of the Lamanite remnant do not necessarily possess the same haplogroups as prototypical Jews, but they do share similar physical and dispositional traits with them since they are related.

Early European settlers have connected indigenous North Americans to Semites. In historian Samuel Smith's 1765 *The Colonial History of New Jersey,* he spoke of the common cultural practices between North American Indian tribes in New Jersey and the Jews,

> They wore ear-rings and . . . jewels[,] bracelets on their arms and legs[,] rings on their fingers[,] necklaces made of highly polished shells found in their rivers and on their coasts. . . . They use shells and turkey spurs round the tops of their moccasins, to tinkle like little bells, as they walk. Isaiah proves this to have been the custom of the Jewish women, or something much like it. 'In that day, says the prophet, the Lord will take away the bravery of their tinkling ornaments about their feet, and their cauls, and their round tires like the moon. The chains and the bracelets and the mufflers. The bonnets and the ornaments of the legs, and the head-bands, and the tablets, and the car rings; the rings and the nose jewels.' [Isaiah 3:18] They religiously observed certain feasts, and feasts very similar to those enjoined on the Hebrews, by Moses[.] . . . In short, . . . both English and Spaniards, were struck with their general likeness to the Jews. The Indians in New Jersey, about 1681, are described, as persons . . . rather resembling a Jew[.] . . . The Indians have so degenerated, that they cannot at this time give any tolerable account of the origin of their religious rites, ceremonies and customs, although religiously attached to them as the commands of the great spirit to their forefathers.[578]

Abraham Mordecai, a historian, the author of *History of American Indians,* and the founder of Montgomery, Alabama, was "an intelligent Jew, who dwelt 50 years [among] the Creek Nation" of North America.[579] Mordecai strongly believed that the Indians were originally of his people, the Jews.[580]

Thomas Thorowgood of the Westminster Assembly of Divines, who was appointed to restructure the Church of England, also connected the North American Indian tribes to the Jews. He said,

> The Indians do themselves relate things of their Ancestors suteable [sic] to what we read in the Bible . . . They constantly and strictly separate their women in a little wigwam by themselves in their feminine seasons [.] . . . [T]hey hold that Nanawitnawit (a God overhead) made the Heavens and the Earth. . . . The rites, fashions, ceremonies, and opinions of the Americans are in many things agreeable to the custom of the Jewes [sic], not only prophane [sic] and common usages, but such as he called solemn and sacred.[581]

Thomas Jefferson, a highly intelligent man and the third President of the United States, was not easily impressed by many people. And yet, in a 1785 letter to French Major General François Jean de Beauvoir, he praised the intellectual abilities of North American Indians. Jefferson remarked, "I believe the Indian then to be in body and mind equal to the whiteman."[582] In effect, Jefferson's acknowledgement of the intellectual abilities of North American Indians connected them to other highly intelligent ethnicities, which includes Semitic groups.

Jefferson believed that the North American Indians he had met had descended from Turks and Tartars (two groups with Middle Eastern origins) that had arrived in America anciently.[583]

John Filson, a Kentucky historian, one of the founders of Cincinnati, Ohio, and Jefferson's contemporary, also believed that North American Indians had descended from Old World lineages. Filson claimed that indigenous North Americans were originally Phoenicians (Canaanites)—a Middle Eastern group that spoke Paleo-Hebrew.[584] All three of these groups—the Turks, Tartars, and Phoenicians—have always predominantly possessed Semitic ancestry owing to the fact that many of them originally came from the Middle East.

Bartolomé de las Casas, an Iberian Jewish historian (1484-1566)

Bartolomé de las Casas (featured above), an Iberian Jewish historian who lived in Spain, was known as the "Protector of the Indians." He spent approximately fifty years of his life actively fighting slavery and the cruel colonial abuse of indigenous peoples. He did his best to persuade the Spanish court to approve a more humane policy of colonization and was known among his countrymen as an honest man, which built him a reputation and gained him some influence at Court.[585] Las Casas wrote about the physical and cultural connections he had made between Semitic people, like himself, and some of the natives he had encountered in the New World. He was perhaps the first Jew, or the first individual in the modern era for that matter, to claim that American Indians were Israelites. He even went so far as to speculate that they were of the lost tribes of Israel.

William Penn (1644-1718), the founder of the Province of Pennsylvania, also connected North American Indian tribes to Hebrew groups such as the Jews. He once remarked that "[T]hey offer their first Fruits they have a Kind of Tabernacles; they are said to lay their Altar upon twelve Stones; their Mournings a year, Customs of Women, with many things that do not now occur."[586]

Circumcision in North America

The very specific practice of foreskin removal (circumcision) was practiced in the New World by natives of North America. This practice seems very out of place, since the religious rite is fairly specific to the Hebrews, especially the Jews.[587] Virtually no other group in the world, including East Asians, Siberians, and non-Jewish Europeans, has practiced this unique religious rite. North American Indians do not appear to have acquired the practice from anyone in East Asia or Siberia. There is a small possibility that the rite may have been learned from Jewish proselytes who traveled to America anciently, but no known accounts exist that make this claim. If this is not the case, then it was likely passed down by Book of Mormon peoples. No other known explanations seem plausible. The odds of North American Indians independently creating such a specific, unpractical, and painful practice—completely unconnected to the Hebrew ritual—seem close to nil.[588]

Mariano Eduardo de Rivero y Ustariz (1798-1857), a prominent Peruvian scientist, and Johann Jakob von Tschudi (1818-1889), a Swiss naturalist and explorer, have said concerning circumcision in North America in their 1857 work *Peruvian Antiquities,*

> Like the Jews, the [American] Indians . . . divide the year into four seasons, corresponding with the Jewish festivals. . . . In some parts of North America circumcision is practised [.] . . . There is . . . much analogy between the Hebrews and Indians[.] . . . The [American] Indians likewise abstain from the blood of animals, as also from fish without scales; they consider divers quadrupeds unclean[.].[589]

New England Puritan minister Cotton Mather (1663-1728), in the course of a yarn of extraordinary letters to the Royal Society in London, drew attention to the existence "in Connecticut, of a tribe of Indians which practiced circumcision."[590] Mather must have been quite surprised by this Hebrew-like practice found among North American Indians.

Similarly, when the Machapunga—a small Algonquian-speaking tribe in North America—were discovered in 1701, they were also found to be practicing circumcision.[591] The number of North American Indian tribes who practiced circumcision at this time authenticates the idea that the rite was passed on from generation to generation for many years, maybe even from Book of Mormon times.

More Evidence to Connect North American Indians and Jews

Red Shirt, an Oglala Sioux (1845-1925)
& **Gregory Pincus,** a Jew (1903-1967)

Select North American Indian tribes, especially ones from the Great Lakes region and the Great Plains, share physical commonalities with Jews. The Sioux, for example, often have Semitic-looking physical characteristics. Featured above is Red Shirt, an Oglala Sioux, and also Dr. Gregory Goodwin Pincus, an American biologist and researcher with Jewish genetics. Both men featured above possessed prominent long and thin noses, high nasal bridges, and long rectangular-shaped faces—all common physical traits among Semitic groups.

The physical characteristics these men both possessed are virtually absent from East Asians, Siberians, and indigenous Mesoamericans. Because of Red Shirt's Semitic visage, he appears to have belonged to the main Lamanite remnant and the genetic lineage of Joseph of Egypt or possibly Judah.

Hehaka Isnala, an Oglala Lakota Sioux (1899)[592]
& a **Kurdish Jew** (1973)[593]

Featured above is Hehaka Isnala—an Oglala Lakota Sioux man—and a Jewish woman from Kurdistan.[dt] Kurds are an ethnic Iranian group in the Middle East who often possess Semitic physical features and common Semitic genetics. Both Hehaka Isnala and the female Kurdish Jew possessed prototypical Semitic noses, high nasal bridges, and long and rectangular-shaped faces.[594]

---

dt   Kurdistan or Greater Kurdistan, is located in Upper Mesopotamia and the Zagros Mountains (and other nearby regions). It is the homeland of the Kurds—an ethnic group in the Middle East. The Kurds are one of the indigenous people of the Mesopotamian plains and the highlands in what are now south-eastern Turkey, north-eastern Syria, northern Iraq, north-western Iran and South Western Armenia..

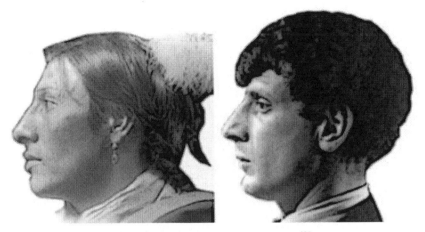

Two Bulls, a Dakota Sioux (1895)⁵⁹⁵
& **Jonathan Miller,** a Jew (b. 1934)⁵⁹⁶

Two Bulls, a Dakota Sioux, and Jonathan Miller, a man with Jewish genetics who grew up in London in a well-connected Jewish family, are featured above. Both men manifest common Semitic physical features: long and prominent noses, high nasal bridges, dolichocephalic (long headed) heads, and long rectangular-shaped faces.

The Sioux often fit William Penn's description of North American Indians: "Their eye . . . [is] not unlike a straight-looked Jew",⁵⁹⁷ and "the noses of many of them have as much of the Roman."⁵⁹⁸ The number of large Roman noses (which are really just Semitic noses) and high nasal bridges found among the Sioux is worth noting. Because of the Semitic visages of many Sioux, it appears that many of them belong to the main Lamanite remnant.

Daniel **Day-Lewis**, a Jew (b. 1957)[599]
& a **Huron** man (Wyandot)[600]

Jewish-American actor Daniel Day-Lewis and a Huron Indian of North America are featured above. This Huron man, if an actual Huron, belonged to the same tribe as Kihue (who is a doppelgänger of Yuya of Egypt). Both Daniel Day-Lewis and the Huron Indian share similar physical features and similar dispositional traits, which are extremely uncommon among East Asians and Siberians. These two men possess(ed) similar Semitic profiles including high nasal bridges, long and thin aquiline noses, and long and rectangular-shaped faces.

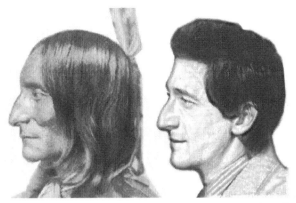

Black Crow, a Sicangu Sioux (1872) & **Adrien Brody**, a Jew (b. 1973)[601]

Black Crow, a Sicangu Sioux, and Adrien Brody, an American actor with Jewish genetics (who is known for the size of his nose), are featured above.[602] Black Crow possessed similar Semitic physical features to Brody, which could mean he belonged to the main Lamanite remnant.

Black Crow, a Sicangu Sioux (1872) & **Adrien Brody**, a Jew (b. 1973)[603]

Both Black Crow and Adrien Brody (featured above) display high nasal bridges, long and thin aquiline noses, and long and rectangular-shaped faces.

Esh-sup-pee-me-shish, a Crow (1854 -1922)[604]
& **Oded Fehr**, a Jew (b. 1970)[605]

Esh-sup-pee-me-shish, a Crow Indian, and Oded Fehr, an Israeli film and television actor now based in the US, are featured above. Esh-sup-pee-me-shish was a Crow scout for George Armstrong Custer's 7th Cavalry during the 1876 campaign against the Sioux and Northern Cheyenne. He survived the Battle of the Little Bighorn.

The appearance and dispositional traits of these two men are extremely rare among East Asian and Siberian populations, but most common among Caucasian males, especially masculine Semitic Caucasian males. Both men display high nasal bridges, long and thin aquiline noses, and long and rectangular-shaped faces.

David Duchovny, a Jew (b. 1960)[606]
& **Little Wound**, a Sioux[607]

Featured above is David Duchovny, an American actor with Jewish genetics, and also Little Wound, a Sioux Indian. According to a 2003 New York Times article, Duchovny's father, Amram, was Jewish.[608] From the photographs featured above, we can clearly see that Duchovny and Little Wound share similar physical characteristics. Both men display high nasal bridges, long aquiline noses, long-headed (dolichocephalic) skulls, and long and rectangular-shaped faces, which of course are common physical traits of Semites from all across the globe.

Because of the Semitic-looking physical characteristics of Little Wound, it appears that he may have belonged to the main Lamanite remnant through the House of Joseph of Egypt or Judah. Little Wound did not look like the prototypical Native Americans of today who often have East Asian and Siberian physical traits.

Crow's Heart, a Mandan[609]
& **Josephus**, a Jewish historian (37 - 100 AD)[610]

Featured above is Crow's Heart, a Mandan Indian,[du] and also Josephus, an ancient Jewish historian. Both men had convex noses, similar dispositional traits, and prototypical Semitic features. Both of them possessed high nasal bridges, long aquiline noses, dolichocephalic heads, and long and rectangular-shaped faces. Many of the dispositional traits and phenotypes of Crow's Heart and Josephus are quite uncommon among East Asian and Siberian populations.[611]

Crow's Heart was a good warrior. At the age of 19, he led the Old Wolves—a group of Mandan—during a war party. Similar to the first Nephi of the Book of Mormon, he was "large of stature" by a young age. The physical features and dispositional traits of Crow's Heart mirror those of Nephi of old. Crow's Heart also has the size, stature, and traits of the 4th century prophet-historian Mor-

---

du The Mandan are the same Mandan tribe featured in *Chapter 7: North American Indian Depopulation* who were virtually wiped off the face of the planet by smallpox.

mon, who abridged the Book of Mormon. Josephus possessed the dispositional traits of the majority of famous Semitic historians. Although it is virtually impossible to know for sure that Crow's Heart belonged to a Semitic lineage, his Semitic-looking physical traits and large stature indicate that he may have belonged to the main Lamanite remnant.

Crow's Heart, a Mandan & **Shimon Peres** of Israel (b. 1923)

Crow's Heart also resembles Shimon Peres, a Jew who became the 9th President of the State of Israel (both featured above). These two men shared similar dispositional traits and the prototypical physical features of Semitic peoples. They both possessed similar high nasal bridges, noses, cheekbones, long faces, sloping foreheads, and head shapes. Because of the Semitic-looking phenotype and dispositional traits of Crow's Heart, he may have belonged to the main Lamanite remnant through the genetic lineage of Joseph of Egypt or Judah.

Lauren Bacall, a Jew,[612] & **Ah-Weh-Eyu**, "Pretty Flower," a Seneca (1914)[613]

Featured above is Lauren Bacall, a blond "bombshell" and former actress, and also Pretty Flower, a Seneca Indian of the Iroquois Confederacy. Bacall had Jewish ancestry and Pretty Flower looks as if she possessed Middle Eastern ancestry as well. Both women shared similar phenotypes and similar dispositional traits. They both possessed a high nasal bridge, which is an uncommon trait among most women of the world. High nasal bridges are almost completely non-existent among East Asians and Siberians. East Asian and Siberian women tend to have low to non-existent nasal bridges and small noses. Because of Pretty Flower's appearance, she could have belonged to the main Lamanite remnant either through the genetic lineage of Joseph of Egypt or Judah.

Once can see that all the pictures of Semitic-looking North American Indians and Jews in this chapter demonstrate how probable it is that select tribes of North American Indians are actually unrecognized Semites. The physical traits of these Semitic-looking North American Indians also authenticate the idea that out of all indigenous peoples of the Americas, they are the most likely to belong to the main Lamanite remnant. Indigenous North Americans not only look the part of Semites, but they also possess the dispositional traits commonly found among Semitic groups.

13

Ancient Artifacts Dating Back to Book of Mormon Times

We can probably safely infer that Lehi and his party showed physical features in the normal range for people in Palestine in his day. (People of that area haven't changed much right up to the modern times, for that matter.) We have skeletons and art representations from early times plus data on living descendants to guide us. . . . Mountainous regions just north of the Near Eastern centers could have furnished genes producing a slightly more rugged build and a more prominent, beaked nose.
—**John L. Sorenson, Latter-day Saint anthropologist (1996)**[614]

The appearance of ancient artifacts, effigies, and masks discovered in North America that date back to Book of Mormon times provide further evidence for the existence of Israelite visages among indigenous North Americans. In nearly all of the ancient North American artwork that dates back to the time of the Book of Mormon, the faces depicted appear exactly the way well-informed Latter-day Saints should expect them to appear: like individuals from the ancient Near East. Many of these representations seem to fit perfectly into the Book of Mormon narrative.

Although it is possible that other ethnicities admixed with Book of Mormon peoples during the time frame it covers, the Book of Mormon is primarily about Israelites and Middle-Easterners who settled in the New World, so these are the types of faces we should expect to see represented in the art of this time period. What follow are pictures and descriptions of ancient artifacts that date back to Book of Mormon times to help us locate Book of Mormon peoples.

A **South Arabian** stone effigy (300 BC-100 AD),[615]
a **Hopewell** effigy, Kentucky, (200 BC-1 AD)[616]
& an early **Delaware Indian** effigy, Pennsylvania[617]

An observable continuity exists between ancient Middle Eastern effigies, ancient indigenous North American effigies (that date back to the time of the Book of Mormon), and early North American Indian effigies. All the effigies featured above possess high nasal bridges, long and thin noses, and long and narrow faces—all physical traits one would expect to find among Book of Mormon peoples. The ancient stone effigy on the left belongs to an ancient South Arabian culture. Arabs, as previously discussed, anciently interacted and admixed with the tribe of Manasseh (Lehi's tribe). The middle effigy was discovered in Kentucky and dates back to the time frame of the Book of Mormon. The effigy on the right was crafted by early Delaware (Lenni Lenape) Indians—the same group of "red men" who were visited by Elder Oliver Cowdery and informed that they were the descendants of Book of Mormon peoples.

These three artifacts represent the physical traits of the groups who made them. They all possess Semitic physical features: long and narrow faces, long and thin noses, and high nasal bridges. None of these ancient effigies are East Asian or Siberian in appearance. Flat and wide faces, flat and wide noses, and low to non-existent nasal bridges cannot be found on these ancient effigies.

The early Delaware and Hopewell effigies are almost identical. Not surprisingly, the Hopewell effigy dates back to the time of the Book of Mormon, and the Delaware are an Algonquian-speaking people with high percentages of Haplogroup R1 (Y-DNA). The Delaware were one of the tribes visited by the first Latter-day Saint missionaries on their missions to the Lamanites (1830-1831). Elder Oliver Cowdery also connected the Delaware Indians to the ancient inhabitants of North America when he said to the Delaware tribe, "This book [the Book of Mormon] was written in the language of the forefathers of the red men."[618] Because the Hopewell effigy that dates back to Book of Mormon times so closely resembles the early Delaware effigy, it is not to difficult to imagine a virtually seamless thread of Semitic-looking peoples among these two groups extending from ancient North America to the Northeastern and Great Plains regions of the United States.

Hopewell human head effigy (3 views), Middle Woodland Period (1 AD-500 AD)[619]

Other ancient effigies that date back to the time of the Book of Mormon also display Semitic physical characteristics. Featured above are three different views of an ancient human head effigy discovered in an Ohio Hopewell Mound in Ross County. This particular effigy dates back to the Middle Woodland Period (approximately 1 AD-500 AD). It is very possible that this effigy memorializes the visage of an actual Nephite or Lamanite since it possesses prototypical Semitic facial features and fits within the correct time frame. The effigy has a high nasal bridge, a long and thin nose, and a long rectangular-shaped face. It does not possess a wide and flat face, epicanthic eye folds, or a low nasal bridge: traits correlated with East Asian genetic origins. The individual it depicts most likely originated in the Old World.[dv]

---

dv To learn more about New World connections to the Old World, refer to *Appendix 4: Old World Symbols in the New World.*

An **ancient death mask** (2 views), Wisconsin (1 AD-400 AD)[620]

Featured above is another ancient Semitic-looking effigy discovered in North America. It also dates back to the time of the Book of Mormon and is similar in appearance to the effigy featured previously (although this person's face was not as long). This two-thousand-year-old burial mound effigy was discovered near the banks of Wisconsin's Red Cedar River. From the effigy's visage, it is clear that the individual depicted had many non-East Asian physical traits; nothing about this ancient effigy gives the indication that this man's genetics were East Asian or Siberian in origin. The artifact has prototypical Semitic physical features: a high nasal bridge, a long and thin nose, and a rectangular-shaped face. The Semitic appearance of this effigy indicates that he too could have been a Nephite or a Lamanite from the Book of Mormon.

A **Woodland Period human face effigy**, Ohio (200 BC-400 AD)[621]

Featured above is another ancient effigy from North America that dates back to the time of the Book of Mormon. This artifact was made around 200 BC-400 AD and possesses prototypical Semitic physical features: a high nasal bridge, a long and thin nose, and a long and narrow face.

An **Adena human effigy pipe**, Ohio (100 BC-100 AD)[622]

An ancient Adena human effigy pipe from North America—also dating back to the time of the Book of Mormon—is featured above. Common to the North American effigies of this time period, this one also has prototypical Semitic physical features: a high nasal bridge, a long and thin nose, and (what appears to be) a narrow face. Once again, from this vantage point, the ancient artifact does not appear to possess any prototypical East Asian or Siberian physical traits.

An ancient Illinois **Hopewell burial effigy** (200 BC-400 AD)[623]

A female effigy holding a baby that dates back to the time of the Book of Mormon is featured above. This ancient North American burial effigy was discovered in Mound 8 of the Knight Mound group in Illinois. This figurine from the Woodland Period (200 BC-400 AD) features faces (a mother's face and a child's face) that are commonly found among Semites but not among East Asians and Siberians. This mother has a high nasal bridge, a long and narrow nose, and a long and narrow face. If this mother and child both possessed prototypically East Asian or Siberian facial features, they most likely would have had flat and wide faces, small and wide noses, and epicanthic eye folds.

After Book of Mormon Times

An **Ohio Intrusive Mound Culture** effigy (abt. 700 AD-900 AD)[624]

Ancient effigies in North America continued to resemble Semites even after the time frame of the Book of Mormon, although the number of East Asian and Siberian-looking effigies began to increase steadily after the Book of Mormon timeline ended. The Ohio Intrusive Mound Culture, an ancient North American group, for example, created an artifact (featured above) dating back to about 700 AD-900 AD that possesses prototypical Semitic physical features. It features a face which is long and narrow, a nose which is long and thin, and possibly a high nasal bridge. Its physical characteristics resemble both those of the Hopewell effigy discovered in Kentucky and the early Delaware Indian effigy found in Pennsylvania that were previously featured in this chapter. Because of the appearance of this ancient effigy, it may depict an individual from the Lamanite remnant.

An **Effigy from the Cahokia Mounds**, Late Woodland Period (500 AD-1000 AD)[625]

The ancient artifact featured above belongs to a period after the Book of Mormon. It comes from the Late Woodland Period (500 AD-1000 AD) and was discovered in a Cahokia mound in Illinois. The figurine possesses a high nasal bridge, a long and thin nose, and a long and narrow face. This artifact does not display common East Asian and Siberian physical characteristics (i.e. a flat and wide face, flat and wide nose, and epicanthic eye folds). The man depicted by this figurine looks like a number of modern-day Native Americans from the Great Lakes and Great Plains regions, but appears more Semitic than Far Eastern. If this ancient figurine was created mostly true to form, it may represent a member of the Lamanite remnant.

Mississippian Culture copper repoussé, Spiro, Oklahoma (800 AD-1600 AD)[626]

The ancient artifact featured above was created by the Mississippian Culture—an ancient North American group.[627] It was discovered in a Spiro Mound in Oklahoma.[628] This North American copper repoussé depicts the face of a man who could have belonged to the Lamanite remnant.[629] Although this copper repoussé only represents the profile of a man, it is still very telling of the typical physical attributes of the people who created it. The effigy lacks prototypical physical traits of East Asians and Siberians, but possesses a long face with an aquiline nose, a high nasal bridge, and eyes without epicanthic folds, which means it is most likely the face of a man who had Old World origins.[dw]

---

dw To learn more about New World connections to the Old World, refer to *Appendix 4: Old*

Early Depictions of North American Indians

A **Timucua Indian** archer (1564)[630]

Many of the earliest depictions of North American Indians match ancient effigies and figurines that date back to the time of the Book of Mormon. In 1564, Jacques Le Moyne de Morgues made etchings of indigenous North Americans during his journey in the New World.[631] The early indigenous Floridians that Le Moyne encountered in the New World were usually tall with light brown skin, muscular, and well-built. The archer (featured above) made by Le Moyne is wearing his hair bundled on top of his head: a common practice of the Timucua intended to make them look taller than most people of the time period. This Timucuan Indian appears to be a Semite because of his characteristic high nasal bridge, long and thin

---

World Symbols in the New World.

nose, and long rectangular-shaped face. The man depicted in this etching does not resemble prototypical East Asians or Siberians.

A **Timucua Indian** (1564)[632]

Other etchings of indigenous North Americans made by Jacques Le Moyne de Morgues in 1564 also display the common physical characteristics of Semites. This Timucuan Indian (featured above) possessed a high nasal bridge, a long and narrow face, and a long and aquiline nose. He does not have epicanthic eye folds, a wide and flat face, or any other physical characteristics most commonly found among East Asians and Siberians.

A **Roanoke Indian** (1585)[633]

Featured above is a 1585 reproduction of a John White etching made of a Roanoke Indian of Virginia.[634] In 1585, White, an English artist and early pioneer of English efforts to settle North America, set sail on an expedition with English sailor Richard Grenville to the shores of present-day North Carolina. This etching of a Roanoke Indian appears Semitic since it possesses prototypical Semitic physical features. Because of the Semitic-looking appearance of this particular Roanoke Indian, he may have belonged to the Lamanite remnant.

A **Susquehannock** Indian chief (1612)[635]

Though almost completely forgotten today, the Susquehannock were one of the most imposing tribes of the mid-Atlantic region during the time that Europeans explored the New World. Featured above is an engraving of a Susquehannock Indian chief that is found on English Captain John Smith's 1612 map of the Chesapeake Region.[636] The depiction was engraved by William Hole,[637] an English artist, etcher, illustrator, and engraver who was known for his historical, industrial, and biblical scenes.

Little was known about the Susquehannock at the time of first contact. They lived some distance inland from the coast, which meant that early Europeans rarely visited their villages. Before they could be studied in depth, epidemics and wars with the Iroquois in 1675 wiped them out.[dx] The descriptions and etchings made of the Susquehannock give one the impression that they were very unique when compared to many modern-day Native Americans.

From August 1-7, 1608, Captain John Smith and his company made contact with some Susquehannocks, whom they called "a mightie [mighty] people."[638] When John Smith arranged a meeting with the formidable Iroquois-speaking Susquehannock, sixty representatives of this giant-like people came down from their village by Octorara Creek to present the new settlers with gifts. They proffered them some venison, tobacco pipes, baskets, targets, and bows and arrows.[639] During John Smith's first meeting with the Susquehannock in 1608, he was especially impressed with their size, deep voices, and the variety of their weapons. Their height must have been exceptional, because the Swedes (who were also often large in stature) who met them also commented on how tall and well-built they were.[640] The Susquehannock were imposing figures with "large statures" like Nephi, Mormon, or Moroni. Because of their size and Semitic-looking features, it is possible that many of them belonged to the Lamanite remnant.

According to John Smith, the Susquehannock Indian chief who was depicted on Smith's 1612 map of the Chesapeake was

---

dx   Hostilities ending in casualties and European diseases reduced the Susquehannock (Conestoga) population from an estimated 5,800 in 1647 to about 250 in 1698.

the greatest of them his hayre [hair], the one side was long and the other shorn close with a ridge over his crowne [crown] like a cock[']s combe [comb] . . . The calfe [calf] of whose leg was ¾ a yard around and all the rest of his limbes [limbs] so answerable to that proportion that he seemed the goodliest man we ever beheld![641]

John Smith further described these Indians of the Chesapeake:

> 60 of those Susquehannocks came to us[.] . . . [S]uch great and well proportioned men are seldome seene [seldom seen], for they seemed like giants to the English[.] . . . [T]hese are the strangest people of all those countries both in language and attire; for their language it may well beseeme [beseem] their proportions, sounding from them as a voice in a vault. Their attire is the skinnes of beares and woolves [skins of bears and wolves], some have cassocks[dy] made of beares heades and skinnes [bear's heads and skins][.] . . . The halfe [half] sleeves coming to the elbows were the heades of beares [heads of bears] and the arms through the open mouth, with paws hanging at their noses. One had the head of a woolf [wolf] hanging in a chain for a jewell [jewel]; his tobacco pipe three-quarters of a yard long prettily carved with a bird, a deer or some such device, at a great end. . . . [O]ne had the heade of a woolf [head of a wolf] hanging from a chain for a jewell [jewel] . . . with a club suitable to his greatness sufficient to beat out one's brains.[642]

These Susquehannock Indians dominated the expansive region between the Potomac River in northern Virginia to southern New York.[643] They seemed to be a bellicose and warlike people who often fought against the Algonquian-speaking Delaware (Lenni Lenape) Indians—the same tribe that would later be visited by Latter-day Saint missionaries during the years 1830-1831.

---

dy A cassock is a full-length garment.

An **Armouchiquois man** by Samuel de Champlain (1613)[644]

Another lesser-known North American Indian tribe is the Armouchiquois. The Armouchiquois once inhabited many parts of the New England area.[645] Featured above is a 1613 etching of an Armouchiquois man done by French explorer Samuel de Champlain. Champlain, who was a French cartographer and draftsman, among other things, was known as "The Father of New France (Canada)." Upon first glance, the Armouchiquois man (featured above), does not appear to be a prototypical North American Indian. He actually looks more like a "hipster" from modern times. If this etching was made true to form and was patterned after an actual Armouchiquois man, then this Semitic-looking North American Indian may have belonged to the Lamanite remnant.

In 1605, when French explorer Samuel de Champlain visited a large native village at the mouth of the Saco River, his Algonquian-speaking Etchemin guides called the people "Armouchiquois" and referred to the village as "Chouacoit." Chouacoit was a large, permanent, palisaded settlement, which could have been similar to one of the many fortified cities spoken of in the Book of Mormon.[646]

By 1616, the village of the Armouchiquois had been hit so hard by disease that most of its inhabitants became sick and died. By 1631, the Armouchiquois village was empty.

Samoset, an Abenaki, artwork made in 1653, the year of his death (depicting 1620)[647]

Featured above is Samoset, meaning "He Who Walks Over Much" (1590-1653). Samoset was a sagamore (lesser chief), ambassador, and interpreter of the Abenaki people. From this 1653 depiction of Samoset, he appears to be a descendant of the Lamanites of old with his shorn head, Semitic physical features, and partially naked body. If this depiction of Samoset is representative of the appearance of the Abenaki people in general, then they were very Middle Eastern-looking.

In 1620, Samoset entered William Bradford's pilgrim settlement at Plymouth, Massachusetts.[648] This group who greeted Samoset had come to America aboard the Mayflower. In history books, Samoset is known as one of the North American Indians who significantly helped these pilgrims during their first winter, and yet he was still viewed by them as a wild "savage."[dz]

Samoset fostered goodwill and trade with the pilgrims. He appears to have enjoyed his interactions with these travelers who had emigrated from Europe to the New World. Alexander Young's *Chronicles of the Pilgrim Fathers* describes a time when the pilgrims fed Samoset, "He asked [for] some beer, but we gave him strong water [distilled liquor], and biscuit, and butter, and cheese, and pudding, and a piece of mallard; all which he liked well."[649]

Samoset introduced the white men to Squanto (also known as "Tisquantum"), an emissary of the great Wampanoag chief, Massasoit, who facilitated the long-term peace between the two civilizations. In later years, Samoset signed the first land sale transaction with the colonists.

---

dz   Samoset had learned to speak English from fishermen who had previously visited his coastal territory.

Great Sun, a Natchez chief in winter clothing (1758)[650]

Featured above is "Great Sun," a Natchez chief with Semitic-looking physical features. Great Sun's tribe originated in the Mississippi region of the United States[651] and was quite interesting to the French, who inhabited parts of North America shortly after John Cabot entered North America for the first time in 1497. The Natchez had holy priests and built temples. Their society was stratified into four layers: the "Suns," "Nobles," "Honorables," and "Stinkards" (a name for their common folk).[652] Their absolute sovereign, the "Great

Sun," was lifted up and carried from place to place on a litter (a type of elevated conveyance). This fascination of the French with the Natchez was reflected in the writings of French writer and historian François-René de Chateaubriand as well as many earlier authors including Antoine-Simon Le Page du Pratz, Dumont de Montigny, and Jean-Bernard Bossu. These Frenchmen wrote at length about this atypical Indian society, often focusing on one of the most tragic events in the French colonization of North America: the Natchez Massacre.[ea]

Around 1730, after several brutal wars with the French, the Natchez were defeated and scattered.[653] Most Natchez survivors were sold by the French into slavery in the West Indies, while others took refuge among various North American Indian tribes such as the Muskogean Chickasaw and Creek, and the Iroquoian-speaking Cherokee.

Chief Tachnechdorus, Mingo tribe[654]

Chief Tachnechdorus of the Mingo tribe of North America is featured above.[655] The Mingo Indians were a small group that were

---

ea The "Natchez Massacre," also known as the "Natchez revolt," was an attack in 1729 by the Natchez people on French colonists near present-day Natchez, Mississippi.

related to the Iroquois living in West Virginia and surrounding areas. This portrayal of Tachnechdorus (circa approximately 1723-1780) is a photographic reproduction of a print.[eb] His prototypical Semitic physical traits make him a likely member of the Lamanite remnant.

A **Mi'kmaq woman** from Whykokamagh, Nova Scotia (1840-1846)[656]

The Mi'kmaq are another group of North American Indians who frequently possess Semitic physical characteristics. Featured above is an old painting of a Mi'kmaq woman from Whykokamagh, Nova Scotia (1840-1846). This woman has a long and thin nose, a

---

eb   Initially, Tachnechdorus encouraged his people not to attack whites who settled in the Ohio country. However, after family members were killed by settlers in 1774, he avenged their deaths, in his estimation, by raiding villages in what is now western Pennsylvania. While his allies, the Shawnee, attempted to make peace with the settlers, Tachnechdorus fueled his vendetta by continuing to be hostile towards whites until his death (around 1780).

high nasal bridge, and (what appears to be) a long, narrow face. She lacks common physical traits found among East Asians and Siberians, but closely resembles females from the Middle East.

Similar to the Egyptians, the Mi'kmaq created a system of writing that uses hieroglyphics and hieratic characters. In fact, many characters from the Mi'kmaq alphabet resemble the Egyptian demotic characters copied down by Joseph Smith from the golden plates to create the Anthon Transcript.[ec] Biologist and epigrapher Barry Fell is one academic who has argued that Mi'kmaq hieroglyphic writing was not only Egyptian in origin but also pre-Columbian.[657]

The Mi'kmaq seem connected to the Old World in other ways. Semitic visages among the Mi'kmaq correspond with the high percentages of Haplogroup R1 (Y-DNA)[658] that have been discovered among the Mi'kmaq (and other Algonquin tribes).[659] Because of the common Semitic appearance of many Mi'kmaq Indians of North America and their hieroglyphic system of writing, it is likely that many Mi'kmaq belong to the Lamanite remnant.

Each picture, photograph, effigy, and figurine featured in this chapter dating back to the time of the Book of Mormon until modern times demonstrates how the Semitic appearance and DNA of the Lamanite remnant has been passed on from generation to generation over the centuries. Semitic-looking indigenous North Americans entered the New World when the Jaredites, Mulekites, and Lehi set foot on the American continent.[660] Semitic-looking

---

ec The "Anthon Transcript," also known as the "Caractors [sic] document," is a small piece of paper on which Joseph Smith, Jr. wrote several lines of characters. According to Joseph Smith, these characters were from the golden plates and represent the reformed Egyptian writing that was on the plates.

peoples were also in the Americas when the prophet-historian Moroni sealed up the golden plates and hid them in the Hill Cumorah, the time Joseph Smith dug up the golden plates, and they still exist today and are easily recognizable (See Mormon 8:14; Moroni 10:2). The large amount of Semitic-looking ancient artifacts, early etchings, and early pictures of indigenous North Americans makes a great case for establishing their place among the main Lamanite remnant. Of all the indigenous groups of the Americas, these particular North Americans appear to be the most connected to the Old World.

14

East Asian Influences in Mesoamerica

All [American] Indians Are *Not* the Descendants of Lehi[.] ... Students of the Book of Mormon should be cautioned against the error of supposing that all the American Indians are the descendants of Lehi, Mulek, and their companions, and that their languages and dialects, their social organizations, religious conceptions and practices, traditions, etc., are all traceable to those Hebrew sources.
—**Janne M. Sjödahl, editor of the LDS Church's** *Millennial Star* **in Liverpool, England (1927)**[661]

Ancient Mesoamerican artifacts, effigies, and masks discovered in Mesoamerica that date back to Book of Mormon times almost all appear East Asian. Seldom do archaeologists on digs in Mesoamerica discover effigies, masks, figurines, murals, and artifacts that represent the common facial features of Semitic peoples. Instead, the faces of sculptures on ancient temples in Mesoamerica, Mayan murals, and Olmec colossal heads almost all possess flat faces, flat noses, and epicanthic eye folds (hooded eyes).⁶⁶²

This chapter features a selection of Mesoamerican effigies, figurines, murals, and artifacts from the time of the Book of Mormon and beyond. The illustrations in this chapter reveal the evident lack of resemblance between ancient and modern Semitic peoples and indigenous Mesoamericans, indicating that they belong to groups from the Far East rather than the main Lamanite remnant.

Ancient **Mayan** effigy, **Lamanai,** Belize, Late Preclassic Period (400 BC-100 AD)⁶⁶³

The large ancient effigy featured above was carved out of stone and is found at Lamanai, Belize, a Maya-named city, on a Mayan temple. It dates back to the Late Preclassic Period of the Maya (400 BC-100 AD) and fits within the time frame of the Book of Mormon. One would imagine that, because of the time it was made, and due to the fact that Lamanai really does sound like a Book of Mormon city, the face of this ancient artifact would have been built by Lamanites. However, it does not represent people who originated in the Middle East (people like Lehi and his caravan). Its high cheek bones, flat and wide face, low nasal bridge, and epicanthic eye folds all point to the ancient Far East. Because of the appearance of this ancient effigy, and the thousands of other effigies just like it from ancient Mesoamerican cities dating back to the time frame of the Book of Mormon, the vast majority of indigenous Mesoamericans do not appear to belong to the Lamanite remnant through the genetic lineage of Joseph of Egypt, but instead look like they descended from East Asians and Siberians.

Ancient **Mayan** effigy, Preclassic Period (1800 BC-250 AD)

The ancient Mayan effigy featured above was discovered in Mesoamerica. It is believed to have come from the Preclassic Period in Maya history (1800 BC-250 AD). This effigy is quite similar in appearance to the Lamanai, Belize, effigy featured previously. Both effigies have high cheek bones, flat faces with small, flat, and wide noses, low nasal bridges, and hooded eyes (epicanthic eye folds). The physical features of this ancient Mayan effigy give no indication that it has prototypical Middle-Eastern physical features or connections.

It is unclear whether this ancient Mayan effigy belongs to a time period of the Jaredites, Mulekites or the Nephites and Lamanites, but even if it does fit within the Book of Mormon time frame, it does not appear to have anything to do with the Semitic peoples of the Book of Mormon, particularly the groups who traveled from the Old World to the Americas. From its appearance alone, it looks like it came from the ancient Far East.

Olmec figurines, La Venta site, Tabasco, Mexico (1200 BC-400 BC)[664]

The cover of Latter-day Saint Assistant Church Historian Davis Bitton's book *Mormons, Scripture, and the Ancient World* features a picture of these Olmec figurines highlighted above.[665] This connection appears strange since most of the Olmec are extremely East Asian and Siberian in appearance. Granted, these figurines featured above appear to date back to the time period of the Jaredites in the Book of Mormon,[666] but they all possess flat faces, epicanthic eye folds, and low nasal bridges. These are not the kinds of physical characteristics one might expect to find among the Jaredites. From what Book of Mormon accounts teach us, the Jaredites were not from the ancient Far East.

An **Olmec** man, Guerrero, Mexico (1000 BC)

An effigy is featured above of an Olmec man from Guerrero, Mexico, dating back to the time frame of the Book of Mormon. His appearance is common among the Olmec, which makes him a very unlikely candidate for a Jaredite.

Ancient Mayan glyphs, **Palenque,** Mexico (226 BC-799 AD)[667]

The majority of the Olmec peoples look just like the Mayan figures on the glyphs featured above. Almost all artwork from ancient Palenque, Mexico, depicts effigies with East Asian and Siberian-looking physical features. Virtually nothing about these glyphs appear to belong in the Book of Mormon.

Mayan effigy, Palenque ruins, Chiapas, Mexico (226 BC-799 AD)[668]

According to a number of LDS scholars, Joseph Smith purportedly claimed in an unsigned *Times and Seasons* editorial that the ancient city of Palenque, Mexico, was built by the Nephites. The unsigned *Times and Seasons* editorial dated September 15, 1842, reads: "[T]hese wonderful ruins of Palenque are among the mighty works of the Nephites."[669] Beyond the questionable authorship of the editorial, ancient effigies from Palenque, Mexico, like the one featured above, point to indigenous Mesoamericans with East Asian and Siberian ancestry as the builders of Palenque. Most Palenque effigies appear to have nothing to do with Semitic cultures such as the ones found in the Book of Mormon. The Mayan effigy featured above possessed the common look of Palenque, Mexico, at the time of the Book of Mormon, and yet it does not have the "look" of any Middle Eastern groups of the Old World. The ancient artifact does not look at all like it could have been made by a Nephite from the Book of Mormon.

A **Mayan** effigy, Palenque ruins, Chiapas, Mexico (226 BC-799 AD)[670]

Another effigy from Palenque, Mexico, that might date back to the time frame of the Book of Mormon is shown above. It also possesses East Asian physical features: a flat face, a low nasal bridge, and epicanthic eye folds. Virtually nothing about the ancient artifacts of Palenque point to the ancient Near East or the Semitic Nephites.

A **Maya** Figurine from **Palenque**, Mexico (226 BC-799 AD)[671]

The Palenque figurine featured here possibly dates back to the time frame of the Book of Mormon. Once again, it looks like almost all other ancient Mesoamerican effigies with its flat face, low nasal bridge, and epicanthic eye folds (hooded eyes).

Mayan Stela E, **Quiriguá**, Guatemala (200 AD-900 AD)[672]

A number of LDS scholars purport that Joseph Smith placed the city of Zarahemla in Quiriguá, Guatemala. An unsigned *Times and Seasons* editorial dated October 1, 1842, reads: "[T]he ruins of Quirigua are those of Zarahemla[.]"[673] Again, the actual source and author of this article is unknown and only assumed to be Joseph Smith. Zarahemla was a city with rich Hebrew traditions, and yet the ancient effigy featured above has the eyes of a prototypical person from East Asia or Siberia. It was during the Hebrew Festival of the Booths that "every family . . . pitched their tents round about the temple" (Mosiah 2:5-6) in the land Zarahemla to hear King Benjamin give his speech—a speech full of Semitic literary devices (including chiasmus)—so why would the people of that land look like the ancient Mayan featured above? These indigenous Mesoamericans likely had little to nothing to do with Hebrew festivals.[674]

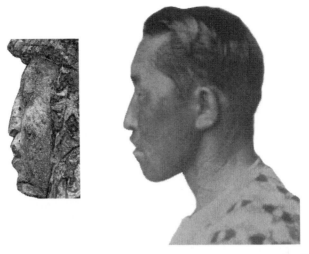

Mayan **Stela H**, Copán, Honduras, Classic Period (751 AD)[675]
& a **Siberian Orochi**

Although a small number of the faces of ancient indigenous Mesoamerican effigies appear to belong to the Old World, these effigies have by and large been East Asian and Siberian in appearance (as would be expected since DNA and appearance are linked together, heritable, and recognizable from generation to generation). An ancient Mayan effigy discovered at Copán, Honduras, is featured above on the left. This Mayan effigy, known as Stela H, demonstrates the fact that ancient Mesoamericans predominantly appeared East Asian and Siberian (rather than Middle Eastern) in 751 AD, a time after the Book of Mormon time frame. It is no coincidence that Stela H looks very similar to the Orochi man from Siberia featured above on the right.

Mural painting, Early Classic **Maya** (250 AD-600 AD)

Discovered in 2004, the mural shown above depicts everyday Mayan life. It appears to date to the time of the Book of Mormon, or shortly after the book ends. The heavy East Asian influence of the Maya mural featured above is difficult to ignore. The physical features depicted in the mural point to the ancient Far East. The ancient Middle East is nowhere to be found.

Maya Vase K1185, Guatemala, Late Classic Period (600 AD-900 AD)[676] & a painting from the **Eastern Han Dynasty,** China (25 AD-220 AD)[677]

Ancient depictions of the Maya that resemble the one featured above on the left closely resemble depictions of the ancient Chinese like the one featured above on the right. Both figures above depict men with prototypical East Asian and Siberian-looking physical features: flat faces, low nasal bridges, and epicanthic eye folds (hooded eyes).[678] Virtually all ancient and modern-day Maya appear East Asian and Siberian, like the photograph of the Mayan scribe featured above on the left. Only a few depictions of ancient Mesoamericans from the time of the Book of Mormon even remotely resemble Semites with physical characteristics common to the tribe of Joseph of Egypt.

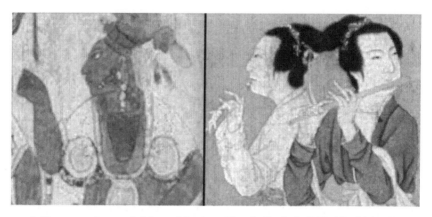

A **Bonampak** mural, Maya, Mexico, Classic Period, (250 AD-900 AD) & a **Chinese** painting, Five Dynasties (907 AD-960 AD)[679]

A Bonampak mural discovered in Chiapas, Mexico, that dates back to the Classic Period of the Maya (250 AD - 900 AD), is displayed above on the left. Featured on the right is a detailed view of a painting of some Chinese women from the Five Dynasties (907 - 960 AD). Almost all ancient Mayan murals from Mesoamerica feature hundreds of individuals just like the ones shown here. All figures above have similar physical characteristics: flat faces, low nasal bridges, and epicanthic eye folds.

As we can clearly see, all of the individuals featured above have a similar appearance. Using what appears to be an East Asian artistic style, the pictures are of East Asian-looking individuals. The hand positions of both the Mayan man and the Chinese women, which seems to be a common style of hand positioning in ancient Chinese art, are also overtly similar.

Mayan figurine (600 AD-800 AD)

The ancient Mayan figurine featured above could pass for an individual from China. It possesses prototypical physical features of the Chinese: a flat face, low nasal bridge, and epicanthic eye folds. These features are so common among the Olmec and Maya that a museum curator could easily set up an ancient Chinese exhibit comprised of thousands of Olmec and Mayan artifacts and fool laymen into believing that they are Chinese. Since this is the reality of Mesoamerica, it is extremely difficult to insert believable Jaredites, Mulekites, Lamanites, and Nephites anywhere into ancient and modern Mesoamerican history.

Chinese Terracotta Warrior armor (300 BC-250 BC),[680]
Mayan warrior, Mexico (approximately 700 AD)
& an East Asian **Buddha** statue

Obvious Hindu-Buddhist influence can be found in ancient Mayan culture. Concerning this connection, Robert Heine Geldern, an ethno-archaeologist, has remarked:

> The influences of the Hindu-Buddhist culture of southeast Asia in Mexico and particularly, among the Maya, are incredibly strong, and they have already disturbed some Americanists who don't like to see them but cannot deny them. . . . Ships that could cross the Indian Ocean were able to cross the Pacific too. Moreover, these ships were really larger and probably more sea-worthy than those of Columbus and Magellan.[681]

Featured above in the center is an ancient Mayan Warrior wearing armor that appears to resemble the armor of an ancient Chinese Terracotta warrior. On both sides of the Mayan Warrior are ancient Chinese artifacts. It is interesting to see the Mayan Warrior making the same Buddhist hand sign of the featured Buddha statue.[ed]

---

ed The hand sign is the eighth Buddha Mudra (i.e. ritual gesture) known as "Vitarka," which can be translated as "application of thought."

A **Mayan Maize god**, Copán Ruinas, Honduras (600 AD-800 AD)[682]
& an ancient **Chinese Buddha**[683]

A Mayan Maize god from Copán, Honduras, is featured above to the left, and an ancient Chinese Buddha is featured above to the right.[684] These two figurines both possess common East Asian physical features, which indicates that the two figures are somehow connected. The matching hand positions of these ancient figurines also indicate connection between the Maya and the Chinese. The hand positioning displayed by the Chinese figurine is the first Buddhist Mudra[ee] known as Abhaya, which means "no fear." It is highly unlikely that the style, hand positioning, and appearance of these two ancient artifacts were created completely independent of one another. Hence the Mayan figurine is also displaying the first Buddhist Mudra. Again, the Mayan Maize god looks nothing like a prototypical Middle Easterner.

---

ee   Mudras are positions of the body that have some kind of influence on the energies of the body, or your mood. In the Theravada tradition, the word "Vitarka" is defined as the mental factor that mounts or directs the mind towards an object.

Maya, Late Preclassic Period (400 BC-100 AD)

Featured above is another example of an ancient Mayan figurine. It comes from the Late Preclassic Period (400 BC - 100 AD), an era that fits within the Book of Mormon time frame. This ancient artifact appears to lack any semblance of common Near Eastern physical traits or garb; it does not appear to belong to the Book of Mormon at all. Like the previous artifacts, it has prototypical East Asian and Siberian physical features: a flat face, a small and flat nose, a low nasal bridge, and epicanthic eye folds. As stated previously, although a tiny minority of ancient effigies from Mesoamerica resemble ancient Middle-Easterners, most ancient Mesoamerican effigies are like this one and possess common Far Eastern physical traits.

Olmec Jade masks (1200 BC-600 BC)

The masks shown above so closely resemble Chinese jade masks that it is difficult to demonstrate that independent creation of these masks occurred with no influence of the ancient Chinese. The Chinese have used jade stone to carve effigies for thousands of years. Because these jade masks look so similar to those from China, it appears that the ancient Chinese may have brought their ideas, skills, and heritable physical traits with them to the New World. It is quite improbable that the Olmec and the Chinese are not connected. These traits and features make it even more improbable that the Olmec are Jaredites since their behavior and appearance are so reminiscent of the Far East.

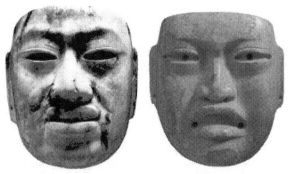

An **Olmec** mask, Rio Pesquero, Veracruz, Mexico (900 BC-600 BC)
& an **Olmec** Jade mask (1000 BC-600 BC)

Featured above are two Olmec Jade masks that may date back to the time of the Jaredites of the Book of Mormon. Note the almond-shaped eyes, flat-faces, and flat and wide-noses which again point to China rather than the Middle East.

An ancient **Kentucky artifact** (200 BC-1 AD),[685]
an ancient **South Arabian artifact** (200 BC-300 BC)
& an ancient **Olmec effigy**, Mexico (1200 BC-400 BC)[686]

For ease of observation and contrast, a prototypical-looking ancient Olmec effigy (featured above in the middle) is shown side-by-side with an average ancient Middle-Eastern effigy (featured above to the right), and an average ancient North American effigy (featured above to the left). The Olmec effigy does not resemble the other two.

Aztec jade mask depicting the god Xipe Totec

Like Far Easterners and the Olmec, the ancient Aztecs carved effigies out of jade. The East Asian and Siberian-like physical features shown here are quite evident. It is once again obvious here that all three of the major groups of ancient Mesoamerica—the Olmec, Maya, and Aztec—predominantly possess physical features of the Far East.

Teotihuacán figurine, Obsidian, Mexico (200 BC-600 BC)

An ancient Teotihuacán figurine that dates back to the time of the Book of Mormon is featured above. It fits right in with all the other East Asian and Siberian-looking artifacts from ancient Mesoamerica.

A **Chinese Terracotta warrior** (210 BC)
& a **Totonac** man, Mexico (600 AD-900 AD)

The Totonac are yet another group of ancient indigenous Mesoamericans with prototypical East Asian and Siberian physical features. They resided in the eastern coastal and mountainous regions of Mexico at the time of the Spanish arrival in the year 1519. Although known to be debated among archaeologists, the earliest historical literature on Mexican indigenous groups has suggested that Teotihuacán (a city often associated with the Book of Mormon by modern-day LDS scholars) was the first major Totonacan city.[687] Featured above on the right is a Totonoc man who—though he has no mustache—strongly resembles an ancient Chinese Terracotta warrior.

Both figures featured above possess prototypical physical features of East Asians and Siberians: flat faces, low nasal bridges, and epicanthic eye folds. Nothing about this ancient Totonac man's appearance leads one to believe that he represents a man who originated in the Middle East or belonged to the Lamanite remnant.

Maya mural painting, Late Preclassic (100 BC-1 BC), Petén, Guatemala[688] & a **Tzendal Mayan** man of Chiapas, Mexico (1902)

A strong physical trait continuity exists among the Maya, and other indigenous Mesoamericans, which spans from ancient times to modern-day Mesoamerica. Ancient Olmecs look like modern-day Olmecs; ancient Maya look like modern-day Maya; ancient Aztecs look like modern-day Aztecs. The general appearance of all these groups does not appear to have changed much over thousands of years.

From the pictures above, we can clearly see that ancient and current Maya commonly possessed flat faces, almond-shaped eyes, small noses, and epicanthic eye folds. Because this ancient Mayan man has so many physical features common to East Asians and Siberians, it is highly likely he possessed the common DNA of East Asia or possibly Siberia. His appearance provides no evidence that he possessed prototypical Semitic DNA. Thus it also appears he is not related to Book of Mormon peoples, including the Lamanite remnant.

An indigenous **Mesoamerican** from Mexico
& an **Olmec** jade mask (1000 BC-600 BC)

Ancient Olmec visages point to an East Asian and Siberian origin rather than a Jaredite origin. Featured above is a modern-day Mesoamerican from Mexico, and also an ancient Olmec jade mask. Both faces appear flat, have epicanthic eye folds, and low nasal bridges. Their lack of Semitic features make it highly improbable that theses two faces depict Jaredites.

An **Olmec** man & an **Olmec** effigy, La Venta, Tabasco, Mexico (1200 BC-900 BC)[689]

Modern-day Olmec look almost identical to the ancient Olmec featured on colossal effigies from places like San Lorenzo Tenochtitlán and Veracruz, Mexico. Both faces featured above appear prototypically East Asian and Siberian, including wider noses and fuller lips.

An ancient **Olmec** figurine, Middle Formative Period (1200 BC-400 BC) & an **indigenous Mesoamerican** (possibly Olmec)

The ancient Olmec, a group many LDS scholars connect to the Jaredites of the Book of Mormon, look almost exactly like modern-day Olmec. Featured above are the head of an ancient Olmec figurine and a modern-day Mesoamerican who look similar.

A **modern-day Tzotzil Mayan**, Chamula, Chiapas, Mexico, an **Olmec** effigy, La Venta site, Mexico (1200 BC-400 BC) & a **Filipino** man

Mayans from Chiapas, Mexico, look like the colossal Olmec effigies from the La Venta site in Tabasco, Mexico. Some of the colossal Olmec heads look like certain Filipinos. These three faces (featured above) possessed prototypical East Asian physical traits.

K'inich Janaab' Pakal (3 angles), a Mayan ruler (603 AD-683 AD)

Featured above is K'inich Janaab' Pakal (603 AD-683 AD), a ruler of the Maya polity of Palenque—a city that Mesoamerican model enthusiasts say was a Nephite city—with a false nose. For whatever reason, some ancient Mayans like K'inich Janaab' Pakal used false noses that made their nasal bridges resemble the high nasal bridges of Semitic peoples. One can tell K'inich Janaab' Pakal has a false nose because it extends all the way up his forehead. Although K'inich Janaab' Pakal ruled after the time of the Book of Mormon, his visage is representative of what Mayans looked like for thousands of years. Without a false nose, his face would be flat. It is possible that Phoenicians or other Middle Easterners interacted with the Maya and influenced the Maya to create false noses, but groups like these appear to be so infrequently represented in effigies, artifacts, and murals that their impact is minor—not the kind of Semitic influence one would expect in a "Book of Mormon region."

A **Mayan** man, Campeche, Mexico (700 AD)

False noses can be found on ancient indigenous Mesoamerican effigies besides Mayan ruler K'inich Janaab' Pakal. Featured above is an ancient Mayan effigy of a man from Campeche, Mexico, dating back to 700 AD, which is a few hundred years after the Book of Mormon time frame. It is unclear where the idea for false noses with high nasal bridges came from, but it is clear that the Mayan practice caught on. Similar to Mayan ruler K'inich Janaab' Pakal, this Mayan effigy (featured above) possesses prototypical East Asian physical features: a flat face, high cheek bones, oval eyes, and epicanthic eye folds.

Because so many Mayan figures on murals, effigies, and figurines at Quiriguá, Palenque, Bonampak, Lamanai, Copán, and Campeche appear East Asian and Siberian (rather than Semitic from the seed of Joseph of Egypt), it is difficult for one to believe the Maya are the Nephites and Lamanites of the Book of Mormon.

Indigenous Mesoamerican, Mexico (100 BC-250 AD)[690]

Featured above is an ancient indigenous Mesoamerican figurine from the Proto-Classic Period (an era that fits within the Book of Mormon time frame). This ancient artifact possesses prototypical East Asian physical features. Its flat face, high cheek bones, oval eyes, and epicanthic eye folds indicate that it most likely has no phenotypic connection to the Middle East or to Book of Mormon peoples. It also does not appear to be a representation of a member of the Lamanite remnant.

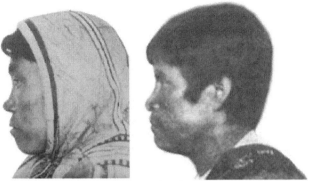

A **Siberian Nganasan** man & a **Totonac** of Mexico[691]

The genetic connection between East Asians, Siberians, and indigenous peoples of the Americas is obvious when the groups are shown together. When a Siberian Nganasan man is compared side by side with the Totonac man (featured above),[692] the two men have similar physical features found among individuals who originated in the Far East.

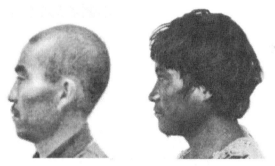

An **East Asian** man & an **Otomi** man from Huixquilucan, Mexico[693]

Featured above is a man with East Asian ancestry and also an Otomi man from Huixquilucan, Mexico.[694] The Otomi are an indigenous ethnic group that inhabit the central altiplano (i.e. the Mexican

Plateau) region of Mexico. Both men have many prototypical East Asian and Siberian physical features including low nasal bridges, flat faces and noses, straight black hair, and epicanthic eye folds. Because the two men featured above appear so similar, and non-Semitic, it is highly unlikely that they belong to the remnant of Lamanites of the Book of Mormon.

An **Olmec** head (two views), San Lorenzo Tenochtitlán, Mexico (1200 BC-400 BC)[695]

Many colossal Olmec heads from San Lorenzo Tenochtitlán, Mexico, that date back to the time of the Jaredites in the Book of Mormon have flat faces (featured above). None of these heads possess prototypical traits of Middle Easterners.

Two views of an **Olmec** effigy (1200 BC-400 AD)

Featured above are two views of an Olmec figurine that dates back to the time of the Jaredites of the Book of Mormon. This ancient figurine possesses prototypical East Asian and Siberian physical traits: a flat face, epicanthic eye folds, and a low nasal bridge. As we can see from this Olmec effigy (and all the other effigies, pictures, photographs, and figurines of indigenous Mesoamericans featured in this chapter), indigenous Mesoamericans almost all appear East Asian.

The Title Page of the Book of Mormon, which was written by the prophet-historian Mormon, teaches us that the book was primarily written to the Lamanite remnant, and yet it was not until 1874—thirty years after Joseph Smith's death—that missionaries were called on missions to Mexico. Guatemalans would have to wait until 1947 for a Latter-day Saint mission to be set up in their homeland. It took over one hundred years after the death of Joseph Smith for the gospel to be preached in a land that, according to Mesoamerican model enthusiasts, is supposed to be the heart of Lamanite civilization. In contrast, during some of the earliest missions of the Church (1830-1831), the Prophet Joseph Smith sent missionaries to the natives of the Great Lakes and Great Plains regions of the United States with the directive to inform these tribes that they had descended from Book of Mormon peoples.

One would think that the overwhelming numbers of East Asian-looking ancient Mesoamerican artifacts, the millions of East Asian-looking modern indigenous Mesoamericans living today, and the late entry into Mesoamerican countries for missionary work would have detoured LDS scholars from believing Mayans were the main Lamanite remnant, but they did not. The general consensus among Latter-day Saint scholars today is that Mesoamerica is the primary setting of the Book of Mormon.⁶⁹⁶ This overall unanimity can be puzzling, especially since the vast majority of effigies, paintings, pictures, murals, and figurines depicting indigenous Mesoamericans during the Book of Mormon time frame appear prototypically East Asian. How could the majority of these scholars believe that almost all Book of Mormon cities were filled with hundreds of thousands of Israelite descendants that looked little to nothing like literal members of the House of Israel? Again, and in summary, by and large, ancient indigenous Mesoamerican skulls resemble the skulls of ancient East Asians and Siberians, the DNA of ancient indigenous Mesoamericans is most similar to East Asians and Siberians. The vast majority of effigies, paintings, pictures, murals, and figurines depicting Mesoamericans during the Book of Mormon time frame appear prototypically East Asian.ᵉᶠ

How could the majority of these scholars believe that almost all Book of Mormon cities were filled with hundreds of thousands of Israelite descendants that looked little to nothing like literal members of the House of Israel? Again, and in summary, by and large, ancient indigenous Mesoamerican skulls resemble the skulls of ancient East Asians and Siberians, the DNA of ancient indigenous Mesoamericans is most similar to East Asians and Siberians, and the vast majority of effigies, paintings, pictures,

---

ef   Flat faces, hooded eyes, and flat noses with low nasal bridges infrequently come from actual Israelite bloodlines, and yet these are the most commonly found physical traits among ancient and modern-day indigenous Mesoamericans. Book of Mormon peoples had "multiplied exceedingly" and so the great dearth of Semitic-looking people on effigies, paintings, pictures, murals, and figurines in ancient Mesoamerican cities show the opposite. A great multitude of King Benjamin's people, too many to number, existed because they had multiplied exceedingly and waxed strong (Mosiah 2:2, 7-8). The Nephites were very numerous, scattered everywhere (Mosiah 27:6). The Lamanites and Amlicites were almost as numerous as the sands of the sea, too numerous to number (Alma 2:27, 35). The Nephites multiplied and waxed strong (Alma 50:18).

murals, and figurines depicting Mesoamericans during the Book of Mormon time frame appear prototypically East Asian.

Lehi and his caravan, the Jaredites, and the Mulekites all arrived in the Americas at separate times anciently, but they all originally came from the Middle East region and therefore almost undoubtedly possessed physical characteristics common to the area. It is a stretch to believe that the majority of them possessed low nasal bridges, wide noses, and round faces. It is even more of a stretch to believe that East Asian and Siberian-looking groups admixed so heavily with the Semitic-looking Book of Mormon peoples in a hypothetical Mesoamerican setting that Semitic bloodlines and visages do not show up anywhere in that region's history.

The great ruins of Mesoamerica are impressive. However, after observing known ancient Israelite edifices, Mesoamerican ruins do not appear to belong in the same category as those typically built by Israelites.[eg] Avi Steinberg, a Jewish author, has written concerning the subject, "The Israelites were far from great builders. Their structures were never impressive."[697] Those who read the Book of Mormon and think of indigenous Mesoamericans are often highly influenced by, as Hugh Nibley put it, "[T]he mighty ruins of Babylon"[698] in Mesoamerica. Many Latter-day Saints are "'[N]ot interested' in the drab and commonplace remains of our

---

eg Latter-day Saint scholar Hugh Nibley once remarked concerning the massive structures of Mesoamerica: "The great monuments [of Mesoamerica] do not represent what the Nephites stood for; rather they stand for what their descendants, mixed with the blood of their brethren, descended to." Hugh Nibley, *The Prophetic Book of Mormon*, pp. 272-273. The pyramid-shaped structures of Mesoamerica most closely resemble the ziggurats (step pyramids) of the non-believer Babylonians, not the Israelites. Israelites for the most part, were a people of ideas, not necessarily stone workers.

lowly [United States] Indians"⁶⁹⁹ even though, as Nibley so aptly stated, "[I]n all the Book of Mormon we look in vain for anything that promises majestic ruins."⁷⁰⁰

Pyramids belonged to the Egyptians, Assyrians, and Babylonians, but not the Israelites. Hugh Nibley remarked about how the massive amount of stone ruins in Mesoamerica do not really belong in the Book of Mormon: "In view of the nature of their civilization one should not be puzzled if the Nephites had left us no ruins at all. . . . [T]he ancients almost never built of stone."⁷⁰¹ Nibley also expressed the idea that "Book of Mormon archaeologists [who almost all believe the Book of Mormon primarily transpired in Mesoamerica] have often been disappointed . . . because they have consistently looked for the wrong things."⁷⁰² To them, he suggested: "We must stop looking for the wrong things."⁷⁰³

In summary, although many indigenous Mesoamericans have been adopted into the seed of Joseph of Egypt over the years via baptism, the great majority of indigenous Mesoamericans do not appear to comprise the main Lamanite remnant. One has to search assiduously to find vestiges of common ancient Middle Eastern and Semitic physical features on ancient artifacts dating back to Book of Mormon times. As demonstrated in this chapter, almost all Mayan, Olmec, and Aztec figurines, effigies and murals depict individuals with East Asian and Siberian ancestry. This implies that the ancestors of Book of Mormon peoples were not mostly indigenous Mesoamericans, but in fact lived further north.

15

The Destiny of the Lamanite Remnant

But before the great day of the Lord shall come, . . . the Lamanites shall blossom as the rose[.] —**Doctrine and Covenants 49:24**

This day the Lord has shown to me . . . You will gather the Redmen to their center from their scattered and dispersed situation, to become the strong arm of Jehovah[.] —**Joseph Smith, Jr., Latter-day Saint Prophet (1844)** [704]

We have forgotten the . . . sons of Joseph [of Egypt] . . . We have forgotten . . . the great redemption of Israel. . . . How do you think their [fore]fathers would feel to see their descendants, without knowledge of God and their ancestors? —**Orson Pratt, Latter-day Saint Apostle (1855)** [705]

Although God loves all of His children, He sets aside His greatest blessings for the offspring who keep covenants. Nephi saw in vision the blessings provided by God to those who are covenant-keepers: "I, Nephi beheld the power of the Lamb of God, that it descended . . . upon the covenant people of the Lord, . . . and they were armed with righteousness and with the power of God in great glory" (1 Nephi 14:14). Since the Lamanites made covenants with God, their descendants will receive great blessings if they renew the covenants their Lamanite forefathers made. This is their destiny. In an early revelation received by Joseph Smith in 1828, he was instructed by God that the golden plates were preserved so the "Lamanites might come to the knowledge of their fathers, and . . . know the promises of the Lord[.]" (Doctrine and Covenants 3:18-20).[eh]

Elder Orson Pratt appears to have understood the destiny of the Lamanite remnant as envisioned by the Prophet Joseph Smith. Elder Pratt also had ideas about actions that Latter-day Saints can take to assist the Lamanite remnant in accomplishing God's purposes for His people. In 1875, Elder Pratt informed members of the Church in Salt Lake City that they had flattered themselves into thinking they alone will build up the kingdom of God. Latter-day Saints, he said, are mere "helpers" who are meant to cooperate with the Lamanite remnant (the remnants of Joseph) to accomplish the great work of God and to return with them to the city of Zion (the New Jerusalem) located in Jackson County, Missouri. To the Saints in Salt Lake City, Elder Pratt added more on the subject:

> [H]ere is a work for us, and we have no need to pray the Father to return us to Jackson County until that work is done. We can pray to

---

[eh] Latter-day Saint scholar Hugh Nibley recognized the Hopi Indians of northeastern Arizona as members of the Lamanite remnant after noticing connections between the Hopi, Latter-day Saint temple rituals, and rituals found in the Book of Mormon. According to Nibley's son-in-law, Boyd Petersen, Nibley saw "Parallels . . . between the language of the Mormon temple ceremony and the Hopi myths of origin in Frank Water's *Book of the Hopi*." Nibley once wrote that the parallels between the Latter-day Saint endowment and the rites of the Hopi "come closest of all as far as I have been able to discover—and where did they get theirs?" What is more, Nibley knew that "Zarahemla means 'red city,' but what attracts me about that is that the Hopis say that their people came from the 'great Red City'". Nibley, Hugh, Teachings of the Book of Mormon: Semester 1, Lecture 26. According to The *Book of the Hopi*: "The Hopis in these villages had long anticipated the coming of their lost white brother, Pahána."

the Father, in the name of Jesus, to convert these Indian tribes around us, and bring them to a knowledge of the truth, that they may fulfill the things contained in the Book of Mormon. And then when we do return, taking them with us, that they shall be instructed not only in relation to their [fore]fathers and the Gospel contained in the record of their [fore]fathers[.] . . . [A]nd then, after having received . . . information and instruction, we shall have the privilege of helping them to build the New Jerusalem. The Lord says—'They,' the Gentiles, who believe in the Book of Mormon, 'shall assist my people, the remnant of Jacob, that they may build a city, which shall be called the New Jerusalem.' [F]or the Lord will have respect unto them, because they are of the blood of Israel, and the promises of their [fore]fathers extends unto them. . . . Do not misunderstand me, do not think that all the Lamanite tribes are going to be converted . . . before we can return to Jackson County. . . . [I]t will only be a remnant. . . . [T]here is another great work which we have got to do in connection with these remnants of Jacob whom we shall assist in building the city. . . . [T]he . . . shepherd will lead Joseph as a flock, and he will stir up his strength and will save the house of Joseph [of Egypt]. But it will be in his own time and way . . . a remnant will be converted.[706]

As Elder Pratt asserted, Latter-day Saints are to assist the Lamanite remnant in building the city of Zion (the New Jerusalem) after they receive information about the covenant their forefathers made with God. Elder Pratt also seems cognizant that not all of the tribes who belong to the Lamanite remnant were going to be converted and assume their leadership roles before Latter-day Saints were to return to Jackson County.[707] Endeavors like that take time, and may never take place if Latter-day Saints have little to no desire to inform the Lamanite remnant of the covenants made by their forefathers with God, or if the Lamanite remnant choose not to renew the covenant of their forefathers.[708]

As previously discussed in *Chapter 2: Did Joseph Smith Change His Mind?*, the Prophet Joseph Smith was committed to fulfilling Book of Mormon prophecy with the Lamanite remnant in North America. He was keenly aware of the promises God made with Book of Mormon peoples that could be extended to their descendants if they chose to follow the teachings of Jesus Christ and received baptism into His Church. Just three days prior to Joseph Smith's death, he demonstrated his commitment to the Lamanite remnant of North America during his last discourse to the Saints near the Nauvoo House in Illinois.[ei] According to this reminiscent account of the event by Latter-day Saint William B. Pace, the Prophet admonished members of the Church to inform "the red men from the west" and "free men" of the US to go west and "gather the red men to their center from their scattered and dispersed situation to become the strong arm of Jehovah, who will be a strong bulwark of protection from your foes."[709]

In 1855, Elder Wilford Woodruff corroborated Joseph Smith's 1844 statement about "red men" (as quoted by William B. Pace) when he spoke in Provo, Utah, about the Lamanite remnant. Elder Woodruff references the physical aspect of the 1844 promise of protection[710] made by Joseph Smith the prophet:

> I can tell you the Lamanites of these mountains will yet be a shield to this people if we do right, and if we will not do our duty, our necks are ready for the halter or the knife; yes, you will find that our necks will be ready for the knives of our enemies, if we do not look to these poor degraded natives.[711]

---

ei   The Nauvoo House was hotel that would be "a delightful habitation for man, and a resting place for the weary traveler" (D&C 124:60).

Our Duty to Our Red Brethren

As etched on the golden plates, the Book of Mormon was primarily written to the main Lamanite remnant in order that they might accept Jesus Christ as their Savior, renew the covenant made with their forefathers, and become baptized into Jesus Christ's church. Elder Wilford Woodruff seems to have understood Joseph Smith's vision for scattered Israel among the American Indian tribes and their duty to them. In 1855, Elder Woodruff remarked:

> I want to know now, if the Mormons can really and truly realize our true position with regard to the Lamanites (the American Indians), or do you consider them inferior? It is necessary to do our duty to our red brethren. When the Priesthood was restored, I knew then that if it were not for the Israelites (the American Indians), the Gentiles might go to hell and be damned. The Lord would not take much pains with us anyhow, were it not for the promised seed (the American Indians). Instead of them being inferior to us in birthright, they are superior in birthright; and they stand first in many instances, with regard to the promises of God in particular.[712]

In this quote, Elder Woodruff referred to the Lamanite remnant as "red brethren," which is a direct reference to indigenous North Americans, especially Algonquian-speaking groups such as the Delaware (Lenni Lenape). The main Lamanite remnant, according to Elder Woodruff, are "superior in birthright" and need to be informed of the promises of God so they can fulfill prophecies made in both the Book of Mormon and the Doctrine and Covenants. In 1873, Wilford Woodruff reiterated a prophecy recorded in Doctrine and Covenants 49:24, and added his own thoughts on the prophecy:

The Lamanites will blossom as the rose on the mountains. I am willing to say here that, though I believe this, when I see the power of the nation destroying them from the face of the earth, the fulfillment of that prophecy is perhaps harder for me to believe than any revelation of God that I ever read. It looks as though there would not be enough left to receive the Gospel; but notwithstanding this dark picture, every word that God has ever said of them will have its fulfillment, and they, by and by, will receive the Gospel. It will be a day of God's power among them, and a nation will be born in a day. Their chiefs will be filled with the power of God and receive the Gospel, and they will go forth and build the [N]ew Jerusalem, and we shall help them. They are branches of the house of Israel, and when the fullness of the Gentiles has come in and the work ceases among them, then it will go in power to the seed of Abraham.[713]

Like Joseph Smith, Elders Woodruff and Pratt both knew of the importance of Zion (the New Jerusalem) and it seems they were aware that Latter-day Saints are meant to assist the main Lamanite remnant in establishing Zion (the New Jerusalem).

Besides Elders Pratt and Woodruff, a number of other General Authorities also understood the destiny of the Lamanite remnant. In October of 1882 in the *Millennial Star* (an official publication of the LDS Church), President John Taylor wrote about the urgency of the work among the Lamanite remnant:

> The work of the Lord among the Lamanites must not be postponed, if we desire to retain the approval of God. . . . [T]he same devoted effort, the same care in instructing, the same organization of priesthood must be introduced and maintained among the house of Lehi as amongst those of Israel gathered from gentile nations. . . . [T]reat them exactly, in these respects, as we would and do treat our white brethren.[714]

During General Conference in 1947, Elder Spencer W. Kimball uttered a vehement message stating that "The Lamanites must rise in majesty and power."[715] Elder Kimball used the word "must" in conjunction with the possible destiny of the Lamanite remnant. The word "must" is a strong statement that displays his passion and love for these indigenous peoples. Although Lamanite descendants still have the free will to fulfill the promises of the Lord, God will not force them to fulfill the promises He made with their forefathers. We as Latter-day Saints, however, can do our part to provide our "red brethren" with the opportunity to choose.

In the same vein, President Kimball later said in 1975 concerning the Lamanite remnant:

> The Lamanites must rise again in dignity and strength to fully join their brethren and sisters of the household of God in carrying forth his work in preparation for that day when the Lord Jesus Christ will return to lead his people, when the millennium will be ushered in, when the earth will be renewed and receive its paradisiacal glory and its lands be united and become one land. For the prophets have said, 'The remnant of the house of Joseph shall be built upon this land; and it shall be a land of their inheritance; and they shall build up a holy city unto the Lord, like unto the Jerusalem of old; and they shall no more be confounded, until the end come when the earth shall pass away' (Ether 13:8).[716]

North American Indian Conversion

Relatively few indigenous North American Indians have been baptized and sealed in the temple since the establishment of the first Latter-day Saint missions to the Lamanites of North

America. In February of 1875 Elder Wilford Woodruff performed the first marriage sealing of Sagwitch and Mogagah, and of Ohetocump (also known as James Laman) and Minnie, two Shoshone Indian couples in the Endowment House—the name of the Salt Lake Temple prior to completion—on the temple block in Salt Lake City, Utah. Sagwitch's son would later be called on a mission and Sagwitch's grandson would become the first LDS bishop among Native Americans. In 1892, during a conference in St. George, Utah, President Wilford Woodruff spoke about the sealings he had performed for members of the Lamanite remnant:

> I am satisfied that although we have done a little for the Lamanites, we have got to do a great deal more. I sealed the first Lamanite-ish man and woman together that ever were sealed in this dispensation. It was in the Endowment House, and quite a number of brethren and sisters were present. The man's name was Laman. The day will come when these Lamanites, with the dark skin that rests upon them, will enter into these Temples of the Lord in these mountains and do a great deal of work. They will come to an understanding of the redemption of the dead. They will have wisdom given unto them. They will have light and truth given unto them, and the spirit of their forefathers will be manifest unto them. I am thankful that I am able to see these Lamanites here. The Prophet of God saw what would come to pass, and he told the truth.[717]

Answers About the Lamanite remnant

This book was set up with the goal to answer the following questions: "Did Joseph Smith know who the main Lamanite remnant were?", "Are all indigenous Americans the descendants of

Joseph of Egypt and also related to Book of Mormon peoples?", "How can one tell the difference between which natives are and which ones are not?", and "Does DNA evidence exist that links Joseph of Egypt to indigenous Americans?"

We began by exploring Joseph Smith's claims regarding the identity of Book of Mormon peoples. The Prophet specified a number of times throughout his life that specific North American Indian tribes belonged to the Lamanite remnant, beginning with the Sauk & Fox tribes in 1841. Concerning their Book of Mormon ancestry, Joseph Smith claimed that the "Great Spirit" told him that the forefathers of indigenous North Americans wrote the Book of Mormon. Joseph Smith was consistent in this claim throughout his entire life, even after the unofficial/unsigned *Times and Seasons* articles were released in the controversial year of 1842—a year that many LDS scholars currently believe the Prophet began to shift his focus from North America to Mesoamerica.[ej] Days before his death in 1844, Joseph solidified his belief that indigenous North Americans (red men) were the main Lamanite remnant when he said shortly before passing: "This day the Lord has shown to me[:] . . . You will gather the Redmen to their center from their scattered and dispersed situation, to become the strong arm of Jehovah[.]"[718]

In addition, the Prophet Joseph Smith provided us with clues to identify the main Lamanite remnant in scripture. In the

---

ej Because of a limited geography for the Book of Mormon, and due to the fact the Prophet informed North American Indian tribes in the last years of his life that their forefathers wrote the Book of Mormon, we can safely assume he did not change his mind about indigenous North Americans as the main Lamanite remnant.

Doctrine and Covenants, he specifically sent missionaries to the "borders of the Lamanites" (D&C 28:14) and the "wilderness among the Lamanites" (D&C 32:2), which were all located in the continental United States of America. In other places, Joseph Smith referred to the "plains of the Nephites" and "Indians (Lamanites) within the territorial limits of the United States" and an "old Nephite altar" that were all located in North America. It is clear from these examples that the Prophet believed at least a few North American Indian tribes belonged to the genetic lineage of Joseph of Egypt.

The fact that the Church began in Seneca County, New York—a county named after the first Indian tribe visited by missionaries sent by Joseph Smith on the first Lamanite-specific mission seems to corroborate the idea that the Prophet believed indigenous North Americans belonged to the Lamanite remnant. This solid connection appears to be a divine correlation between the Prophet's birthplace, the location of the organization of the Church, the location of where the golden plates were buried, and the locations of the American Indian tribes to which missionaries visited during Joseph Smith's lifetime. A limited geography for the Book of Mormon plus divine correlation between where Joseph Smith lived and where the Hill Cumorah is located, leads one to agree with the Prophet's belief that a number of North American Indian tribes and the Lamanite remnant are one and the same.

With a knowledge that genetic markers are consistent from generation to generation, we explored the DNA and physical traits of Semites and looked to identify where those markers appear in the Americas. It is clear that the vast majority of Mesoamerican DNA haplogroups and traits came from East Asia and Siberia. In contrast, we are able to pinpoint numerous individuals from North American Indian tribes with Semitic ancestry, demonstrating that they are descendants of Semitic groups, and more specifically belong to the tribe of Joseph of Egypt. We also observed numerous artifacts and images that indicate a strong correlation between North American tribes and their Semitic ancestors. This type of solid evidence that links to Semitic groups is virtually non-existent in Mesoamerica. Between the DNA, artifacts, skulls, and physical traits it becomes apparent that Joseph Smith was correct in stating that the red men (North American Indians) did in fact descend directly from Book of Mormon peoples.

While we may never have sufficient evidence to prove beyond doubt the location of the Book of Mormon, it is worth our time to consider the identity of the Lamanite remnant and how we can help God to fulfill His promise to them. This book has focused on the idea that select groups of North American Indians are the main Lamanite remnant. Of course readers can form their own opinions on the matter, however, this book invites readers to consider the evidence that seems to favor many indigenous North Americans as descendants of the Lamanites. Regardless of which group of indigenous Americans readers believe is the main Lamanite remnant, it is important for us to come together as one heart and one mind without any contention (Moses 7:18), understanding that the main purpose of identifying the Lamanite remnant is to gather Israel and establish Zion. We all have the opportunity to follow the admonition contained in Doctrine and Covenants Section 100 to become a committed society of "pure people" who serve God in righteousness (D&C 100:13,16).

Appendix 1

Genetic Markers Expansion

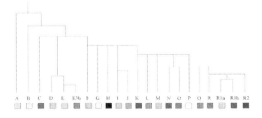

Semitic DNA can be found among Kurds,[719] Kurdish Jews,[720] Sephardic Jews,[721] Ashkenazi Jews,[722] the Bedouin,[723] Palestinians,[724] Yemeni,[725] Albanians,[726] Macedonians,[727] and Italians (especially Sicilians from Italy),[728] Greeks (especially Greek Cypriots),[729] Bulgarians,[730] French,[731] Germans,[732] Austrians, Belgians, Bosnians, Croatians, Czechs, Danish, Estonians, Finns, Hungarians, Icelanders, Latvians, Lithuanians, Netherlanders, Norwegians, Portuguese, Romanians, Russians, Serbians, Slovakians, Spaniards, Swedes, Swiss, Ukrainians, and countries of the United Kingdom. By no coincidence, many of these groups share similar physical traits. For example, many Sicilians from Italy not only look like, act like, and have the same temperament traits as Jews, but they also share almost the exact same sets of genetic markers.

Common Semitic Paternal DNA

Among Semites, a number of common genetic markers exist. Common paternal lineages (Y-DNA) among Semites include Haplogroups J, E, T, G, I, and R1, R1b, and R1a. Geneticists have discovered that the Jews mostly belong to Haplogroups J,[733] E,[734] T,[735] and G.[736] Joseph of Egypt's descendants appear to come from many of these lineages, but especially from Haplogroups I, R1, R1b, and R1a (most Scandinavians and Brits belong to these four groupings). Ephraimite lineages especially tend to come from Europe and belong to these four haplogroups.

Paternal Lineage 'R1b'

Barbara Arredi and colleagues were the first to point out that the distribution of R1b STR variance in Europe forms a cline from east to west, which is more consistent with an entry into Europe from Western Asia [the Middle East and the Near East] with the spread of farming.
—**Barbara Arredi, E. S. Poloni, and C. Tyler-Smith (2007)**[737]

The territory of Ephraim, which was part of the Kingdom of Israel, was conquered by the Assyrians, and the tribe was exiled. They became "lost."[738] It appears from genetic data that Ephraimites became "lost" in regions of Europe. Paternal lineage Haplogroup R1b, which is related to Haplogroup R1 (a paternal lineage featured in *Chapter 10: The Genetic Markers of Semites, East Asians, and Siberians*) appears to belong to the tribe of Ephraim.

The Prophet Joseph Smith wrote a letter to Isaac Galland,[739] a merchant and doctor, about the Ephraimites he believed were located in Europe:

Some of the Twelve and others have already started for Europe [in

September 1839], and the remainder of that mission we expect will go now in a few days. ... The work of the Lord rolls on in a very pleasing manner, both in this and in the old country. In England many hundreds have of late been added to our numbers; but so, even so, it must be, for 'Ephraim he hath mixed himself among the people' [Hosea 7:8]. And the Savior He hath said, 'My sheep hear my voice' [John 10:27].[740]

Joseph Smith quoted from the Book of Hosea in the Bible to indicate that the seed of Joseph of Egypt was in England.[741] Brigham Young appears to have espoused this idea as well. Young seemed convinced that Ephraimite bloodlines were abundantly found in the British Isles.[ek] As a successful missionary in England, Young most likely believed that the Ephraimite ancestry of much of Great Britain played a role in his success. Young wrote:

> We landed in the spring of 1840, as strangers in a strange land and penniless, but through the mercy of God we have gained many friends, established churches in almost every noted town and city in the kingdom of Great Britain, baptized between seven and eight thousand.[742]

According to DNA statistics, nearly 70% of males tested in England belonged to paternal lineage Haplogroup R1b (a common paternal lineage among certain Middle Easterners).[743] Prominent figures in world history such as Abraham Lincoln,[el] John Adams, and John Quincy Adams—who appear to have possessed "lost"

[ek] A recent study states that most Britons are direct descendants of farmers who moved west from the Near East 10,000 years ago. After analyzing the DNA of more than 2,000 men, researchers at Leicester University say they have compelling evidence that four out of five white Europeans can trace their roots to lands that form part of modern-day Iraq and Syria.

[el] Jewish Union Rabbi Isaac Wise, who knew Abraham Lincoln, was convinced Lincoln was Semitic when he said in 1865, "Abraham Lincoln . . . possessed the common features of the Hebrew race both in countenance and features." *Abraham Lincoln and the Jews*, p. 5, Isaac Markers (1909).

Israelite blood because of their Semitic-looking physical traits—all belonged to Haplogroup R1b.[744]

The highest percentages of paternal lineage Haplogroup R1b are often found within countries in the United Kingdom. Of the Welsh males who participated in a DNA study, 92.3% belonged to Haplogroup R1b.[745] In the early days of the Church, many of the Welsh were receptive to the Latter-day Saint message. They joined the LDS Church in droves. Under Latter-day Saint Elder Dan Jones's leadership, LDS missionaries serving in Wales between 1845 and 1848 baptized approximately 3,600 people. Because of how prevalent the idea about Ephraimites in the British Isles was among Latter-day Saints at the time, Elder Jones may have considered attributing much of his success to the idea that Ephraimites were commonly found among the Welsh people.

Other prominent groups that belong to Haplogroup R1b:

Group of People	% of males tested who belonged to Haplogroup R1b
Basques (of France and Spain)[746]	87.1
Baskirs (Perm) of Russia[747]	86
Irish[748]	85.4
Scottish[749]	77.1
Shewa Arab[750]	40
Assyrians (of Iran)[751]	32.3
Siwa Berber (of Egypt)[752]	28
Kurdish Jews (of Iran)[753]	20.2
Arabs (of Jordan)[754]	17.8
Ashkenazi Jews[755]	15

This table is significant because it connects two groups with many Israelites among them who belong to paternal lineage R1b: European Israelites and Israelites from the Middle East.

Common Siberian and East Asian Maternal DNA

The most common female lineages (mtDNA)[em] of American Indians also stem from Siberian and East Asian haplogroups: A, B, C, and D. (Haplogroup X is also found among American Indians, but it most likely did not originate in Siberia or East Asia.) A great majority of those who possess these maternal lineages do not appear to belong to the Lamanite remnant because their phenotypes and DNA are mostly Siberian and East Asian—rarely Semitic.

Lehi and his caravan were largely, if not all, Semites with common Semitic phenotypes of the time. The Book of Mormon does not give any indication that the Nephites, Lamanites, Mulekites, and Jaredites had flat faces, flat and wide noses, and epicanthic eye folds—all common phenotypes of East Asians and Siberians. Highly religious Semites in ancient times were discouraged by their leaders from marrying foreigners, and primarily married their own people. This limited amount of admixture among East Asians living at the time of Lehi. Laman and Lemuel (and even Nephi) might have had descendants who admixed with East Asians and Siberians, but it seems most likely that these men possessed pureblooded Semitic lineages which persisted through Book of Mormon times. The prophet Mormon emphatically declared, "I am Mormon, and a pure descendant of

---

em   Although mitochondrial DNA is used to track maternal lineages, it can also be tested in males as well as females.

Lehi. I have reason to bless my God and my Savior Jesus Christ, that he brought our fathers out of the land of Jerusalem" (3 Nephi 5:20).

Maternal Lineage 'A'

Siberian **Chukchi woman**, Yerchin River (1901),
a modern Lacandon **Mayan** woman, Mexico (1952)
& an ancient **Mayan** woman effigy, Guatemala (300 AD-800 AD)[756]

Members of Haplogroup A—a maternal lineage—follow the common Siberian and East Asian phenotypic patterns of paternal lineage Haplogroup Q.[757] Featured above is a Siberian woman and also two prototypical Mayan females. One of the women is a modern-day Maya, whereas the other woman is a Mayan dignitary featured on an ancient effigy which dates back to the approximate time of the Book of Mormon. The physical similarities between these three women is quite noticeable. They all have short and flat faces, epicanthic eye folds, and low nasal bridges—all common physical traits among East Asians and Siberians.

Haplogroup A appears to have originated in and spread from the Far East.[758] Certain Siberian lineages are known to belong to

Haplogroup A in high percentages. The Siberian Chukchi are one of these groups. Of the Siberian Chukchi whose mitochondrial DNA was tested, approximately 70% belonged to Haplogroup A.[759] This percentage is similar to the numbers found among the modern-day Maya of Mesoamerica. Of the 27 modern-day Mexican Mayans of the Yucatan whose mitochondrial DNA was tested, 55.6% belonged to Haplogroup A.[760] Nearly 90% of the ancient Mayans of Xcaret in Mexico (Yucatan, Quintana Roo) whose mitochondrial DNA was tested belonged to Haplogroup A.[761] These high percentages of Haplogroup A and the common East Asian and Siberian phenotypes, found among Siberian Chukchi peoples and the Maya (ancient and modern) point to similar Northeast Asian origins: Siberia.

Similar to their Mayan relatives, Aztec women of Mesoamerica who have had their DNA tested also mostly belong to Haplogroup A. Of the 43 Aztec women of Tlatelolco (Mexico) whose mitochondrial DNA was tested, 65.2% belonged to Haplogroup A.[762] This pattern mirrors the haplogroups of the Maya and the Siberian Chukchi. This data collected from these three groups indicates that the great majority of indigenous Mesoamericans who belong to Haplogroup A not only come from East Asia and Siberia, but have no known ties to the Middle East or Semites. This is why most of the Maya do not appear to belong to the Lamanite remnant.

A **Greenlandic Inuit** woman
& a **Siberian Chukchi** woman

Featured above is a Greenlandic Inuit woman and a Siberian Chukchi woman. Greenlandic Inuit females largely belong to Haplogroup A as well. The East Asian and Siberian physical features of these two women—jet black hair, high cheekbones, flat faces, wide and flat noses, and epicanthic eye folds—indicate their likely common genetic origin.

The physical similarities of the two women featured above correlate with groups of genetic markers found among East Asians and Siberians. Of the 385 Greenlandic Inuit whose mitochondrial DNA was tested, 96.1% belonged to Haplogroup A. That is nearly all Greenlandic Inuit women! Of the Siberian Chukchi whose mitochondrial DNA was tested, nearly 70% were Haplogroup A.[763] These numbers indicate that the women featured above most likely belong to Haplogroup A. The data and visages of these two women also indicate that they most likely do not belong to the Lamanite remnant.

Maternal Lineage 'B'

A **Yao** woman, Tiantouzhai, Longji Terraces, China & a **Maya** woman

Haplogroup B—a maternal lineage—is frequently found in southeastern Asia and in the Americas. Featured above is a Dingban Yao woman of China, and also a Maya woman from Mesoamerica. Because of the visages of the women featured above, they may have belonged to Haplogroup B. Of the 10 Dingban Yao of China (Mengla, Yunnan) whose mitochondrial DNA was tested, 60% belonged to Haplogroup B.[764] Of the 27 modern-day Mexican Maya of the Yucatan whose mitochondrial DNA was tested, 33.1% were Haplogroup B.[765]

Groups in Southeast Asia, including Filipinos, belong to Haplogroup B in high percentages. Of the 64 Filipinos whose mitochondrial DNA was tested, 42.2% belonged to Haplogroup B.[766] Because of the common visages of the Chinese, Maya, and Filipinos, it is easy to see that individuals who are Haplogroup B are primarily East Asian in appearance. Due to the lack of prototypical Semitic physical traits among these groups, they do not appear to have physical or genetic ties to Middle Eastern groups, especially Semites.

Maternal Lineage 'C'

An East Siberian **Yukagir** woman & a **Mayan** woman, Valladolid, Yucatán, Mexico

The maternal lineage Haplogroup C is commonly found in Northeast Asia and in the Americas.[767] Featured above is an East Siberian Yukagir woman and a Mayan woman from Mexico. Both women possessed flat faces, wide noses, and epicanthic eye folds—all common physical traits among East Asians and Siberians. The prevalence of Haplogroup C occurs at above 50% among Siberian Yukaghirs of northeastern Asia, central Siberian Yakuts, and Siberian Evenks, as well as East-Sayan Tofalars. Of the 9 ancient Maya of Copán, Honduras, whose mitochondrial DNA was tested, 89.9% were Haplogroup C.[768] Of the 27 modern-day Mexican Mayans of the Yucatan whose mitochondrial DNA was tested, 6.2% were Haplogroup C.[769] Some Aztecs belong to this particular haplogroup as well. Of the 43 Aztec women of Tlatelolco (Mexico) whose mitochondrial DNA was tested, 4.3% belonged to Haplogroup C.[770] These groups that possess percentages of Haplogroup C also have prototypical East Asian appearances with no known physical or genetic ties to the Middle East or Semites. The data collected on these groups, and the common East Asian and Siberian visages found among them, fail to indicate that they belong to the Lamanite remnant.

Maternal Lineage 'D'

An **Aleut** woman

Haplogroup D, a maternal lineage, is found in high percentages in Northeast Asia (which includes Siberia).[771] Featured above is an Aleut woman who possessed prototypical East Asian and Siberian physical traits: a flat face, jet black hair, epicanthic eye folds, and a low nasal bridge. Of the 36 Aleuts of the Commander Islands of North America whose mitochondrial DNA was tested, 100% belonged to Haplogroup D. That is every single Aleut who was tested! Of the 163 Aleuts of the Aleutian Islands of North America (near Alaska) whose mitochondrial DNA was tested, 65.6% of them were Haplogroup D.[772] Because of the high percentages of maternal lineage 'D' among the Aleut, it is highly likely the Aleut woman featured above belongs to 'D.'

Some Aztecs also belong to Haplogroup D. Of the 43 Aztec women of Tlatelolco (Mexico) whose mitochondrial DNA was tested, 17.4% belonged to 'D.'[773] These groups all share Northeast Asian haplogroups and possess prototypical East Asian appearances with no known physical or genetic ties to the Middle East or Semites.

Common Semitic Maternal DNA

It is probable that groups of Joseph of Egypt's descendants can be found among prototypical Semite Haplogroups: K,[774] H,[775] I,[776] N1b,[777] T,[778] U,[779] V,[780] W,[781] X, and J1.[782]

Maternal Lineage 'X'

A **Druze** woman, Isfiya, northern Israel (about 40 miles from the Sea of Galilee)[783] & **Falling Star,** an Abenaki of the Algonquin peoples (1900)[784]

The maternal lineage Haplogroup X (mtDNA) is found among Semites and appears to originate in the Middle East. Featured above is an Israeli Druze woman and Falling Star, an Algonquin woman of the Abenaki tribe. Genetically speaking, the Druze are similar to Bedouin Arabs, who have historically had close ties to the tribe of Manasseh. Both of these women share common Semitic appearances: long and thin noses with high nasal bridges indicating they might belong to maternal lineage 'X.' The Semitic visages of the Israeli Druze woman and the Algonquin woman are extremely rare in East Asia, Siberia, and Mesoamerica.

One study of Israeli Druze found that of the 41 Druze from the Galilee heights of northern Israel who had their DNA tested, 95.1% were Haplogroup X.[785] Since this figure is by far the highest percentage of X on earth, Israel is most likely where Haplogroup X originated.

Significant amounts of 'X' have also been found among Algonquin tribes of North America. Within Algonquin tribes, especially the Ojibwe and Mi'kmaq, Haplogroup X has been found in percentages of 40-50%.[786] Of the Sioux and Iroquois who had their DNA tested, 15% of each of these tribes possessed the genetic markers of Haplogroup X.[787] Haplogroup X is also found among other indigenous North American groups, such as the Nuu-Chah-Nulth (13.3%), Navajo (11-13%), Dogrib (7%), and Yakima (5%).[788]

Concerning Haplogroup X, William Marder and Paul Tice—authors of *Indians in the Americas: The Untold Story*—have said:

> In the Americas, the X group has so far been found . . . in larger numbers among the Ojibway (Ojibwe, Chippewa), Iroquois and Oneida tribes, and in ancient remains in Illinois near Ohio and near the Great Lakes. The geneticists are certain haplogroup X type is very ancient and not a result of intermarriages.[789]

Because Haplogroup X is an ancient maternal lineage that apparently first appeared in the New World over 10,000 years ago, it seems to have arrived long before the Book of Mormon time frame. However, the fact that this haplogroup was discovered in the Americas much earlier than the time period of the Book of Mormon does not necessarily mean that Haplogroup X did not arrive in the New World via Lehi and his caravan. It is possible that this particular haplogroup could have been brought to the Americas at several different time periods.

A **Druze** woman, Isfiya, northern Israel (about 40 miles from the Sea of Galilee), **Falling Star,** an Abenaki of the Algonquin peoples (1900) & a **Gypsy** woman

The Middle Eastern origin of maternal lineage Haplogroup X becomes more apparent when other groups with this maternal lineage are analyzed.[790] Featured above is a Druze woman, who is situated next to an Algonquin woman and a Gypsy woman. All three women possessed similar non-East Asian phenotypes. Each of them had a common Middle Eastern look, and could have been related to Semites. Of the Roma Gypsies whose DNA was tested, 7.6% were Haplogroup X.[791] Even though some scholars believe that the Gypsies originally came from South Asia (specifically

350

India), it is possible that they actually originated in the Middle East since many of them share similar haplogroups with Semites.

North American Indians who belong to Haplogroup X appear to share some genetics and physical traits with Gypsies. This possible connection seems to go beyond phenotypes and genotypes. William Penn connected a common practice shared by North American Indians and Gypsies when he wrote,

> The NATIVES [are of] complexion black, but by design, as the gypsies[.] . . . They grease themselves with bear's fat clarified, and using no defense against sun or weather, their skins must needs be swarthy [(dark-skinned)]. . . . I find the Indians of the like countenance with the Jews.[792]

North American Indians who would blacken their faces with "bear's fat clarified," reminded Penn of the Gypsies—a group long remembered for being expert metalworkers. Fine metallurgists existed among ancient indigenous North Americans dating back to Book of Mormon times. It is possible that this "blacksmith look" in North America among the natives was passed down to from generation to generation.[793]

The maternal lineage Haplogroup X, which is one of the only possible genetic links that indigenous American peoples have to the Middle East, is virtually non-existent, or absent, among indigenous Mesoamericans. Subgroups (subclades) of Haplogroup X—X2a and X2g—are found in North America but are yet to be discovered among native Latin Americans.[794]

Mitochondrial DNA tested among living and deceased indigenous Mesoamericans:

Group of People	% of indigenous Mesoamericans tested who belonged to Haplogroup X
Aztecs of Tlatelolco (Mexico)[795]	0
Ancient Maya (of Xcaret, Mexico)[796]	0
Ancient Maya (of Copán, Honduras)[797]	0
Modern-day Maya (Yucatan, Mexico)[798]	0

Indigenous Mesoamericans are almost completely devoid of Haplogroup X, which means they have fewer genetic ties to the ancient Middle East than indigenous North Americans.[799] The lack of Haplogroup X among indigenous Mesoamericans makes them appear less Middle Eastern than indigenous North Americans.

The maternal lineage Haplogroup X2 is almost entirely absent from Siberia, which means it most likely did not originate there. Conversely, Haplogroup X2e (a subgroup of X2) is present in North American Indian DNA, demonstrating that North American Indians did not originate in Siberia, but more likely came from the Middle East.[800]

Particular concentrations of maternal lineage Haplogroup X2 appear in Georgia of the Caucasus (8%), the Orkney Islands in Scotland (7.2%), and Lebanon (5.8%). This haplogroup can also be found among the Turks (4.4%), Greek Cypriots (6.7%), mainland Greeks and Cretans (4.4%), and Abazins of Russia (6.3%).[801] The 1000 year old skeleton called the "Norwich Anglo-Saxon" belongs to Haplogroup X.[802] Nine percent of ancient Basque maternal lineages whose DNA was tested belonged to Haplogroup X.[803] Common genetic connections between all of these groups point to a Middle Eastern origin of Haplogroup X.

Ancient people known as the Adena (also referred to as "Alleghans") lived in North America during the time of the Book of Mormon. This time period correlates with the early Woodland Period (1000 BC-200 BC). Of the ancient American Indian bones tested, 1 in 16 of the Adena tested (6.2%) were either Haplogroup X or other haplogroups common to indigenous North Americans. Among other indigenous North Americans whose DNA was tested (and which dates back to the time of the Book of Mormon), 2.1% of them were Haplogroup X.[804] Some indigenous North Americans belonging to Haplogroup X came before Book of Mormon peoples, including individuals found in an ancient Florida bog, and an ancient man discovered in Washington known as "Kennewick Man."[805]

Appendix 2

The Joseph of Egypt Connection and "Believing Blood"

The Book of Mormon is a record of the forefathers of our western tribes of Indians[.] . . . By it we learn that our western tribes of Indians, are descendants from the Joseph that was sold into Egypt[.] —**Joseph Smith, Jr., Latter-day Saint Prophet (1833)**

To the Red Men of America . . . You are a branch of the house of Israel. You are descended from the Jews, or rather more generally, from the tribe of Joseph, which Joseph was a great prophet and ruler in Egypt. —**Parley P. Pratt, Latter-day Saint Apostle (1851)**[806]

Joseph of Egypt—the great-grandson of Abraham—is an extremely important figure in the Mormon Church. The Bible narrative of Joseph of Egypt's life describes how he utilized certain inborn temperament traits that he had inherited from his great-grandfather Abraham's DNA, including high levels of agreeableness and conscientiousness, to curry favor with Pharaoh and also become a great political and religious leader.[807] These traits appear to have stemmed from the spiritual gifts that he possessed.

The Doctrine and Covenants discusses the idea that, although God loves all of His children, not all of His offspring are born with the necessary genetics and spiritual maturity to possess all spiritual gifts. Doctrine and Covenants 46:11 reads: "For all have not every gift given unto them; for there are many gifts, and to every man is given a gift by the Spirit of God."[808] Nevertheless, just because everyone in the world does not possess identical spiritual gifts and heritable traits, it does not mean God loves them less. The diversity of traits among humans has nothing to do with preferential treatment from a Higher Power.[en] That is to say, we do not need to possess traits such as agreeableness and high conscientiousness to be righteous and favored by God. From the scriptures, we learn that proper behavior and the acquisition of a good character matter far more to God than the specific genetics of individuals.

In 2005, Ana Sacau, a Professor of Psychology at the University of Fernando Pessoa, and her team performed a study that found individuals who are born with the heritable temperament traits of agreeableness and conscientiousness[eo] are more often credulous (i.e. believers) and consistently religious than those who lack these traits.[809] Vassilis Saroglou, a leading expert in personality and religious psychology research from the Université Catholique de Louvain, and his team discovered a similar phenomena in 2009. Saroglou and his team discovered while comparing individuals who consistently considered themselves to be religious with those who

---

en This diversity stems from varying genetic codes, and as these codes are passed down through certain genetic lineages, so are the combinations of their accompanying temperament traits.
eo Conscientiousness is the personality trait of being thorough, careful, and vigilant and agreeableness is the tendency to be compassionate and cooperative rather than suspicious and antagonistic towards others. Conscientious types are principled and governed by conscience.

seldom to never consider themselves consistently religious that "[R]eligiosity is generally linked to agreeableness and conscientiousness."[810]

As indicated by the scriptures and events in world history, Joseph of Egypt's bloodline has always been full of religious individuals with a natural penchant for religiosity. Because of how genetic principles of heredity work, not all members of Joseph of Egypt's bloodline are born with this specific combination of traits, but they are commonly found among his descendants. This combination of traits often leads to religiosity, which can be considered a form of "believing blood."

"Believing Blood"

The term "believing blood" has been used from time to time by Latter-day Saint brethren to describe traits believed to be associated with the literal bloodline of Abraham.[ep] Certain brethren recognized that certain descendants of Abraham inherit traits that increase their ability to believe in, recognize, and live according to the truth.[811] In 1985, Elder Bruce R. McConkie, in his book *A New Witness for the Articles of Faith*, described the term in the following way:

> What then is believing blood? It is the blood that flows in the veins

---

ep   Although it is often taboo to speak of genetics and heredity in our modern world, it is unwise to pretend that heritable traits do not exist or to downplay the importance of genetics in the scriptures and early Latter-day Saint ideologies. Despite the opinions of the world, or even perspectives of select groups of skeptical Latter-day Saints, genetics are an important factor in encouraging the creation of certain types of religious communities. Therefore, knowledge of these genetic lineages and their attendant spiritual gifts provides a very useful tool to Latter-day Saints who desire to create a community where the fruits of "believing blood"—such as faith, covenant-keeping, and consistent religiousness—are prominent.

of those who are the literal seed of Abraham–not that the blood itself believes, but that those born in that lineage have both the right and a special spiritual capacity to recognize, receive, and believe the truth. The term is simply a beautiful, a poetic, and a symbolic way of referring to the seed of Abraham to whom the promises were made. It identifies those who developed in pre-existence the talent to recognize the truth and to desire righteousness.[812]

Despite any possible inaccuracies in Elder McConkie's explanation of where "believing blood" originated, his idea that Abraham's literal descendants are genetically endowed with certain temperament traits that are associated with "a special spiritual capacity to recognize, receive, and believe in truth" appears correct.

President Charles W. Penrose of the First Presidency also understood the genetic aspect of "believing blood" and was known to discuss the idea in General Conference. President Penrose connected the concept to the acceptance of truth in October General Conference 1922:

> The power and ability to receive truth is a great thing—that is a gift which I believe is largely bestowed upon the Latter-day Saints. I believe there is something in our racial connection which has to do with this. It is evident to me that in the last days the Lord has wrought mightily upon the descendants of the house of Israel. . . . I do not mean to say that this is confined to us, but that particularly those who are of the house of Ephraim are ready to receive the word and act according to it as the Lord shall direct.[813]

Many Latter-day Saint prophets and apostles over the years have supported the idea of "believing blood" among Joseph of Egypt's bloodline and for good reason. In 1949 LDS General Conference,

Elder Spencer W. Kimball said, "the Lamanites have believing blood."[814] Kimball also said concerning the Lamanite remnant:

> Intolerant people reproachfully indict these red men . . . [T]he red man will remain loyal and true to the gospel and the Church [.] . . . The Lamanites are firm and steady[.] . . . 'the Lamanites had become, the more part of them, a righteous people insomuch that their righteousness did exceed that of the Nephites, because of their firmness and their steadiness in the faith' (Helaman 6:1).[815]

President James E. Faust of the First Presidency once averred that "[N]o race or class seems superior to any other in spirituality and faithfulness,"[816] and yet he also spoke of "the blessing of having believing blood."[817] President Faust gave a talk that mentions his ancestors as some of the early leaders who helped the Prophet Joseph Smith establish the LDS Church. In the talk, he expressed his gratitude for his genetics: "I am grateful for the believing blood of my forebears as well as those of my wife[.]"[818]

Elder M. Russell Ballard, a descendant of Hyrum Smith, also believes in "believing blood." He is known to have said, "Lehi and Sariah, both [had] believing blood[.]"[819] He further mentioned that

> With the marriage of Joseph Smith[,] Sr. and Lucy Mack, the Lord wove together the believing blood of the Smiths and the believing blood of the Macks. . . . I, of course, am very humble to have such noble and faithful pioneer forefathers, . . . When I contemplate the believing blood that flows through the veins of my family, I realize that we have a great responsibility to do all that we can to bring God's children to a knowledge of the restored gospel of Jesus Christ. . . . One of the remarkable things that every Latter-day Saint needs to be grateful for is the believing blood that flowed in the veins of the Smith family.[820]

"Believing Blood" and Keeping Covenants

Although Joseph Smith was a universalist in many respects,[eq] he was known to focus his attention on specific genetic lineages and heredity of Israelites, especially groups from the lineage of Joseph of Egypt. Joseph Smith must have been aware to some capacity that heredity influences the behavior of individuals. For example, he seems to have known that some people are born with a desire to keep things sacred, while others innately struggle to separate the sacred from the profane. The Book of Mormon speaks of this principle (See Jacob 5), and Joseph could have learned it from the book. Joseph's knowledge of how certain traits are passed on from parent to child is probably why he sent missionaries to find tame lineages rather than wild ones, especially tame lineages among scattered Israel.[821] The Prophet demonstrated that he knew the God of Israel prefers to make covenants and promises with individuals who can keep things sacred. He also apparently knew that Israelites, especially from Joseph of Egypt's bloodline, have many among them who naturally make and honor covenants.

Wild and rebellious individuals often struggle to keep things sacred, which is probably why Joseph Smith did not send early Latter-day Saint missionaries to find and convert them. Because of the genetic predispositions and character traits of the wild and rebellious, they tend to lack self-control and high conscientiousness

---

eq Christian Universalism is a school of Christian theology which includes the belief in the doctrine of universal reconciliation, the view that all human beings will ultimately be restored to a right relationship with God in Heaven and the New Jerusalem. The teachings of the LDS Church in relation to God's children are epitomized by a verse in the second book of Nephi: "[The Lord] denieth none that cometh unto him, black and white, bond and free, male and female; ... all are alike unto God, both Jew and Gentile" (2 Nephi 26:33).

—traits commonly found among Joseph of Egypt's descendants.[822] Joseph LeDoux, an American neuroscientist and author of *The Synaptic Self: How Our Brains Become Who We Are,* has shown in his research that dispositional traits are heritable. He once remarked, "Temperament runs through bloodlines."[823]

People born with low conscientiousness, low self-control, and disagreeableness are unlike Joseph of Egypt: a man who was responsible, agreeable, dutiful, and highly self-controlled. Joseph of Egypt's high conscientiousness, self-control, and agreeableness helped him curry favor with Potiphar and the pharaoh of Egypt. Had Joseph of Egypt not possessed these heritable traits, it is unlikely he would have been as consistent, self-controlled, and submissive as he was to impress pharaoh enough to become a high-ranking official in Egypt.

In his spiritual life, Joseph of Egypt set apart sacred things for special purposes and maintained great self-control throughout his life, even in precarious situations. When Potiphar's wife attempted to seduce him, for example, he never surrendered to her advances. This type of behavior is common among those with good character and high conscientiousness. A 2014 Yale study published in the *Journal of Research in Personality* indicates that high conscientiousness has a strong correlation with self-control and self-discipline.[824] Conversely, in 2001, Psychologist David C. Watson, and his team from MacEwan University in Edmonton, Canada, found that individuals who are low in conscientiousness, and also choose to be wild and rebellious, are usually impulsive and out of control.[825]

The *Doctrine and Covenants* speaks of the genetic and character aspects of rebelliousness:

> Behold, the Lord requireth the willing heart and a willing mind; and the willing and obedient shall eat the good of the land of Zion in these last days. And the rebellious shall be cut off out of the land of Zion, and shall be sent away, and shall not inherit the land. For verily I say that the rebellious are not of the blood of Ephraim, wherefore they shall be plucked out (Doctrine and Covenants 64:34-36).

According to this scripture in the Doctrine and Covenants, there is a genetic aspect to rebelliousness. Ephraim appears to have inherited his father's high conscientiousness and agreeableness, both of which are likely reasons for why he was put in charge of Joseph's lineage, despite the fact that his brother Manasseh was older than him and technically had the birthright. Because of Ephraim's genetic endowment and character traits, as Brigham Young said in 1855, "[I]t is the very lad [Ephraim, son of Joseph of Egypt] on whom father Jacob laid his hands, that will save the house of Israel."[826] Conscientious and agreeable types like Ephraim and his father Joseph of Egypt tend to be reliable individuals and the most compassionate. What better traits can one possess to save the House of Israel?

Lehi's Family and Joseph of Egypt's Heritable Traits

Although individuals from the literal bloodline of Joseph of

Egypt are not all tame and highly conscientious, many individuals in his family tree are. The original Nephi and Lehi of the Book of Mormon, who were both descendants of Joseph of Egypt through Manasseh, each possessed high conscientiousness as indicated by their language patterns in the Book of Mormon. Nephi indicated his lack of wildness and rebelliousness when he uttered, "I will go and do the things which the Lord hath commanded" (1 Nephi 3:7). Conversely, Laman and Lemuel, who also had descended from Joseph of Egypt via Manasseh, both lacked high conscientiousness and agreeableness. Their recorded behaviors in the Book of Mormon denote their wildness. Lehi lamented over the low conscientiousness and rebelliousness of his sons Laman and Lemuel. To Laman, Lehi pleaded: "O that thou mightest be like unto this river, continually running into the fountain of all righteousness!" (1 Nephi 2:9). To Lemuel, Lehi cried: "O that thou mightest be like unto this valley, firm and steadfast, and immovable in keeping the commandments of the Lord!" (1 Nephi 2:10).

It was because of the consistent "rudeness" of Laman and Lemuel that they were rendered unfit to succeed their father as head of the family (2 Nephi 2:1). Specifically, they forfeited their birthright to their younger brother Nephi because of their perpetual wildness and rebelliousness. According to Latter-day Saint scholar John Tvedtnes, the word "rudeness" in this context in the Book of Mormon, does not mean "impoliteness." At the time Joseph Smith translated the Book of Mormon, "rude" meant "wild" or "savage."[827]

Laman and Lemuel lacked the temperament traits of high conscientiousness and high agreeableness, which made it difficult for them to be tame and obedient (especially since they demonstrated little to no desire to be righteous). When Laman and Lemuel's wives had children, their disagreeableness and low conscientiousness were passed on to their descendants, the Lamanites. These genetic traits among their descendants are some of the main reasons why collectively the Lamanites became "a wild, and ferocious, and a blood-thirsty people" (Mosiah 10:12).

In the Book of Mormon, it seems that only when the Lamanites admixed with the Nephites that the heritable traits of high conscientiousness and agreeableness began to be more dominant, thus helping the Lamanites became less wild than the Nephites.[828] The Lamanite high civilization was likely only possible because of the growing amounts of individuals among them with high conscientiousness and agreeableness.

Tame vs. Wild

Studies performed by American psychologist Robert Hogan and his team in 1997 demonstrate that high conscientiousness is associated with obedience and conventional integrity—two traits commonly found in tame and civilized societies.[829] Jacob 5 in the Book of Mormon contains Zenos's allegory of the olive tree, which demonstrates the genetic and character aspects of tameness and wildness.

The God of Israel wants all of His children to receive salvation and exaltation, but He knows that certain genetic lineages make it difficult for some of His children to keep promises and covenants. He knows that wild branches (genetic lineages) naturally bear wild fruit and that tame branches (i.e. genetic lineages) naturally bear tame fruit (See Jacob 5). The God of Israel is keenly aware of the extreme difficulty for those who are born into wild branches to produce tame fruit.[830] He does all he can to help everyone bear tame and good fruit no matter if they are born tame or wild (See Jacob 5:27).

Naturally tame individuals are more tractable and submissive than "wild" individuals. One definition of "tame" is "to be rendered useful."[831] To be tame is to not be wild or in a savage state. Conversely, to be wild is to be unpolished, uncultivated, barbarous, rude, and boorish. Studies indicate that tame individuals with high conscientiousness are the most prone to keep promises and covenants.[832] It is why naturally conscientious types often receive admirable descriptors such as hard-working, reliable, and persevering.[833] High conscientiousness predicts happier marriages and longer lives, since conscientious people are less likely to do things that aggravate marriages or get themselves killed.[834] Research performed in 2005 by Brent W. Roberts, a Professor of Psychology at the University of Illinois, and his team, indicates that conscientiousness is represented by positive traits such as "industriousness, order, responsibility . . . and virtue."[835] Conscientiousness has been shown to be negatively associated with many adverse health behaviors (such as overeating).[836] The trait has been positively linked to rational strategies of decision-making, and negatively linked to irrational strategies.[837] In addition, conscientiousness is involved in the coordination of self-regulatory traits.[838] On average, conscientious types are the most dependable and industrious people on the planet.

The highly heritable trait of conscientiousness[839] has always been found in large percentages in the US, the UK, Canada, Germany, Scandinavia, and Israel—all areas Joseph Smith knew had Joseph of Egypt's seed. It is probably one of the reasons why the first missionaries in this dispensation were sent to many of these locations.[840] Latter-day Saint missionaries were sent to parts of the US and Canada, and also Lamanite lands of North America beginning in 1830,[841] the United Kingdom in 1837,[842] Australia in 1840, and Germany, the Netherlands,[843] Turkey,[844] and Israel/Palestine all in 1841.[845]

With a clear understanding of the science behind "believing blood," and also a comprehension of the idea that many of the seed of Abraham—the descendants of Isaac, Jacob (Israel), and Joseph of Egypt—were provided at birth with a certain combination of temperament traits that led to belief and religiosity, we get closer to recognizing the main Lamanite remnant.

Appendix 3

Egyptians, American Indians, and Feathers

Ma'at wearing a feather in a headband,[846]
& **Princess Red feather,** Wampanoag tribe[847]

Lehi and his caravan were influenced by the Egyptians since Egypt was such a dominant power in the ancient world. This is probably why the Book of Mormon was written in reformed Egyptian. Lehi must have known the Egyptian language and been familiar with Egyptian culture due to the amount of Egyptian hieratic and demotic (forms of colloquial Egyptian) that were included in the Book of Mormon. The oldest brothers of Nephi—Laman and Lemuel—and their families, may have decided to worship Egyptian gods such as Ma'at (featured above), a single feather-wearing god, after rejecting the God of Israel.[848]

It is possible that ancient Egyptian gods who wore feathers could have been retrieved from the memories of Laman and Lemuel and worshipped by the Lamanites. That is, Laman and Lemuel could have remembered the gods worshipped in Egypt and then chose to follow the way of the Egyptians.

At least five Egyptian gods and goddesses are depicted with feathers in headbands and headdresses. Feathers were also commonly used by ancient Libyans and Nubians. Ma'at (featured in this section) is a woman with an ostrich feather held up with a headband in North American Indian-like fashion. The tall feather, attached by a headband, is the hieroglyph for truth, order, balance, justice and freedom. The reason for the association of the ostrich feather with Ma'at, the goddess of truth, is unknown, as is also the primitive conception which underlies the name, but it is certainly very ancient, and probably dates from pre-dynastic times (generally dated 3100-3000 BC).[849] To enter the afterlife in Egypt, Ma'at weighed the heart of the deceased man against the weight of an ostrich feather. If the man's heart weighed less than the feather, then he was allowed to enter the afterlife.

Shu, another Egyptian god, is generally depicted as a man wearing an ostrich feather headdress with between one and four feathers. The feather Shu used was the same ostrich feather of Ma'at.[850] As the air, Shu was known to be a calming influence and is often portrayed in art as wearing an ostrich feather.

Detail of an **Assyrian relief** (the captive has a feather in a headband)[851]

Featured above is a photograph of an alabaster wall panel relief depicting an Assyrian capture of an Egyptian fortress. The captive on the Assyrian relief is most likely a Nubian who is from a land called Nubia, located in northern Sudan and southern Egypt. The figure has a headband with a feather in it. In a way, the Nubian, who is most likely copying the appearance of an Egyptian god, looks somewhat like an American Indian.

The Temehu Berber tribe of ancient Libya and Egypt wore feathers. Some of the Temehu people were said to be fair-skinned and blue-eyed. They wore single hair locks on each side of the head and pointed beards, and had a headdress of two ostrich plumes. According to some sources, one feather symbolizes "chieftain status," while two feathers are generally worn by everyone else. But in other representations of Libyans we see chieftains with two feathers and their subjects wearing only one feather. Also Libyan goddesses of ancient Libya were portrayed by the ancient Egyptians with one feather, like the Libyan Goddess Ament, the consort of Libyan Amen or Amon.

Anqet, goddess of fertility (with a feathered headdress)[er]
& **Rain-in-the-Face**, a Sioux[852]

Anqet (Anuket or Anukis), known as the huntress goddess, can be recognized by her distinctive feathered headdress (featured above).[853] Most often Anqet is portrayed as a woman holding a papyrus scepter and an ankh and wearing a tall headdress made of ostrich feathers or reeds.[854] Also featured above is Rain-in-the-Face, a Lakota Sioux with a feathered headdress. It is possible that the feathered headdress idea came from Egyptian culture and was passed down from Laman and Lemuel to their descendants.

Amentet, the Personification of the West, was depicted as wearing the Standard of the West. The Standard of the West is usually represented by a half circle sitting on top of two poles of

---

er   Anqet was a goddess of the whole Aswan area—of the islands in the Nile and of the First Cataract—and also a goddess of Nubia, which was the land to the south of Egypt.

uneven length, the longer of which is tied to her head by a headband. Often a hawk or an ostrich feather is seen sitting on top of the standard. Hathor, the "Lady of the West," is often depicted as wearing this headdress. In this form, she is known as Hathor-Amentet.[855]

The Egyptian god Amen (Amun) is usually depicted as a man wearing a headdress with two tall plumes rising from a short crown. As Amen-Ra (Amun-Ra), the sun disk is added between the plumes, showing his connection to the sun. Horus, one of the oldest and most significant deities in ancient Egyptian religion, is also seen wearing the headdress of Amen. Amen carries two high plumes atop his head since plumes are a sign for wind or air.[856]

Appendix 4

Old World Symbols in the New World

Triskelion[857]

In a tructed stone sarcophagus in which the face of the skeleton was looking toward the setting sun, a beautiful shell ornament was found resting upon the breastbone of the skeleton. This shell ornament is 4.4 inches in diameter, and it is ornamented on its concave surface with a small circle in the center and four concentric bands, differently figured, in relief. The first band is filled up by a triple volute [triskelion][.]. . . This ornament, on its concave figured surface, has been covered with red paint, much of which is still visible[.] . . . This ornament when found lay upon the breastbone with the concave surface uppermost, as if it had been worn in this position suspended around the neck, as the two holes for the thong or string were in that portion of the border which pointed directly to the chin or central portion of the jaw of the skeleton. The marks of the thong by which it was suspended are manifest upon both the anterior and posterior surfaces, and, in addition to this, the paint is worn off from the circular space bounded below by the two holes. —**Joseph Jones, of a Tennessee archaeological dig (1876)**[858]

During the 4 ½ years I spent researching and writing this book, I made a number of interesting and unexpected discoveries while attempting to tie the Old World to the New World. Because the material didn't fit in the body of this book, it was put into the appendices. The following are examples of the discoveries I found in scientific journals and books about ancient America.

The late Jewish-American scholar Cyrus H. Gordon[859] heavily studied Near Eastern cultures and ancient languages. He was convinced that Jews, Phoenicians, and other Semitic groups made transatlantic voyages in ancient times, which ultimately influenced both North and South America.[860] Pre-Columbian artifacts discovered all over the Americas appear to support his claims. Some of the ancient artifacts that corroborate Gordon's claims can be found in North America and date back to the time of the Book of Mormon.

Old World symbols in North America seem to further tie the Book of Mormon to indigenous North Americans. As extensive scholarship has shown, the Book of Mormon lines up perfectly with ancient Near Eastern, Egyptian, Arabian, and Mediterranean languages, writing styles, and cultural practices. Latter-day Saint polymath and scholar Hugh Nibley compared Joseph Smith's translation of the Book of Mormon to archery, stating that it hit multiple "bulls-eyes."[861] The names, places, phrases, and idioms contained in the Book of Mormon match up perfectly with Old World languages, writing styles, and cultural practices.[862] The following are examples of Old World symbols that appear in North America.

Hamesh

Jewish hamesh with inscribed Hebrew words: Tzedek, Shutafut, Shalom[es]

Carved into a number of Hopewell, Adena, and Mississippian artifacts that were discovered in North America are ancient symbols from the Old World. Included in these discoveries is the ancient Middle Eastern symbol known as a "hamesh" in Hebrew, "hamsa" in Arabic, and "humsa" in Hindi (featured above). The term hamesh is the Hebrew word for the number "five" (חמש), which represents the five fingers of a hand.[et] This ancient symbol is known to mean "God's protective hand," and the "all seeing eye of mercy." The hamesh is often fashioned into palm-shaped amulets, which for many years have remained popular throughout the Middle East and North Africa. These amulets are commonly used in jewelry and wall hangings even today!

---

es   This picture is of a Jewish hamesh with the inscribed words: Tzedek, Shutafut, Shalom, which is Hebrew for "Justice/Righteousness, Partnership, Peace."
et   "Chamesh" or "hamesh" in Hebrew means "five" (חמש).

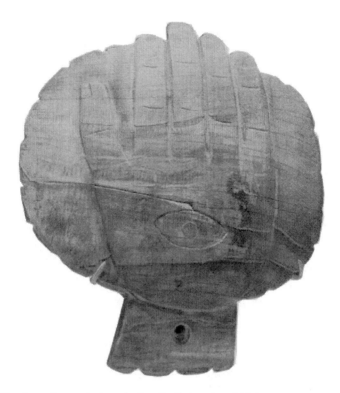

An upside-down **hamesh** (eye-in-hand), Spiro site, Oklahoma (about 800 AD)[863]

The Old World custom of hanging a hamesh (hamsa) around the neck as jewelery appears to have made its way to ancient North America. An ancient artifact featuring a hamesh was discovered in North America, more specifically Oklahoma, with a hamesh symbol on it (featured in this section). The hamesh ("eye-on-hand") symbol featured in this section on the Oklahoma pendant predates Christianity and Islam.[eu] This ancient symbol is thought to combine two senses: an eye added to the palm of a hand includes sight (ophthalmoception) and touch (tactioception).[864] Because this ancient

---

eu   In Islam, it is known as "the hand of Fatima," so named to commemorate Muhammad's daughter Fatima Zahra. Levantine Christians call it "the hand of Mary," for the mother of Jesus, but the symbol is much older than Fatima and Mary. To some, the symbol represents God's "protective hand."

symbol is so unmistakable, the chance of independent creation of the symbol both in the Middle East and America is close to zero. Odds favor an ancient connection between the Middle East and America.

Sandstone "Rattlesnake Disk" with hamesh, Moundville, Alabama (about 1277 AD)[865]

A ceremonial object (featured above), known as "Rattlesnake Disk" for its design, was also discovered in Moundville, Alabama. It contains the hand-and-eye (hamesh) symbol.

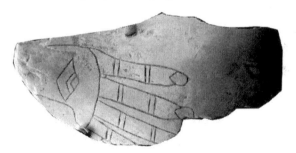

An ancient North American hamesh engraved shell cup[866]

Another ancient hamesh symbol was discovered in North America on a shell cup (featured above). This hamesh was carved on a shell and features long fingernails on the hand. The hand most likely represents a shaman's hand because of the length of the fingernails.

Sheet-copper pendant from Moundville, Alabama, with hamesh (about 1000 AD)[867] & a drawing of the **copper pendant** from Moundville, Alabama, with a hamesh

Another Pre-Columbian artifact with the ancient hamesh symbol was discovered in North America (featured above). This artifact from Moundville, Alabama, is made of sheet-copper and contains a sun symbol (on the top) and a hamesh symbol reaching down from the sun. According to legend, the Choctaw Indians of North America once believed that the sun watched them with its great blazing eye, and so long as the eye was on them they were all right, but if the eye was not observing them they believed they were doomed.[868]

Cross

Oklahoma Spiro Mound double-hand shell gorget pendant (about 800 AD)[869]

Featured above is an ancient double-hand shell gorget pendant that dates back to about 800 AD. Interestingly, the two hands of this pendant have crosses on the palms. The cross symbols found on this artifact can possibly be traced back to the Old World. This particular artifact was found in a Spiro Mound, which is one of many important pre-Columbian Caddoan Mississippian Culture mounds located in present-day eastern Oklahoma. The crosses on this artifact may be a Pre-Columbian Christian reference, although their actual meaning is unknown. However, the average Christian cannot help but make a mental connection between this ancient artifact and the crucifixion scars of Jesus Christ.

Ohio Hopewell mica hand (100 BC-500 AD)[870]
& an **Etruscan bronze hand**, Italy (650 BC)[871]

A sheet-mica hand (featured above on the left) was discovered in a Hopewell burial mound (Mound 25) in Ross County, Ohio. This artifact dates back to approximately 100 BC-500 AD, which fits into the Book of Mormon timeline. This indigenous North American "hand" possibly represents a group of stars still known today among the Plains Indians as the "Hand Constellation."[872] Somewhat similar to this Hopewell hand is an Etruscan bronze hand (featured above on the right), which was discovered in Italy. The Etruscan bronze hand dates back to 650 BC.

Cernunnos

Wooden mask with antlers, Oklahoma[873]
& **Cernunnos**, Denmark (200 BC-300 AD)[874]

A wooden mask with antlers (featured above) was discovered in the Spiro Mounds of Oklahoma. The visage of this ancient North American artifact resembles a figure that was discovered in Europe and is generally identified as the ancient Celtic god Cernunnos. This ancient Cernunnos figurine (featured above on the right) dates back to approximately 200 BC-300 AD—a time period that coincides with the Book of Mormon time frame. Cernunnos-like motifs discovered in North America lead one to believe that a connection exists between ancient America and ancient Europe.

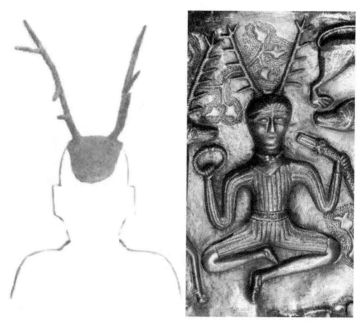

Ohio Hopewell copper antlers (300 BC - 500 AD)
& an **Antlered figure**, Gundestrup Cauldron, Denmark (200 BC-300 AD)[875]

Other Cernunnos-like artifacts have been discovered in North America. In an Ohio Hopewell burial mound copper antlers were discovered. They appear to be made for a man to wear on top of his head (featured above). These copper antlers date back to about 300 BC - 500 AD, which fits into the Book of Mormon timeline. Also featured above is Cernunnos from an ancient silver cauldron found in Denmark dating back to 200 BC - 300 AD. Cernunnos is holding a torc. The torc is often found among Scythian (Cimmerian), Illyrian, Thracias, and Celtic cultures of the European Iron Age (700 BC - 200 AD). The Cernunnos is holding a torc in his right hand and the ram-horned serpent in his left. He is attended by a stag on his right and a boar on his left.

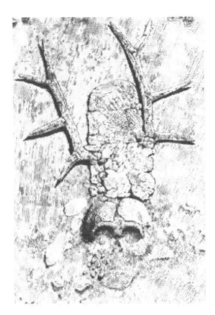

Ohio Hopewell skull with copper horns (300 BC-500 AD)[876]

Another set of Hopewell copper antlers was discovered in North America. This set of antlers (featured above) were found on the skull of a man who lived during the Hopewell time period, about 300 BC-500 AD, which fits into the Book of Mormon timeline. Although these antlers do not necessarily tie the Hopewell to Book of Mormon peoples, they do link the Old World with the New World in Pre-Columbian times. These artifacts could rewrite history because antlers like these have been found in the Old World at a time period long before Christopher Columbus and even before Leif Ericson.

Swastika

A **Hopewell Double Serpent Swastika,** Mississippi Valley (200 BC-200 AD)[877]

Five swastika crosses were discovered in Hopewell mounds in Ross County, Ohio. Featured above is one of them. It was fashioned out of stone in the shape of two serpents and dates back to approximately 200 BC-500 AD, which fits within the Book of Mormon timeline. It is likely that the shape for this double serpent swastika originally came from the Middle East or Europe. However, to date, no definitive statement has been made by scholars about how the shape for this swastika symbol made its way to the New World.

Long before German Nazis appropriated the swastika for their own uses, it was a well-known ancient Aryan symbol[878] that carried a positive meaning.[879] Contrary to popular belief, the swastika did not originate in East Asia, although it is widely used there even today. Germans from 1930-40 never would have used the symbol to represent the Third Reich in Nazi Germany if it was anything less than Aryan.[880]

In Thomas Wilson's 1896 book, *The Swastika: The Earliest Known Symbol and its Migrations*, this former curator of the Department of Prehistoric Anthropology in the US National Museum describes the ancient North American mound swastikas,

> The Swastika of the ancient mound builders of Ohio and Tennessee is similar in every respect, except material, to that of the modern Navajo and Pueblo Indian. Yet the Swastikas of Mississippi and Tennessee belong to the oldest civilization we know in America. . . . the Swastika had an existence in America prior to any historic knowledge we have of communication between the two hemispheres. . . . Some authorities are of the opinion that it was an Aryan symbol and used by the Aryan peoples before their dispersion through Asia and Europe.[881]

The Book of Mormon never mentions this ancient Old World symbol. However, since this ancient symbol dates back to the time frame of the Book of Mormon, it might have been brought from the Middle East by Lehi and his caravan, or even the Mulekite caravan, to the American continent. Whether it was Book of Mormon peoples who brought the symbol to the New World or not, some group from the Old World appears to have brought the ancient symbol with them to the continent long before Columbus made his way to the Americas. It does not seem too likely that the swastika symbol was created independently in the Old World and in the Americas.

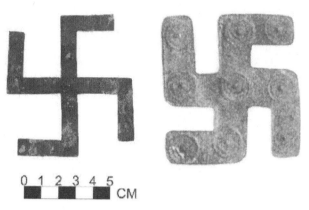

An **Ohio Hopewell copper swastika** artifact (100 BC-500 AD)[882]
& a **Fylfot brooch swastika**, found in modern-day Germany (400 AD-500 AD)

A number of swastikas have been found in ancient Hopewell archeological sites in North America. Featured above are two sheet-metal swastikas that date back to approximately the same time period —an era that fits within the Book of Mormon time frame. The swastika featured on the left was discovered in an ancient Hopewell earthwork and ceremonial center, which is a part of the Hopewell Mound Group in Ohio. It dates back to about 100 BC-500 AD. The swastika featured on the right was discovered in modern-day Germany and dates back to approximately 400 AD-500 AD. The two swastikas have different designs but share a similar form.

Because of the distinct shape and meaning behind the Old World symbol, the swastikas discovered in Hopewell archeological sites most likely came from the ancient Middle East or Europe. No known evidence points to the fact that the Hopewell developed the design independently from the Old World. No other explanation for how the symbol made it to the New World seems plausible at this point.

Hohokam village ruins (approx. 400 AD) near Phoenix, Arizona & a **Swastika** on Arizona Hohokam pottery (approx. 400 AD)[883]

The Hohokam are a group of ancient indigenous Americans who lived in an area now known as Arizona. They ensconced in the area from about 300 AD to 1500 AD, a time that partially overlaps the Book of Mormon time frame.[884] Villages were built by the Hohokam close to streams in order to farm the region's rich bottomlands. The Hohokam decorated pottery and other artifacts with the ancient Aryan swastika symbol (featured above).

An **Ancient Greek bowl,** Artemis (or Diana, Isis in Egypt) with swastikas (700 BC)

Swastikas were commonly used in the Old World. Ancient Greece, for example, used them often in their artwork, but referred to them as "tetraskelions" or "gammadions." Featured above is an ancient Greek bowl depicting Artemis with tetraskelions (swastikas). The Greeks, who have possessed many Semitic genetic lineages for thousands of years, were known to be heavily influenced by Hebrew and Aryan peoples. They most likely borrowed the swastika symbol from Middle Eastern culture.

An **Etruscan gold pendant** with swastika symbols, Bolsena, Italy (700 BC-650 BC)

Ancient Etruscans of Europe used the swastika symbol to adorn valuable items. The Etruscans of ancient Etruria (modern-day Italy) belonged to a civilization that reached its height around 500 BC. The Etruscan gold pendant (featured above) has four swastikas on it. Ancient Etruscans, who were in many ways similar to the Greeks, possessed a number of Semitic genetic lineages and often shared physical characteristics of Semites.[885] They were known for being heavily influenced by Hebrew and Aryan peoples. The swastika they used to decorate various items most likely came from Middle Eastern and Aryan cultures.

A **meander swastika** on a Byzantine church floor, Shavei-Zion, Israel (360 AD-636 AD)[886]

Long before the swastika became a symbol associated with anti-Semitism, it was actually used by religious Semitic sects in Israel. Featured above is portion of a mosaic floor of a Byzantine church in Shavei-Zion, Israel, with a meander swastika pattern (360 AD-636 AD).

Meander swastikas on a Byzantine church floor, Shavei-Zion, Israel (360 AD-636 AD)[887]

Featured above are numerous meander swastikas on the floor of a Byzantine church in Israel. These meander swastikas date back to approximately 360 AD - 636 AD.

A **meander swastika**, Katzrin Synagogue, Israel (500 AD-600 AD)[888]

Featured above is another meander swastika pattern carved into a stone from the Katzrin Synagogue that is located in Israel. This stone carving dates back to approximately 500 AD-600 AD.

Triskelion

Engraved shell disk, three-branched volute (triskelion), Tennessee[889]

The triskelion is another Old World symbol that made its way to the New World. Featured above is an ancient North American ornament with a triskelion located in its center. Triskelion (also triskele) is a Greek word for "three-branched." This ornament was found in a stone grave at the summit of a mound on the banks of the Cumberland River, opposite Knoxville, Tennessee. The triple spiral (or triskele) is a Celtic and pre-Celtic symbol located on numerous Irish Megalithic and Neolithic sites, most notably inside the Newgrange passage tomb, on the entrance stone, and on a few of the curbstones surrounding the mound. The meaning of the ancient triskelion (triskele) symbol is unclear. However, the spirals of the triskelion could represent "the cycle of life" or "the sun."

Dr. Joseph Jones, of New Orleans, described the artifact (featured in this section) and where it was found:

> In a carefully constructed stone sarcophagus in which the face of the skeleton was looking toward the setting sun, a beautiful shell ornament was found resting upon the breastbone of the skeleton. This shell ornament is . . . ornamented on its concave surface with . . . a triple volute [triskelion][.] . . . This ornament, on its concave figured surface, has been covered with red paint, much of which is still visible . . . This ornament when found lay upon the breastbone . . . as if it had been worn in this position suspended around the neck.[890]

Different forms of the triskelion made up of three symmetrically arranged spirals, triangles, or bent human legs show up in all kinds of Old World cultures. In ancient Celtic culture, the design was very popular on jewelry and coat of arms. The symbol can be found even today in various cultures. It is possible the ancient symbol was created independently in the Old World and in ancient North America. However, it appears most likely that the Old World symbol was brought to the New World.

Celtic bronze disc **triskele**[891]
& shell disk with **triskelion**, Tennessee mound[892]

Featured above are two ancient triskelions. The one featured on the left is from ancient Europe, and the one featured on the right is from ancient North America. The triskelions found in the center of these ancient artifacts closely resemble one another. During the European Iron Age, ancient Celtic art of the La Tène Culture included triskeles (triskelions). The ancient Celtic triskele symbol (featured above on the left) appears in ancient Europe on one of the earliest known astronomical calendars in Ireland at the famous megalithic tomb of Newgrange built around 3200 BC. The ancient triskelion from Tennessee (featured above on the right) appears to have been made by a Mound Builder society that could have dated back to Book of Mormon times.

The existence of ancient triskeles (triskelions) in North America appears to change US History and lead to many questions. Although answers about how triskeles (triskelions) from ancient cultures traveled around the world are unavailable, one thing seems fairly certain: this ancient symbol found its way to the New World.

Ancient Rainbow Cup coin (200 BC-100 BC), obverse (left), reverse (right)[893]

An ancient "Rainbow cup"[894] coin with a Celtic triskele (triskelion) dating back to about 200 BC-100 BC is featured above. Coins like this one are the result of a combination from Greek, Roman, and Celtic cultures. Their weight corresponded to the Greek stater, while the design was created by Celts and reverted back to Roman coins.

Triskelion, Nashville (1100 AD-1450 AD)[895]
& **Triskelion**, Spain (1200 BC-550 BC)[896]

Featured above are two triskelions: one is from North America, and the other from Europe. Galicia, the part of Spain where the triskelion on the right was found, is in the northwest part of the Iberian peninsula. It appears that the triskelion belongs to the Castro Culture. Gallaecians are Northwestern Hispano-Celts from Spain, and many of them possess Semitic paternal and maternal lineages.

Other Relics

A **Gallaecian warrior's head** wearing a torc, a neck ring (1000 BC-2000 BC)

Featured above is an effigy of a Gallaecian warrior with a torc around his neck. His torc is similar to the one in the hand of the Cernunnos (the man with antlers on his head) in this appendix. Torcs are found in the Scythian (Cimmerian, ancient Iranian), Illyrian Thracian, Celtic, and similar cultures of the European Iron Age from around the 8th century BC to the 3rd century AD. Notice his high nasal bridge, long nose, and long face; all common traits of Semites.

Chief Joseph's **Assyrian cuneiform tablet**[897]

According to a few sources, when Chief Joseph of the Nez Percé[898] surrendered to the US Army in 1877, he was carrying an authentic Assyrian cuneiform tablet in his medicine bag. General Nelson Appleton Miles of the US Army commandeered the tablet, and according to one source, it ended up archived at the West Point Museum in Virginia. The tablet was later translated by Robert Biggs, a professor of Assyriology at the Oriental Institute of the University of Chicago. After translating the tablet, Professor Biggs discovered that it was a sales receipt for a lamb dating back to 2042 BC.[ev]

No one today is exactly sure how an authentic Assyrian cuneiform tablet ended up in Chief Joseph's medicine bag, but some have ideas about how he acquired it. Mary Gindling, for example, of Helium's *History Mysteries* wrote about Chief Joseph's cuneiform tablet:

> The chief said that the tablet had been passed down in his family for many generations, and that they had inherited it from their white ancestors. Chief Joseph said that white men had come among his ancestors long ago.[899]

Some sources claim that Chief Joseph asserted that white men had come among his ancestors long ago and had taught his people many truths. If this statement is true, then it appears that the cuneiform tablet connects the Old World to the Americas.

---

ev   The cuneiform tablet reads, "Nalu received 1 lamb from Abbashaga on the 11th day of the month of the festival of An, in the year Enmahgalanna was installed as high priestess of Nanna."

Chief Joseph with a possible Assyrian Star of Asshur, Nez Percé Indian & an **Assyrian Star of Asshur**

During his life, Chief Joseph (featured above on the left), may have worn the Star of Asshur, an Assyrian symbol (featured above on the right). This could indicate a connection existed between the Assyrians and the Nez Percé Indians of North America. If Chief Joseph wore an Assyrian symbol, it would make sense that he would have had an ancient cuneiform tablet in his possession when he surrendered to a US army in 1877. Asshur was a son of Shem, which makes him a Semite. Asshur was the father of the Assyrians and the country of Assyria was named after him.

Another cuneiform tablet known as the "Hearn Cuneiform Tablet" was discovered in 1963 in Georgia by Mrs. Joe Hearn. It is comparable to the ancient tablet of Chief Joseph. The "Hearn Cuneiform Tablet" has the Sumerian language on it and dates back to approximately 2040 BC.

These four appendices have been supplemental materials to aid you, the reader, in identifying the Lamanite remnant for yourself. Joseph Smith had a special place in his heart for the Lamanites, and it is my hope that the information presented in this book has created such a space in your heart as well.

Index

A

Adair, James 71-72, 178-179
Aztec xix, 73, 197, 214, 311, 313, 324, 343, 346-347, 352

B

Book of Mormon geography xvii-xviii, xxv, xxvi, 27 30-31, 41, 47, 51-53, 55

C

Cherokee xiv, 16, 149-150, 180-182, 201, 289
Chief Apaquachawba 2, 5, 8

D

Delaware (Lenni Lenape) Indians 15-16, 35-36, 97, 144, 147, 149, 156, 168-171, 174, 245, 270-271, 276, 284, 329

E

Egyptian xxviii, 51, 80-96, 98-100, 124, 197, 201, 291, 324, 367-371, 374

F

Florida bog 353

G

Golden plates xxvi, 4, 9, 17, 28, 43, 291-292, 326, 329, 334

H

Handsome Lake 57-60, 62-64, 66, 156, 163
Hill Cumorah 28, 31, 37-38, 40, 52, 54, 61, 70, 292, 334

I

Iroquois xix, xxi, xxxiv, 16, 26, 35, 57-60, 62, 64-65, 97, 143, 146, 157-163, 174, 201, 232-233, 266, 283, 290, 349

J

Jaredite 215, 291 296-297, 305, 309-310, 314, 320, 320, 323, 341
Joseph of Egypt xi, xv-xvi, xviii, xix-xx, xxvii-xxx, xxxiii, 15, 21, 36, 65-68, 70, 76-77, 79-122, 139, 141, 150, 156, 165, 167, 172-173, 187, 190, 197, 206-209, 212, 215-217, 219-221, 225-226, 230-231, 234, 236, 251, 258, 263, 265-266, 295, 303, 317, 324, 333-335, 338-339, 348, 355, 357-358, 360-363, 366
Judah 110, 113-114, 118, 206, 207-210, 251, 258, 263, 265-266

K

Keokuk 7-8, 166-167

L

Lamanites xvi, xviii-xx, xxv, xxvii-xxviii, xxx-xxxi, xxxiii-xxxiv, 4-5, 8-11, 15-21, 23-44, 52-55, 57, 61-63, 65-67, 69, 72-73, 80-81, 97, 99, 101, 122, 138-141, 149, 155-162, 164-165, 167-168, 172, 176, 180, 182-183, 186, 188, 190-191, 197-198, 200, 2003, 207-208, 210, 215-217, 219-225, 231-232, 234, 237, 244, 251-252, 258-259, 261, 263, 265-267, 273, 276-278, 281, 283, 286-287, 290-292, 294-296, 305, 312-313, 317-318, 320-321, 324-336, 341, 343-344, 346, 359, 364, 366, 368, 397

M

Mandan 97, 141-143, 182-185, 264-265
Maya xix, 73, 197, 294, 296-297, 305, 309-315, 320-321, 324
Mulekite 207, 215, 291, 296, 305, 323, 341, 385
Munsee xxi

N

Nauvoo (Illinois) 7, 66, 328

O

Olmec xix, 73, 197, 294, 296-297, 305, 309-315, 320-321, 324

P

Plains of the Nephites 37-38, 54, 334
Pottawatomie 1-7, 15-16, 21-23, 28, 52-53, 72-73, 164

R

Red Men xxii-xxiii, 20-24, 36, 45, 55, 72, 102, 170-171, 190, 193-194, 271, 328-329, 333, 335, 359

S

Sauk & Fox (Sac & Fox) 7-8, 10-11, 15-16, 22-23, 28, 53, 70, 73, 97, 157, 165-167, 333
Scalping 72-75
Seneca xxviii, 16, 18-19, 26, 28, 33-35, 57-62, 65-66, 156-157, 160, 163, 266, 334

T

Times and Seasons 43-50, 52, 55-56, 298, 300

U

Utah xiv, 328, 332

V

Virginia 281, 284, 395

W

West Virginia 72, 290

X

X, haplogroup (mtDNA)   341, 348-353

Y

Yuya   xxviii, 79-100, 104, 106, 108-109, 117-119, 196-197, 201, 209, 260

Z

Zion's Camp   37
Zarahemla   xxvi, 50, 55-56, 207, 300

End Notes

End Notes

1. *Patriarchal Blessings*, 1:3. Joseph Smith, Sr., blessing, 9 Dec. 1934, Church Historical Department, 1:3-4.
2. Neal A. Maxwell, *A Choice Seer*.
3. William Brandon, *American Heritage: Book of Indians*, p. 8, Introduction by John F. Kennedy (1961). John F. Kennedy, a Catalyst (Out-Curviplex-Circle-Point) in the P2MR approach.
4. A. Irving Hallowell, "The Backwash of the Frontier: The Impact of the Indian on American Culture," in Walker D. Wyman and Clifton B. Kroeber, eds., *The Frontier in Perspective* (Madison: University of Wisconsin Press, 1957), p. 230.
5. Joseph Smith, Jr. *Teachings of the Prophet Joseph Smith*, p. 43.
6. The Munsee natives were Algonquian natives. The Algonquian natives consisted of various tribes that spoke similar languages. The Munsees originally lived in New York and New Jersey, but moved westward as white people forced them from their lands. By the 1720s, the Munsee natives had reached western Pennsylvania. There, missionaries from the Moravian Church attained some success in converting the Munsees to Christianity.
7. Evan T. Pritchard, *Native New Yorkers: The Legacy of the Algonquin People of New York*, p. 46.
8. Europeans such as the Dutch, English, and Swedes.
9. Manhattan was a great location for farming and was once almost a paradise for North American Indians before the Europeans arrived. *Native New Yorkers: The Legacy of the Algonquin People of New York*, p. 35, Evan T. Pritchard. Munsee (also known as Munsee Delaware, Delaware, Ontario Delaware) is an endangered language of the Eastern Algonquian subgroup of the Algonquian language family, itself a branch of the Algic language family. Munsee is one of the two Delaware languages.
10. A memoir of Jacques Cartier, sieur de Limoilou, his voyages to the St. Lawrence, a bibliography and a facsimile of the manuscript of 1534. Baxter, James Phinney, 1831-1921; Roberval, Jean François de La Roque, sieur de, 1501?-1560?; Alfonce, Jean, i.e. Jean Fonteneau, known as, 1483?-1557?
11. ibid.
12. "How Indians Got to Be Red," American Historical Review, p. 102, June, 1997, Nancy Shoemaker, assistant professor of history at the University of Wisconsin. James P. Howley, The Beothucks or Red Indians: The Original Inhabitants of Newfoundland (Toronto, 1915), pp. 2-3, 10, appears to have made the creative leap to John and Sebastian Cabot and Jacques Cartier as the first Europeans to call Indians "red." However, "redskin" was not common usage until the 19th century.
13. Parley P. Pratt, 1851.
14. An Indian's view of Indian Affairs: Chief Joseph, 1879. *Voices of the American Indian Experience*, p. 331, James E. Seelye, Steven Alden Littleton.
15. Joseph Smith Journal, November 9, 1835; cited in Dean C. Jessee, Mark Ashurst-McGee, Richard L. Jensen, eds., The Joseph Smith Papers: Journals Volume 1:1832-1839, pp. 88-89.
16. Christopher Columbus failed to land in the East Indies as he had planned. That is why the natives he found in the New World were obviously not "East Indians." Counter to what many Americans may believe, Columbus did not discover North America. It was Leif Ericson (and others before him). During four separate trips that begun with the famous 1492 voyage of Columbus, he landed on various Caribbean islands that are currently the Bahamas as well as the island later known as Hispaniola. He also explored Mesoamerican and South American coasts but never set foot in North America. Leif Ericson, on the other hand, is documented as the first European to discover the northern part of the American continent in 1000 AD. After him, John Cabot, Italian-born English navigator and explorer, set foot for the first time in North America in 1497 AD. A short time later, French explorer Jacques Cartier, who claimed Canada for France, landed in North America.
17. The term "Native American" applies to North American Indian tribes. In Canada, indigenous groups are known as "First Peoples."
18. In a technical sense, if the term "Indian" were used too broadly when referring to indigenous peoples of America, it might lead to confusion and a fundamental misunderstanding about the origins of some ancient inhabitants of the Americas. If all American Indian tribes, for example, are clustered together and labeled "Indians," or even "Native Americans" for that matter, and treated as a single, homogeneous, and genetically-related group, one might ignore the non-East Asian provenance of some of these tribes.
19. Scientists confirm at least three different times when foreigners made their way to the New

End Notes

World. The term "Indian," as Robert F. Berkhofer, Jr., former professor of history at the University of California, has pointed out, embodied a unitary concept of the native inhabitants of the Americas invented by Europeans. "By classifying all these many peoples as Indians," writes Berkhofer, "whites categorized the variety of cultures and societies as a single entity for the purposes of description and analysis, thereby neglecting or playing down the social and cultural diversity of Native Americans then—and now—for the convenience of simplified understanding." Dan Vogel, *Indian Origins and the Book of Mormon: Religious Solutions from Columbus to Joseph Smith* (1986), pp. 8—9. The reader should be cautioned that Vogel—a former Church member and current atheist—believes that the Book of Mormon is a 19th-century fiction concocted by Joseph Smith. For a review of the strengths and weaknesses of this volume, see Kevin Christensen, "Truth and Method: Reflections on Dan Vogel's Approach to the Book of Mormon (Review of *Indian Origins and the Book of Mormon*)," *FARMS Review* 16/1 (2004): 287–354.

20   Well-known is the fact that many pre-Columbian inhabitants of the Americas possessed East Asian and Siberian origins, however it is likely that few people are aware that recent research has affirmed that a third of the collective North American Indian genome originated in Western Eurasia (which includes Israel). Because the majority of "Native Americans" possess East Asian and Siberian DNA, and many of them look like East Asians and Siberians, they do not appear to be actual members of the Lamanite remnant. Native Americans who possess only East Asian and/or Siberian ancestry, although not genetically of the Twelve Tribes of Israel, can "become" Israelites through adoption. However, these adoptees still retain their East Asian and/or Siberian roots and visages, unless over time they admix with literal Israelites. Some Latter-day Saint brethren of the past believed that the blood and genetics Latter-day Saint converts literally change to become Israelite. However, this does not appear to be the case.

21   Nibley, Hugh W. (2010), "The Book of Mormon: A Minimal Statement", Journal of the Book of Mormon and Other Restoration Scripture (Maxwell Institute) 19 (1): 78–80, retrieved 2010-09-28.

22   The introduction to the Book of Mormon was changed after the Latter-day Saint brethren realized that the erroneous belief had crept into the belief system of Elder Bruce R. McConkie (who wrote the introduction to the Book of Mormon in the 1981 edition). McConkie's phrase was changed from "The Lamanites . . . were the principal ancestors of the American Indians" to "The Lamanites . . . are among the ancestors of the American Indians."

23   Most Native Americans are descended from a small group of migrants that crossed a 'land bridge' between Asia and America during the ice ages 15,000 years ago. Rob Waugh, columnist (2012). Native Americans arrived in THREE great migrations across land bridge from Siberia 15,000 years ago. Rob Waugh, 12 July 2012. "Based on genetics, morphology and geographic location, there is a consensus among scientists that Siberian hunters and gatherers peopled the Americas. Obviously, these migrants brought their genes with them." Michael H. Crawford, British ancient historian (2001). *The Origins of Native Americans: Evidence from Anthropological Genetics*, p. 296, Michael H. Crawford. Concerning the first Americans, Michael H. Crawford, a biological anthropologist at the University of Kansas and author of *The Origins of American Indians: Evidence from Anthropological Genetics*, has said, "Since the eighteenth century, most scientists have been convinced that the first Americans originated from Northeastern Asia. . . . evidence indicates extremely strong biological and cultural affinities between New World and Asian populations and leaves no doubt that the first migrants into the Americas were Asians, probably from Siberia. Stewart, 1973. *The Origins of American Indians: Evidence from anthropological genetics*, p. 3, Michael H. Crawford.

24   Since only Latter-day Saints consider the Book of Mormon to have any ancient historical basis, mainstream scholars do not consider the Lamanites to be a valid category of people.

25   A limited geography model for the Book of Mormon is one of several theories by Latter Day Saint movement scholars that the book's narrative was a historical record of people in a limited geographical region, rather than of the entire Western Hemisphere as believed by some early Latter Day Saints. Coon, W. Vincent "How Exaggerated Settings for the Book of Mormon Came to Pass".

26   Nibley, Hugh W. (2010), "The Book of Mormon: A Minimal Statement", Journal of the Book of Mormon and Other Restoration Scripture (Maxwell Institute) 19 (1): 78–80, retrieved 2010-09-28.

27   Coon, W. Vincent "How Exaggerated Settings for the Book of Mormon Came to Pass."

28   *Religion and Republic: the American Circumstance* (1987), p. 324, Martin E. Marty.

End Notes

29 *History of the Church,* 5:480. Mr. Hitchcock, Letter to John Chambers, in John King to John Chambers, 14 July 1843, Iowa Superintendency, 1838-49, Letters Received by the Office of Indian Affairs, 1824-81, BIA Microfilm #363, 357-60; Helen Mar Whitney, "Life Incidents," Woman's Exponent 9-10 (1880-1882) and "Scenes and Incidents in Nauvoo," Woman's Exponent 11 (1882-83), July 2, 1843. The latter incorrectly cites the Wilford Woodruff journal as its documentary source.

30 Moroni, a Nephite Chief Captain (Alma 46:23–24, 27). Readers of the English Bible might suspect that this terminology was taken from Amos 5:15, where the prophet wrote, "It may be that the Lord God of hosts will be gracious unto the remnant of Joseph." However, close analysis suggests that Amos, as well as those whose words are recorded in the Book of Mormon passages cited above, may have been influenced by an earlier passage of scripture.

31 During the translation process of the Book of Mormon, Joseph Smith received information about the connection between Joseph of Egypt and people discussed in the Book of Mormon. The information must have helped him understand his role in regards to the Lamanite remnant (See 2 Nephi 3:5-22).

32 "The principal nation of the second race fell in battle towards the close of the fourth century. The remnant are the Indians that now inhabit this country [The United States of America]." Joseph Smith, Jr. *History of the Church,* 4:535–41. The Wentworth Letter was originally published in Nauvoo in the *Times and Seasons,* 1 Mar. 1842, and it also appears in *A Comprehensive History of the Church,* 1:55.

33 Parley P. Pratt was baptized in Seneca Lake by Oliver Cowdery on or about September 1, 1830, formally joining the Latter Day Saint church. The same day, he was ordained an elder at the house of Father Whitmer, Seneca County, New York.

34 Ziba Peterson was baptized into the Church of Christ (an early name for The Church of Jesus Christ of Latter-day Saints) by Oliver Cowdery on April 18, 1830 in Seneca Lake, New York.

35 Elder Mark E. Petersen, a more modern-day member of the Quorum of the Twelve Apostles, apparently believed in a Western Hemispheric model. He is quoted as saying: "As the ancient Israelites suffered a dispersion which sprinkled them among all the nations, so the descendants of Laman and Lemuel [sons of Lehi] were sifted over the vast areas of the western hemisphere. They are found from pole to pole." Mark E. Petersen, "Children of Promise," Salt Lake City: Bookcraft, 1981, p. 31.

President Spencer W. Kimball also apparently believed in a Western Hemispheric model. He was known to make many broad-brush statements about the Lamanite remnant. In 1971, for example, he asserted, "Lehi and his family became the ancestors of all of the Indian and Mestizo tribes in North and South and Central America and in the islands of the sea." Spencer W. Kimball, "Of Royal Blood," Ensign, July 1971, p. 7. President Spencer W. Kimball defined "Lamanite" thusly: "The term Lamanite includes all Indians and Indian mixtures, such as the Polynesians, the Guatemalans, the Peruvians, as well as the Sioux, the Apache, the Mohawk, the Navajo, and others." Spencer W. Kimball, "Of Royal Blood," *Ensign,* July 1971, p. 7.

Although some LDS brethren like President Kimball have given their opinions on what a "Lamanite" is defined as, this book only focuses on merely one definition of Lamanite— Lamanites who are genetic Lamanites, or in other words, "[A] people of Israelite origin from the Book of Mormon."*Encyclopedia of Mormonism,* "Lamanite." When the Prophet Joseph Smith spoke of the "Lamanites," he was clearly speaking of the descendants of the people described in the Book of Mormon. The Book of Mormon describes people as being Lamanite either through lineage, or through dissension.

36 A broad definition of the term "Lamanite" is often used by Latter-day Saints, even General Authorities, to include all definitions for the term (i.e. the political, genetic, and general use of "Lamanite"). In many cases, an all-inclusive definition of "Lamanite" has replaced the genetic definition. Because of the broad-brush use of this term by many LDS Church members, it is often unclear whether it is being implementing in a broad sense or in the genetic sense. Also, the ambiguity of the term in the Book of Mormon itself means that discerning which meaning of the term is intended (or understood by the speaker) is no longer a straightforward task.

37 The Lamanites were descendants of Lehi and Ishmael. Lehi was a descendant of Joseph of Egypt (See 1 Nephi 5:14) and Ishmael might have also descended from that same lineage.

38 *Encyclopedia of Mormonism,* "Lamanite." When the Prophet Joseph Smith spoke of the "Lamanites," he was clearly speaking of the descendants of the people described in the Book of

End Notes

Mormon. The Book of Mormon describes people as being Lamanite either through lineage, or through dissension.

39   BH Roberts, *New Witness for God,* Volume 3, p. 40.

40   Joseph Smith, Journal, 9 November 1835. See Scott H. Faulring, ed., An American Prophet's Record: The Diaries and Journals of Joseph Smith (Salt Lake City: Signature Books and Smith Research Associates, 1987), p. 51; Dean C. Jessee, ed., The Papers of Joseph Smith: Volume 2- Journal, 1832-1842 (Salt Lake City: Deseret Book, 1992), p. 70; Dan Vogel, ed., Early Mormon Documents (Salt Lake City: Signature Books, 1996), 1:44.

41   *The Genius of Charles Darwin, God Strikes Back,* part 2/5, Richard Dawkins (2009).

42   *Abraham Lincoln and the Jews,* p. 5, Isaac Markers (1909).

43   *History of the Church,* 6:366; from a discourse given by Joseph Smith on May 12, 1844, in Nauvoo, Illinois; reported by Thomas Bullock.

44   Joseph Smith, Letter to Rochester, New York, newspaper editor N. C. Saxton, January 4, 1833. American Revivalist February 2, 1833.

45   Donald Q. Cannon, Church History Regional Studies BYU Department of Church History and Doctrine, Regional Studies, Illinois, "Zelph Revisited" (1995), pp. 97-109.

46   The Pottawatomie are North American Indians who belong to the Algonquian family. They generally call themselves "Bodéwadmi," meaning "Keepers of the fire." The Pottawatomie originally called themselves "Neshnabé," which means "Original People." The Pottawatomi, also spelled Pottawatomie and Potawatomi (among many variations), are a Native American people of the Great Plains, upper Mississippi River and Western Great Lakes region.

47   *History of the Church,* 5:480. Mr. Hitchcock, Letter to John Chambers, in John King to John Chambers, 14 July 1843, Iowa Superintendency 1838-49, Letters Received by the Office of Indian Affairs, 1824-81, BIA Microfilm #363, 357-60; Helen Mar Whitney, "Life Incidents," Woman's Exponent 9-10 (1880-1882) and "Scenes and Incidents in Nauvoo," Woman's Exponent 11 (1882-83), July 2, 1843. The latter incorrectly cites the Wilford Woodruff journal as its documentary source.

48   ibid.

49   The meeting was primarily set up by the Pottawatomie to convince Joseph Smith to "give them any assistance in case of an outbreak [of war] on the frontier." Joseph told the Pottawatomie during their visit that they "could give them no assistance [in war] ... [for] his hands were tied by the US but ... he could sympathize [with] them." Henry King to John Chambers, 14 July 1843, "Letters Received by the Office of Indian Affairs," K1824-81, Iowa Superintendency 1838-1849, microfilm, 1949, National Archives, Washington DC. Henry King to John Chambers, 14 July 1843, "Letters Received by the Office of Indian Affairs," K1824-81, Iowa Superintendency 1838-1849, microfilm, 1949, National Archives, Washington DC.

50   *History of the Church,* 5:480. Mr. Hitchcock, Letter to John Chambers, in John King to John Chambers, 14 July 1843, Iowa Superintendency , 1838-49, Letters Received by the Office of Indian Affairs, 1824-81, BIA Microfilm #363, 357-60; Helen Mar Whitney, "Life Incidents," Woman's Exponent 9-10 (1880-1882) and "Scenes and Incidents in Nauvoo," Woman's Exponent 11 (1882-83), July 2, 1843. The latter incorrectly cites the Wilford Woodruff journal as its documentary source.

51   *History of the Church,* 5:480. Mr. Hitchcock, Letter to John Chambers, in John King to John Chambers, 14 July 1843, Iowa Superintendency, 1838-49, Letters Received by the Office of Indian Affairs, 1824-81, BIA Microfilm #363, 357-60; Helen Mar Whitney, "Life Incidents," Woman's Exponent 9-10 (1880-1882) and "Scenes and Incidents in Nauvoo," Woman's Exponent 11 (1882-83), July 2, 1843. The latter incorrectly cites the Wilford Woodruff journal as its documentary source.

52   The Book of Mormon was primarily written to the Lamanite remnant. Joseph Smith fulfilled a prophecy by informing the Pottawatomie of their Book of Mormon ancestors and that the Book of Mormon contains the answers to their problems. He also helped them to begin the repentance process by telling them to make peace with one another and to stop killing Indians and white men.

53   The Book of Mormon also states that Joseph of Egypt had obtained a promise of the Lord that in the Latter days his descendants (i.e. the Lamanite remnant) would be brought out of hidden darkness and captivity and become free" (See 2 Nephi 3:5). The Lord led Lehi and his caravan out of the land of Jerusalem that he might raise up unto him "a righteous branch from the fruit of the loins of Joseph [of Egypt]" (See Jacob 2:25). The whole purpose of Lehi leaving Jerusalem before it was destroyed was so that Joseph of Egypt's lineage would be preserved.

End Notes

The Lord even told Joseph of Egypt that he would preserve his bloodline forever (See 2 Nephi 3:16). Because of the covenant Joseph of Egypt's seed made with the Lord, his bloodline is blessed. The Lord reiterated his promise: "for thy seed shall not be destroyed" and they shall hearken unto the words of the book [the Book of Mormon] (See 2 Nephi 3:15).

54   The Lamanite remnant shall "know that they are of the house of Israel, and . . . the covenant people of the Lord; and . . . know and come to the knowledge of their forefathers, and also to the knowledge of the gospel of their Redeemer" (1 Nephi 15: 14). The Book of Mormon echoes this significant prophecy. "[T]he fulness of the gospel of the Messiah," according to Lehi of the Book of Mormon, is prophesied to "come unto the Gentiles, and from the Gentiles unto the remnant of our seed [i.e. The Lamanite remnant]" (1 Nephi 15: 13).

55   Thomas, Robert Murray. Manitou and God: North-American Indian Religions and Christian Culture. Greenwood Publishing Group, 2007.

56   The Pottawatomie letter to Joseph Smith, Jr.: "Father . . . we are like a blind people in this [situation][.] . . . [O]ur chiefs are very uneasy[.] . . . [F]ather is what we want." Summer of 1843.

57   Like other Latter-day Saints, Joseph Smith was part Gentile. The Prophet learned from a revelation he had received in Kirtland, Ohio, that many Latter-day Saints of his day were either full-fledged Gentiles or else part Gentile and part Israelite due to how much Joseph of Egypt's descendants admixed with Gentiles "[C]oncerning the revelations and commandments which thou hast given us, who are identified with the Gentiles." Joseph Smith dedicatory prayer (Kirtland Ohio). Elder Mark E. Petersen of the Quorum of the 12 Apostles, "Who are these modern Gentiles? We are! . . . We are also of Israel, however—mostly of Ephraim—but we are Gentiles too, inasmuch as Ephraim was widely scattered among the Gentiles in ancient times and intermarried with them." *The Great Prologue* (1975), p. 5-6.

58   Joseph Smith apparently had the bloodlines of Ephraim, Judah (possibly), and Gentiles.

59   Both Joseph Smith, Sr. and Brigham Young claimed Joseph Smith, Jr. was an Ephraimite. Brigham Young, *Journal of Discourses* 2:268–69. Even though certain Latter-day Saints are direct descendants of Joseph of Egypt, Brigham Young stated that some of Joseph Smith's progenitors may have come from bloodlines other than that of the tribe of Ephraim. Brigham Young, *Journal of Discourses*, 2:268. Author Hoyt W. Brewster has suggested that the Rod of Jesse in the Bible is Joseph Smith, which, if true, would mean that he is related to Jesus Christ via the house of Judah. *Isaiah Plain and Simple: the Message of Isaiah in the Book of Mormon*, by Hoyt W. Brewster. See chapter 11.

60   Original spelling of some misspelled words were corrected. Joseph Smith to Pottawatomie Indians, August 28, 1843, Joseph Smith Collection, church archives, The Church of Jesus Christ of Latter-day Saints.

61   Original spelling of some misspelled words were corrected. Joseph Smith to Pottawatomie Indians, August 28, 1843, Joseph Smith Collection, church archives, The Church of Jesus Christ of Latter-day Saints.

62   "At the grove President [Joseph] Smith addressed the Indians at some length, upon what the Lord had revealed to him concerning their forefathers, and recited to them the promises contained in the Book of Mormon respecting themselves. ... How their hearts must have glowed as they listened to the prophet relate the story of their forefathers—their rise and fall; and the promises held out to them of redemption from their fallen state!" (B.H. Roberts, A Comprehensive History of The Church of Jesus Christ of Latter-day Saints, 2:88-89, 1957), President John Taylor.

63   A large village of Sauk and (Fox) Meskwaki lived along the Mississippi near what is Nauvoo, established in the late 18th century; this village had as many as 1,000 lodges.

64   Joseph Smith, Jr., *History of the Church of Jesus Christ of Latter-day Saints*, ed. BH Roberts, 2d ed. Rev., 7 vols. (SLC: Deseret News Co., 1932-1951), 4:401-402.

65   "I stood close by the Prophet [Joseph Smith, Jr.] when he was preaching to the Indians in the Grove by the [Nauvoo] Temple. The Holy Spirit lighted up his countenance till it glowed like a halo around him, and his words penetrated the hearts of all who heard him and the Indians looked as solemn as Eternity." "Joseph Smith, the Prophet," *Young Woman's Journal* 16 [December 1905]: 558. Original spelling was altered ("prophet" was capitalized).

66   The Sauk & Fox (also Sac and Fox) have been located in Oklahoma, Kansas, Iowa, and Nebraska.

67   Joseph Smith, *An American Prophet's Record:The Diaries and Journals of Joseph Smith*, edited by Scott Faulring, Significant Mormon Diaries Series No. 1, (Salt Lake City, Utah:

End Notes

Signature Books in association with Smith Research Associates, 1989), p. 482. Aug. 12th, 1841, the *History of the Church* vol. 4, pp. 401-402.

68   Keokuk had facial hair. Although uncommon among many American Indians, goatees and beards did exist, contrary to what you might hear from some scholars about Indians and facial hair.

69   *History of the Church,* 4:401-402; Alexander Neibaur, Journal, 12 August 1841, Special Collections, Harold B. Lee Library.

70   Mormon 7:4-5.

71   See *History of the Church* 1:191.

72   *Joseph Smith Journal,* Nov. 9, 1835; cited in Dean C. Jessee, Mark Ashurst-McGee, Richard L. Jensen, eds., *The Joseph Smith Papers: Journals* Volume 1:1832-1839 (Salt Lake City: Church Historian's Press, 2008), 88-89.

73   History of Joseph Smith, The Latter-day Saints Millennial Star, vol. XV, p.565.

74   Joseph Smith Journal, 23 May, 1844; cited in Joseph Smith, *An American Prophet's Record:The Diaries and Journals of Joseph Smith,* edited by Scott Faulring, Significant Mormon Diaries Series No. 1, (Salt Lake City, Utah: Signature Books in association with Smith Research Associates, 1989), p. 482. HC 6:401-402

75   *History of the Church* 2:450.

76   Salem Gazette, 22 Nov. 1785, Richard Lloyd Anderson, *Joseph Smith's New England Heritage* (1971) pp. 89, 91.

77   On March 16, 1621, the settlers were quite surprised when Samoset walked straight through the middle of the encampment at Plymouth Colony and said, "Welcome Englishmen!" In the English language. Samoset had learned to speak English from fishermen who had previously visited his coastal territory.

78   Quoted in James R. Clark, comp., *Messages of the First Presidency of The Church of Jesus Christ of Latter-day Saints,* 6 vols. (1965–75), 5:279–280.

79   Jessee, Dean C., ed. And comp. *The Personal Writings of Joseph Smith,* Salt Lake City: Deseret Book, 1984.

80   In 1958, Elder Spencer W. Kimball spoke to faculty at BYU concerning the this time in early Church history. Elder Kimball told them that during this collective excitement to preach to the Lamanites, Joseph Smith had a vision of the Lord's work and so the very first thing he did prior to the organization of Church, was to swiftly send Elders Oliver Cowdery, Ziba Peterson, Parley P. Pratt, and Peter Whitmer to the North American Indians. "The very first thing before the Church was organized, Joseph Smith caught the vision of this work. He sent Oliver Cowdery, Ziba Peterson, and Parley P. Pratt and Peter Whitmer to the Indians immediately[.]" Spencer W. Kimball, "The Children of the First Covenant," An address to Seminary and Institute faculty at Brigham Young University, Provo, Utah, June 27, 1958.

81   Elder Cowdery was instructed by revelation through the Prophet Joseph Smith to preach the gospel of Jesus Christ to the Lamanites (See D&C 28:8).

82   Manuscript History, 1838-1856, volume A-1, *Joseph Smith Papers,* p. 60.

83   Moroni could not have been referring to natives of Mesoamerica when he proclaimed to the Prophet that "the covenant which God made with ancient Israel was at hand to be fulfilled because they were apparently an afterthought. Jessee, Dean C., ed. And comp. *The Personal Writings of Joseph Smith,* Salt Lake City: Deseret Book, 1984.

84   Parley P. Pratt, *Autobiography of Parley P. Pratt,* p. 47. The word "travelling" was changed to "traveling."

85   Parley P. Pratt, *Autobiography of Parley P. Pratt,* pp. 56-61. Elder Oliver Cowdery, *Church History* 1:178-183.

86   Parley P. Pratt also remarked concerning the Delaware Indians: "We found several among them who could read, and to them we gave copies of the Book [of Mormon], explaining to them that it was the Book of their forefathers. Some began to rejoice exceedingly, and took great pains to tell the news to others, in their own language. The excitement now reached the frontier settlements in Missouri, and stirred up the jealousy and envy of the Indian agents and sectarian missionaries to the degree that we were soon ordered out of the Indian country as disturbers of the peace; and even threatened with the military in case of non-compliance." *Autobiography of Parley P. Pratt,* pp. 56-61.

87   Mormon 7.

88   "The principal chief [of the Delawares, Chief Kikthawenund] says he believes every word of the Book [of Mormon] & there are many more in the Nation who believes [sic][.]" Oliver

End Notes

Cowdery to Joseph Smith, *Joseph Smith's Letter Books*.

89   The Book of Mormon states that Joseph of Egypt obtained a promise of the Lord that in the Latter days his descendants (i.e. the Lamanite remnant) would be brought out of hidden darkness and captivity and become free" (See 2 Nephi 3:5). So all of this could happen, the Lord led Lehi and his caravan out of the land of Jerusalem that he might raise up unto him "a righteous branch from the fruit of the loins of Joseph [of Egypt]" (See Jacob 2:25). The whole purpose of Lehi leaving Jerusalem before it was destroyed was so that Joseph of Egypt's lineage would be preserved. The Lord informed Joseph of Egypt that he would preserve his bloodline forever (See 2 Nephi 3:16). Because of the covenant Joseph of Egypt's seed made with the Lord, his bloodline is blessed. The Lord reiterated his promise to them: "for thy seed shall not be destroyed" and they shall hearken unto the words of the book [the Book of Mormon] (See 2 Nephi 3:15).

90   Parley P. Pratt, *Autobiography of* Parley P. Pratt, pp. 56-61. Original spelling of some misspelled words were corrected. Joseph Smith to Pottawatomie Indians, August 28, 1843, Joseph Smith Collection, church archives, The Church of Jesus Christ of Latter-day Saints.

91   Brigham Young to Chief Walkara of the Utes, January 22, 1855, as cited in Leonard J. Arrington, *Brigham Young: American Moses* (1985), p. 217.

92   In 1854, Chief Seath'tl of the Seattle tribe lamented concerning the white man's treatment of his people. He said, "I am a red man. . . . The white man's God cannot love his red children." Chief Seath'tl (1788 - 1866) of the Seattle (Duwamish) Indians. *Great Speeches by Native Americans,* p. 118, Robert Blaisdell. Frederic James Grant. History of Seattle, Washington, 1891, pp. 434-436.

93   Although neutrality concerning the term "Red Man" is less prevalent these days, the term is used throughout this book with a neutral tone.

94   Captain John Smith, *The General History of Virginia, New England, and the Summer Isles*, from the Third Book, chapter 2.

95   Potawatomi History, http://www.tolatsga.org/pota.html

96   Oklahoma Quick Facts: Ben's Guide to US Government.

97   D&C 54: 7-8. Note: the term "borders of the Lamanites" is used rather than the term "borders of *some* Lamanites." The use of the article "the" in the term "borders of the Lamanites" might imply that North America is the predominant location of the Book of Mormon if the article "the" was used by the Prophet Joseph Smith to limit the location of the Lamanites to North America.

98   *History of the Church* 2:224-225.

99   Jonathan Dunham was a member of the Seventy—a Latter-day Saint priesthood office in the Melchizedek priesthood. He would later be appointed as brigadier general of the Nauvoo Legion in 1844 and admitted to the Council of Fifty in 1845.

100   The Brotherton Indians, located in Wisconsin, are a North American tribe formed in the early 19th century from communities of several Pequot and Mohegan (Algonquian-speaking) tribes of southern New England and eastern Long Island, New York.

101   The Tuscarora ("Shirt-Wearing People") are a North American people of the Iroquoian-language family, with members today in New York, Canada, and North Carolina.

102   Blessing Given to Jonathan Dunham, 15 July 1837, and Missionary Diaries, 1837 and 1839, Jonathan Dunham Papers, LDS Church Archives. Journal History, 20 December 1836, 9 April 1837, 13 March 1838, and 15 July 1838. Dunham, Diary, 3 May-9 June 1840.

103   William Clayton Diary, 1 Mar. 1845, as reproduced in Andrew F. Ehat, "'It Seems Like Heaven Began on Earth': Joseph Smith and the Constitution of the Kingdom of God," BYU Studies, 20 (Spring 1980): 253–80; and Thomas Burdick to Joseph Smith, 28 Aug. 1840, Joseph Smith Collection, Church Archives.

104   Jonathan Dunham, Diary, 3 May-9 June 1840.

105   Dunham to Hyrum Kellogg, quoted in Thomas Burdick, Letter to unnamed correspondent, 28 August 1840, Joseph Smith Papers, LDS Church Archives

106   For additional data and interpretation on Wight and his later activity, see Davis Bitton, ed., *The Reminiscences and Civil War Letters of Levi Lamoni Wight: Life in a Mormon Splinter Colony on the Texas Frontier* (Salt Lake City: University of Utah Press, 1970); and "Mormons in Texas: The Ill-fated Lyman Wight Colony," *Arizona and the West: A Quarterly Journal of History* 11 (Spring 1969): 5-26.

107   Wilford Woodruff, Diary, 9 November 1840, Woodruff Papers, LDS Church Archives.

108   Nibley, Hugh W (2010), "The Book of Mormon: A Minimal Statement", Journal of the Book

End Notes

of Mormon and Other Restoration Scripture (Maxwell Institute) 19 (1): 78–80, retrieved 2010-09-28.

109   *Mormon's Codex: An Ancient American Book*, p. 20, John L. Sorenson.

110   *Times and Seasons* articles, September 15, 1842, and October 1, 1842. Author(s) uncertain.

111   Joseph Smith, Jr., *Teachings of the Prophet Joseph Smith*, p. 313. *History of the Church*, 5:498–99; from a discourse given by Joseph Smith on July 9, 1843, in Nauvoo, Illinois; reported by Willard Richards.

112   Dan Vogel (editor), Early Mormon Documents (Salt Lake City, Signature Books, 1996–2003), 5 vols, 1:296. citing Lucy Mack Smith, Biographical Sketches of Joseph Smith the Prophet, and His Progenitors for Many Generations (Liverpool, S.W. Richards, 1853), 36-173.

113   Joseph Smith, Jr., 2 Feb. 1833: *American Revivalist.*

114   Joseph Smith, Jr., 1 March 1842: Wentworth letter.

115   In 2003, LDS scholar Matthew Roper surmised that Joseph Smith translated the Book of Mormon, "but apparently did not know the scope of its geography." Matthew Roper, "Nephi's Neighbors: Book of Mormon Peoples and Pre-Columbian Populations," *FARMS Review* 15/2 (2003): 91-128. In 1999, LDS scholar Kenneth W. Godfrey wrote, "Exactly what Joseph Smith believed at different times in his life concerning Book of Mormon geography in general is . . . indeterminable." Kenneth W. Godfrey, "What is the Significance of Zelph In The Study Of Book of Mormon Geography?," *Journal of Book of Mormon Studies* 8/2 (1999): 70–79.

In 1985, prominent LDS scholar John L. Sorenson asserted: "The historical sources give no indication that Moroni's instructions to the young Joseph Smith included geography. Nor did Joseph Smith claim inspiration on the matter. Ideas he later expressed about the location of events reported in the book apparently reflected his own best thinking." John L. Sorenson, *An Ancient American Setting for the Book of Mormon* (Salt Lake City, Utah: Deseret Book Co.; Provo, Utah: Foundation for Ancient Research and Mormon Studies, 1996 [1985]), 1. In other words, if Roper has not changed his mind since 2003, he evidently believes Joseph Smith was unaware that Book of Mormon cities were not sprawled across the entire American continent. Prominent LDS scholar John L. Sorenson espouses a different view from Roper, although both academics believe in a Mesoamerican model for the Book of Mormon. Sorenson asserted in his 2013 book *Mormon's Codex: An Ancient American Book*: "Joseph Smith became convinced in the last years of his life that the lands of the Nephites were in Mesoamerica." John L. Sorenson, *Mormon's Codex*, p. 694. Sorenson apparently does not believe Joseph Smith wholly embraced a Western Hemispheric model or a North American model for the Book of Mormon, but is convinced the Prophet fully accepted a Mesoamerican model in the last years of his life.

Because distinct indigenous peoples are tied to specific geographical locations, Roper is implicitly claiming that Joseph Smith did not know the identity of the primary Lamanite remnant in his day and Sorenson is implicitly claiming that before Joseph Smith was killed he became convinced the Maya of Mesoamerica were the main Lamanite remnant.

116   *History of Joseph Smith,* Lucy Mack Smith, pp. 82–83.

117   Because the Prophet Joseph Smith understood the Lord's promises to the Lamanite remnant, he deliberately moved the Latter-day Saints closer to them. Where were the Remnant located? North America. According to revelations he had received from God, indigenous North Americans were meant to be an integral part of the establishment of Zion and the New Jerusalem. Book of Mormon prophecy states this very thing: the Saints are meant to assist the Israelite bloodline of the Book of Mormon in establishing Zion and the New Jerusalem (See 3 Nephi 21:24). In his lifetime Joseph Smith never traveled to Mesoamerica or South American to be closer to indigenous peoples south of the United States border. He only ever fulfilled Book of Mormon prophecies with indigenous North Americans. Revelation given through Joseph Smith the Prophet to Oliver Cowdery, at Fayette, New York, September 1830. CR D&C 28:14.

118   Revelation given through Joseph Smith the Prophet to Parley P. Pratt and Ziba Peterson, in Manchester, New York, early October 1830. CR D&C 32:2.

119   Covenant of Oliver Cowdery and Others, October 17, 1830, Ohio Star.

120   Parley P. Pratt, *Autobiography of* Parley P. Pratt, pp. 56-61.

121   Revelation given through Joseph Smith the Prophet to Newel Knight, at Kirtland, Ohio, June 10, 1831. CR D&C 54:14.

122   The plains of the Nephites appear to correspond perfectly with the Great Plains of the United

End Notes

States. Joseph Smith, Jr.: "The whole of our journey, in the midst of so large a company of social honest and sincere men, wandering over the plaines [sic] of the Nephites, recounting occasionaly [sic] the history of the Book of Mormon, roving over the mounds of that once beloved people of the Lord, picking up their skulls & their bones, as a proof of its divine authenticity...." Letter from Joseph Smith to his Wife, 1834, Dean C. Jessee, *The Personal Writings of Joseph Smith*, p. 324.

123   Nephihah was the second Nephite Judge (83 - 67 BC). He succeeded Alma the Younger when Alma had surrendered the judgment seat to him to devote more time to missionary work.

124   The term "plains of the Nephites" does not include a proper noun, just a reference to the plains that belonged to the Nephites.

125   According to Latter-day Saint Levi Hancock, Joseph Smith addressed Sylvester Smith on the Zion's Camp march and said, "on the way to Illinois River where we camped . . . this land was called the land of desolation" *Levi Hancock Journal*, 1834.

126   Joseph Smith, Jr. (Jan. 6, 1836). DHC 2:357. *The Journal of Joseph: The Personal History of a Modern Prophet*, p. 101.

127   Joseph Smith, Jr. (Jan. 6, 1836). DHC 2:357. *The Journal of Joseph: The Personal History of a Modern Prophet*, p. 101. The words "Esto Perpetua" can be traced back to the Venetian theologian and mathematician Paolo Sarpi (1552–1623), also known as Fra Paolo. The day before his death he had dictated three replies to questions on affairs of state, and his last words were "Esto perpetua" reportedly in reference to his beloved Venice and translated as "Mayest thou endure forever!"

128   Joseph Smith, Jr. (Jan. 6, 1836). DHC 2:357. *The Journal of Joseph: The Personal History of a Modern Prophet*, p. 101. The words "Esto Perpetua" can be traced back to the Venetian theologian and mathematician Paolo Sarpi (1552–1623), also known as Fra Paolo. The day before his death he had dictated three replies to questions on affairs of state, and his last words were "Esto perpetua" reportedly in reference to his beloved Venice and translated as "Mayest thou endure forever!"

129   Joseph Smith, Jr. (1838), *History of the Church of Jesus Christ of Latter-day Saints*, 3:34-35. George W. Robinson, a scribe of Joseph Smith who was with him at the time, corroborated the story: "We next kept [traveling] up the river mostly in the timber for ten miles, until we came to Colonel Lyman Wright's who lives at the foot of Tower Hill. A name appropriated by President Smith in consequence of the remains of an old Nephitish Altar and Tower where we camped for the Sabbath." Scott H. Faulring ed., *An American Prophet's Record: The Diaries and Journals of Joseph Smith*, SLC: Signature Books, 1989, p. 184.

130   ibid.

131   Wilford Woodruff Journal, 16 September 1841.

132   The book is comprised of two volumes. On November 16, 1841, Joseph Smith dictated a letter to John Bernhisel thanking him for the gift of *Incidents of Travel in Central America, Chiapas, and Yucatan* (1841): "I have read the volumes with the greatest interest & pleasure & must say that of all histories that have been written pertaining to the antiquities of this country it is the most correct [,] luminous & comprihensive [sic]." Personal Writing of Joseph Smith, p. 533. The letter is written in the handwriting of John Taylor. The letter could suggest that Joseph Smith dictated the apostle to write Bernhisel on his behalf.

133   *Personal Writing of Joseph Smith*, Dean C. Jessee, p. 533.

134   Joseph Smith, Jr., 1 March 1842: Wentworth letter.

135   Joseph Smith Jr, 2 May 1842: *Times and Seasons*— Evidence from Kentucky. Joseph Smith, ed., "A Catacomb of Mummies Found in Kentucky," Times and Seasons, 3:781-782 (May 2, 1842).

136   Joseph Smith Jr, 2 May 1842: *Times and Seasons*— Evidence from Kentucky. Joseph Smith, ed., "A Catacomb of Mummies Found in Kentucky," Times and Seasons, 3:781-782 (May 2, 1842). "In our own country, the opening of forests and the discovery of tumuli or mound and fortifications, extending in ranges from the lakes through the valleys of Ohio and Mississippi, mummies in a cave in Kentucky, the inscription on the rock at Dighton, supposed to be in Phoenician characters, and the ruins of walls and a great city in Arkansas and Wisconsin Territory, had suggested wild and wandering ideas in regard to the first peopling of this country, and the strong belief that powerful and populous nations had occupied it and had passed away, whose histories are entirely unknown. The same evidences continue in Texas, and in Mexico they assume a still more definite form." Stephens, *Incidents*, 1:98.

137   Samuel L. Mitchill (1764 - 1831).

End Notes

138   A letter from Dr. Samuel L. Mitchill, of New York to Samuel M. Burnside, Esq., Secretary of the American Antiquarian Society, on North American Antiquities (Aug. 24, 1815). "A Further Contribution to the Study of the Mortuary Customs of the North American Indians," Dr. HC Yarrow (1879 - 1880), Annual Report of the Bureau of Ethnology. It appears that for part of Mitchell's letter he is quoting an individual who had discovered the mummies in Kentucky.

139   John L. Lund, a Latter-day Saint, tour guide, scholar, and author of *Mesoamerica and the Book of Mormon, Is This The Place?*, is thoroughly convinced that the *Book of Mormon* primarily transpired in Mesoamerica. John L. Lund: " I am an advocate for Mesoamerica or Southern Mexico and Central America as the primary American lands of the Book of Mormon." *Joseph Smith and the Geography of the Book of Mormon*, p. 6, John L. Lund. Concerning the subject, Lund has asserted, "The correlations between Mesoamerica and the Book of Mormon are overwhelming. . . . Mesoamerica defines the geographical region of Southern Mexico and includes Central America." John L. Lund prefaced his statement with, "It is important to remember that a correlation doesn't prove anything." *Mesoamerica and the Book of Mormon, Is This The Place?*, p. 5, John L. Lund.

John E. Clark, a Latter-day Saint archaeologist specializing in ancient pottery, is also a proponent of the Mesoamerican model. He has stated concerning Book of Mormon geography, "The one requirement for making comparisons between archaeology and the Book of Mormon is to be in the right place[.] . . . Mesoamerica is the right place." John E. Clark, "Archaeology, Relics, and Book of Mormon Belief," Address given May 25, 2004 in the de Jong Concert Hall at BYU.

John L. Sorenson, an emeritus professor of anthropology at Brigham Young University, is just as, if not more, emphatic than Lund and Clark about Mesoamerica being the central location for Book of Mormon geography. Sorenson surmised in 1985, "[T]he land of Nephi in the broader sense constituted the highlands of southern Guatemala." "The city of Nephi was probably the archaeological site of Kaminaljuyu, which is now incorporated within suburban Guatemala City;" John L. Sorenson, *An Ancient American Setting for the Book of Mormon* (Salt Lake City, Utah : Deseret Book Co.; Provo, Utah: Foundation for Ancient Research and Mormon Studies, 1996 [1985]). Kaminaljuyu is a Pre-Columbian site of the Maya civilization that was primarily occupied from 1500 BC to 1200 AD, which means that Sorenson was implying that Nephites were Mayans. This quote leaves little doubt in the minds of readers about where he believes the Book of Mormon predominantly transpired. Sorenson's ideas, in effect, attempt to connect the Maya to the Nephites and Lamanites of the Book of Mormon. He, and many other Latter-day Saint scholars, also believe that the ancient Olmec are Jaredites from the Book of Mormon. Sorenson states: "The hill Ramah of the Jaredites, which is the same as the hill Cumorah of the Nephites, was where the final extermination of both peoples took place; that hill corresponds in all relevant parameters to Cerro El Vigía in the Tuxtla mountains of south-central Veracruz state, Mexico. . . . A decline of the society in which the Jaredites lived took place over a period of several centuries before their extinction around 600 BC. The Olmec cultural tradition declined and disappeared from the culture history of Mexico at the same time." "The city of Nephi was probably the archaeological site of Kaminaljuyu, which is now incorporated within suburban Guatemala City;" John L. Sorenson, *An Ancient American Setting for the Book of Mormon* (Salt Lake City, Utah : Deseret Book Co.; Provo, Utah: Foundation for Ancient Research and Mormon Studies, 1996 [1985]). Kaminaljuyu is a Pre-Columbian site of the Maya civilization that was primarily occupied from 1500 BC to AD 1200, which means that Sorenson was implying that Nephites were Mayans.

140   Traits of the Mosaic History, Found Among the Aztaeca [Azteca] Nations, *Times and Seasons*, vol. III no. 16 Pg 820.

141   Joseph Smith, *History of the Church of Jesus Christ of Latter-day Saints*, 7 vols. (Salt Lake City: Deseret Book, 1980), 5:44.

142   John E. Page, reply to "'A Disciple,'" Morning Chronicle, Pittsburgh, 1 July 1842, as quoted in Limited Geography and the Book of Mormon: Historical Antecedents and Early Interpretations, Matthew Roper, FARMS Review: Volume – 16, Issue – 2, Pages: 225-76 Provo, Utah: Maxwell Institute, 2004 .

143   John E. Page, "Mormonism Concluded: To 'A Disciple,'" Morning Chronicle, Pittsburgh, 20 July 1842.

144   Stephens, *Incidents*, 2:193–97, 305, 364; ibid., 2:128–29, 280. John Bernhisel's Gift to a

End Notes

Prophet: *Incidents of Travel in Central America* and the Book of Mormon, Matthew Roper, Interpreter: A Journal of Mormon Scripture 16 (2015): 207-253.

145   Joseph Smith (editor), "American Antiquities," Times and Seasons 3 no. 18 (15 July 1842), 858–860.

146   Joseph Smith (editor), "American Antiquities," Times and Seasons 3 no. 18 (15 July 1842), 858–860.

147   Joseph Smith (editor), "American Antiquities," Times and Seasons 3 no. 18 (15 July 1842), 858–860.

148   Joseph Smith (editor), "American Antiquities," Times and Seasons 3 no. 18 (15 July 1842), 858–860. "Mormon" may have been misspelled "Mormen." This error was fixed.

149   "Extract from Stephens' 'Incidents of Travel in Central America'," *Times and Seasons* 3 no. 22 (15 Se "And how was you destroyed? was the inquiry of those efficient antiquarians Messrs. Catherwood and Stephens, the charge d'affairs of these United States, as they sit on the wondrous walls of "Copan," situated near the western extremity of the Bay of Honduras, in the narrowest neck of land between the waters of the Atlantic ocean and the Pacific ocean, the very place where the Book of Mormon located a great city, on the narrow neck of land between the two seas. . . . How was this city, with seven or eight others, which Stephens gives us an account of, destroyed? Read the Book of Mormon, and that will tell the story of their sad disasters" John E. Page, reply to "'A Disciple,'" Morning Chronicle, Pittsburgh, 1 July 1842, as quoted in Limited Geography and the Book of Mormon: Historical Antecedents and Early Interpretations, Matthew Roper, FARMS Review: Volume – 16, Issue – 2, Pages: 225-76 Provo, Utah: Maxwell Institute, 2004. September 1842), 911 - 915.

150   "We are not going to declare positively that the ruins of Quirigua are those of Zarahemla, but when the land and the stones, and the books tell the story so plain, we are of opinion, that it would require more proof than the Jews could bring to prove the disciples stole the body of Jesus from the tomb, to prove that the ruins of the city in question, are not one of those referred to in the Book of Mormon". *Times and Seasons*, October 1, 1842.

151   John Bernhisel's Gift to a Prophet: *Incidents of Travel in Central America* and the Book of Mormon, Matthew Roper, Interpreter: A Journal of Mormon Scripture 16 (2015): 207-253.

152   *Wrestling the Angel: The Foundations of Mormon Thought: Cosmos, God, Humanity*, Terryl L. Givens, pp. 18-19.

153   Orson Pratt, 1832 & 1840.

154   *Times and Seasons*, Sept 15 1841, vol. 2 No. 22 pg. 544-545.

155   In September of 1840, Parley P. Pratt asked if anyone would like proof of the Book of Mormon, just look at "the ruins of cities, towns, military roads, forts, fortifications, mounds, artificial caves, temples, statues, monuments, obelisks, hieroglyphics, sculptured altars, aqueducts, and an endless variety of articles of husbandry, cooking utensils, &c. &c. which are the product of some ancient race, who inhabited that land, and who had risen to a high state of refinement in the arts and sciences, as the relics of their labours prove-as they now lie scattered over a vast extent of North and South America, either on the surface, or buried beneath by the convulsions of nature, or the visitations of the Most High, as recorded in the fore-going extract; and which are frequently discovered and brought to light by antiquarian travellers." (emphasis added). Parley P. Pratt, Millennial Star, September 1840, Vol. 1 No. 4 "Book of Mormon."

In February, 1842, Elder Parley P. Pratt after an article was published titled "Ruins in Central America. Ancient Monument at Copan", he quoted many parts of it, and reviewed Incidents of Travel in Central America, Chiapas, and Yucatan, and added: "We publish the foregoing for the purpose of giving our readers some ideas of the antiquities of the Nephites–of their ancient cities, temples, monuments, towers, fortifications, and inscriptions now in ruin amid the solitude of an almost impenetrable forest; but fourteen hundred years since, in the days of Mormon, they were the abodes of thousands and millions of human beings, and the centre of civil and military operations unsurpassed in any age or country. It is a striking and extraordinary coincidence, that, in the Book of Mormon, commencing page 563, there is an account of many cities as existing among the Nephites on the "narrow neck of land which connected the north country with the south country;" and Mormon names a number of them, which were strongly fortified, and were the theatres of tremendous battles, and that finally the Nephites were destroyed or driven to the northward, from year to year, and their towns and country made most desolate, until the remnant became extinct on the memorable heights of Cumorah (now western New York),–I say it is remarkable that Mr. Smith, in translating the

End Notes

Book of Mormon from 1827 to 1830, should mention the names and circumstances of those towns and fortifications in this very section of country, where a Mr Stephens, ten years afterwards, penetrated a dense forest, till then unexplored by modern travellers, and actually finds the ruins of those very cities mentioned by Mormon. The nameless nation of which he speaks were the Nephites. The lost record for which he mourns is the Book of Mormon. The architects, orators, statesmen, and generals, whose works and monuments he admires, are, Alma, Moroni, Helaman, Nephi, Mormon, and their contemporaries. The very cities whose ruins are in his estimation without a name, are called in the Book of Mormon, "Teancum, Boaz, Jordan, Desolation," &c." Parley P. Pratt, "Ruins in Central America," Millennial Star 2/11 (March 1842): 165.

156 Some LDS brethren have believed in South American models for the Book of Mormon. Other LDS brethren have believed in North American models and also Mesoamerican models. Some LDS brethren have believed that the Book of Mormon took place across the entire American continent. For example: in 1971, President Kimball (who was acting President of the Council of the Twelve Apostles at the time) asserted at the Lamanite Youth Conference in Salt Lake City, "Lehi and his family became the ancestors of all of the Indian and Mestizo tribes in North and South and Central America and in the islands of the sea."Spencer W. Kimball, "Of Royal Blood," Ensign, July 1971, p. 7. In 1981, Elder Mark E. Petersen of the Quorum of the Twelve Apostles claimed in his book *Children of the Promise: The Lamanites Yesterday & Today*: As the ancient Israelites suffered a dispersion which sprinkled them among all the nations, so the descendants of Laman and Lemuel [sons of Lehi] were sifted over the vast areas of the western hemisphere. They are found from pole to pole. Mark E. Petersen, "Children of Promise," Salt Lake City: Bookcraft, 1981, p. 31.

157 Emphasis added to the word "among." New Introduction to the Book of Mormon (2007). According to LDS spokesman Mark Tuttle, the change "takes into account details of Book of Mormon demography which are not known."

158 *History of the Church,* 5:480. Mr. Hitchcock, Letter to John Chambers, in John King to John Chambers, 14 July 1843, Iowa Superintendency 1838-49, Letters Received by the Office of Indian Affairs, 1824-61, BIA Microfilm #363, 357-60; Helen Mar Whitney, "Life Incidents," Woman's Exponent 9-10 (1880 1882) and "Scenes and Incidents in Nauvoo," Woman's Exponent 11 (1882-83), July 2, 1843. The latter incorrectly cites the Wilford Woodruff journal as its documentary source.

159 Joseph Smith Journal, 23 May, 1844; cited in Joseph Smith, *An American Prophet's Record:The Diaries and Journals of Joseph Smith*, edited by Scott Faulring, Significant Mormon Diaries Series No. 1, (Salt Lake City, Utah: Signature Books in association with Smith Research Associates, 1989), p. 482. Aug. 12th, 1841, *History of the Church* vol. 4, pp. 401-402.

160 LDS scholar John L. Sorenson asserted in his 2013 book *Mormon's Codex: An Ancient American Book*: "Joseph Smith became convinced in the last years of his life that the lands of the Nephites were in Mesoamerica." John L. Sorenson, *Mormon's Codex*, p. 694. Sorenson apparently does not believe Joseph Smith wholly embraced a Western Hemispheric model or a North American model for the Book of Mormon, but is convinced the Prophet fully accepted a Mesoamerican model in the last years of his life.

161 Internal Book of Mormon geography, which many LDS scholars currently agree upon, does not add up when a Guatemalan Zarahemla is added to the mix. Since internal Book of Mormon geography focuses on the relationships between lands and other geographic features as explained in the Book of Mormon, independent of where they might be physically located on a real map of the Americas, its foundation is interpretation coupled with logic and reasoning. Hence, a dramatic detour from internal Book of Mormon geography, such as Zarahemla being located thousands of miles away from Cumorah, requires one to question Book of Mormon writers' reckoning of time and distance in the Book of Mormon. Since this is highly improbable, either the North American models or the Mesoamerican models must be incorrect; they cannot simultaneously be true. If one assumes a Mesoamerican setting and does not question the reckoning of Book of Mormon writers, as some scholars have suggested, there must be two Hill Cumorahs.

162 President Anthony W. Ivins of the First Presidency, in Conference Report, April 15-16, 1929.

163 President Anthony W. Ivins of the First Presidency, in Conference Report, April 15-16, 1929.

164 Harold B. Lee, "Loyalty," address to religious educators, 8 July 1966; in *Charge to Religious*

Educators, 2nd ed. (Salt Lake City: Church Educational System and the Church of Jesus Christ of Latter-day Saints, 1982), 65; cited in Dennis B. Horne (ed.), *Determining Doctrine: A Reference Guide for Evaluation Doctrinal Truth* (Roy, Utah: Eborn Books, 2005), 172-173.

165 John A. Widtsoe, "Evidences and Reconciliations: Is Book of Mormon Geography Known?," Improvement Era 53 (July 1950), 547. …out of the studies of faithful Latter-day Saints may yet come a unity of opinion concerning Book of Mormon geography. John A. Widtsoe, foreword to Thomas Stuart Ferguson's Cumorah—Where? (Oakland: Published by the author, 1947), cited by John L. Sorenson, Mormon's Map (Provo, Utah: FARMS, 2000), 7–8. ISBN 0934893489.

166 Around 1918, President Joseph F. Smith: The present associate editor of *The Instructor* was one day in the office of the late President Joseph F. Smith when some brethren were asking him to approve a map showing the exact landing place of Lehi and his company. President Smith declined to officially approve of the map, saying that the Lord had not yet revealed it, and that if it were officially approved and afterwards found to be in error, it would affect the faith of the people.

167 The arguments laid out in the last two chapters have had one main purpose: to help Latter-day Saints become more aware of the overwhelming evidence that creates a presumption in favor of indigenous North Americans as the predominant Lamanite remnant. Unless substantial evidence exists to overturn this view, indigenous North Americans should be assumed as the main Lamanite remnant. Latter-day Saints with enthusiasm for the Mesoamerican model of the Book of Mormon have an enormous hurdle to clear if they wish to dislodge this presumptive position from its current location.

168 *Autobiography of* Parley P. Pratt, pp. 56-61, *Church History* 1:178-183.

169 D&C 30:6.

170 When this first revelation was published in the *Book of Commandments* (Chapter 30), Sidney Rigdon modified the text "among the Lamanites" to read "on the borders by the Lamanites". 8. And now behold I say unto you, that it is not revealed, and no man knoweth where the city shall be built, but it shall be given hereafter. 9. Behold I say unto you, that it shall be on the borders by the Lamanites. Revelation [second quote] given through Joseph Smith the Prophet, in Zion, Jackson County, Missouri, July 20, 1831. D&C 57:1-3.

171 "Handsome Lake and the Great Revival in the West," Anthony F. C. Wallace, *American Quarterly*, Vol. 4, No. 2, Summer 1952, pp. 149-150.

172 Three other glorious and radiant beings then appeared, each holding a bow in one hand and a huckleberry with berries of many colors in the other.

173 *Handsome Lake or The Reformed Drunkard Who Talked with Angels*, Alfred G. Hilbert. *Indians of Eastern Woodlands*, Sally Sheppard. *Iroquois Culture*, Edith Drumm.

174 Handsome Lake was born into the Wolf clan of his mother, as the Iroquois have a matrilineal kinship system, and was named Hadawa'ko ["Shaking Snow"] at birth. He was eventually adopted and raised by the Turtle clan people and later dubbed "Handsome Lake." Handsome Lake, by no coincidence, shares the same temperament type as Joseph of Egypt: Catalyst (Out-Curviplex-Circle-Point). See *Recognizing People* by Alexander T Paulos.

175 Handsome Lake even taught former US president Thomas Jefferson his religious beliefs. Letter from Thomas Jefferson To Brother Handsome Lake, Washington, November 3, 1802.

176 Daguerreotype of Thaonawyuthe, Blacksnake, or "Chainbreaker," Seneca (born between 1737 & 1760, died 1859). He died at Cold Spring, New York, in South Valley Allegany Reservation Dec 26th 1859 at age 117 or 120.

177 Parley P. Pratt was baptized in Seneca Lake by Oliver Cowdery on or about September 1, 1830, formally joining the Latter Day Saint church. The same day, he was ordained an elder at the house of Father Whitmer, Seneca County, New York.

178 Ziba Peterson was baptized into the Church of Christ (an early name for The Church of Jesus Christ of Latter-day Saints) by Oliver Cowdery on April 18, 1830 in Seneca Lake, New York. Ziba was excommunicated from the LDS Church in 1833.

179 Before Elders Pratt and Peterson visited the Seneca, Joseph Smith called Latter-day Saint Elder Peter Whitmer, Jr. to accompany his brother-in-law Oliver Cowdery on a mission to the Lamanites with instructions to "open thy mouth to declare my Gospel" and to "give heed unto the words and advice of thy Brother" who had been given power "to build my Church among thy Brethren the Lamanites". See D&C 30:5-6. Revelation, September 1830-D, JSP (*Joseph Smith Papers*). Even though Elders Whitmer and Cowdery were called first, Elders Pratt and Peterson appear to have visited as missionaries the first Lamanites on record.

180 Parley P. Pratt, ed., *Autobiography of Parley P. Pratt,* Classics in Mormon Literature series

End Notes

(Salt Lake City: Deseret Book Co., 1985), p. 35.
181   Peter Whitmer, Sr.'s home, Fayette, New York.
182   Peter Whitmer, Sr.'s home, Fayette, New York.
183   Neal A. Maxwell, "Encircled in the Arms of His Love." October 2002.
184   Nicholas C. P. Vrooman, "Handsome Lake, Joseph Smith, and the Word," recorded 7 November 1994. "For clarity, I've made my paraphrase and quotation of the story italics."— Lori Taylor, Ph.D., "Joseph Smith in Iroquois Country: The Handsome Lake Story."
185   Ganargua Creek (which became Anderson township), also known as "Mud Creek."
186   Engraving made in 1853 by C Burt after Captain Seth Eastman. Red Jacket wearing the "Peace Medal" given to him by George Washington.
187   Painting of Sagoyewatha (Red Jacket) by Charles Bird King, circa 1828. Original in the Albright-Knox Art Gallery, Buffalo, NY. Red Jacket wearing the "Peace Medal" given to him by George Washington.
188   *Joseph Smith: Rough Stone Rolling*, pp. 37-38, Richard L. Bushman. *Joseph Smith: The Making of a Prophet*, ch. 3, Dan Vogel. Turner, History of the Pioneer Settlement of Phelps and Gorham's Purchase, 214. There are also periodic newspaper notices to "the young people of the village of Palmyra and its vicinity" to attend "a debating school at the school house near Mr. [Benjamin] Billings" (see Western Farmer, 23 Jan. 1822; Palmyra Herald, 26 Feb. 1823). When the debating club began or when it was moved to the school house cannot be determined. Turner's use of "us" indicates that Joseph was involved with the club before Turner left Palmyra in 1820, although he was in nearby Canandaigua until 1822.
189   Red Jacket's Speech (1822), Palmyra, NY. *Republican Advocate*, Batavia: November 15, 1822.
190   *Teachings of the Prophet Joseph Smith*, p. 75.
191   *Teachings of the Prophet Joseph Smith*, p. 75.
192   *Kirtland Council Minute Book*, Kirtland High Council Minutes, MS 3432 CHL. 2 May 1835, p. 115, *Brigham Young Journal*. Spelling corrected. Originally, it read, "[W]ee [We] their [there] saw meney [many] of the seed of Joseph," wrote Young in his journal, "among them ware [were] two Chefts [chiefs,] one a prsbeterin [Presbyterian,] the other a Pagon [pagan]."
193   History of the Church 5:183, 386, 447.
194   Brigham Young. Manuscript History of Brigham Young, 1801-1844. Ed. Eldon Jay Watson. Salt Lake City: Smith Secretarial Service, 1968.
195   DHC 1:312-316, Joseph Smith, Jr. (1833). Letter to Noah C. Saxton, 12 February 1833. Some spelling mistakes are corrected, but the poor grammar has not been fixed.
196   Joseph Smith, Jr., 2 Feb. 1833: *American Revivalist*. Joseph Smith was in New York. Original spelling corrected. The original letter is no longer extant, but Frederick G. Williams copied it into JS's letterbook, probably soon after its composition.
197   Joseph Smith, Jr., 2 Feb. 1833: *American Revivalist*. Joseph Smith was in New York. Original spelling corrected.
198   Millennial Star, Vol. 54 (1852), p. 605. Wilford Woodruff, in Conference Report, Apr. 1898, p. 57; punctuation and capitalization modernized. "Elder's" was changed to "Elders."
199   Millennial Star, Vol. 54 (1852), p. 605-606. Wilford Woodruff, in Conference Report, Apr. 1898, p. 57; punctuation and capitalization modernized.
200   *Myths Encyclopedia*.
201   When Joseph Smith and early LDS brethren personally identified Lamanite remnants living in North America, they never told anyone (or even intimated that) they had anciently migrated from Mesoamerica.
202   James Adair (1709 - 1783).
203   Lord Kingsborough's Mexican Antiquities, Vol. VIII., pp. 356, 358. James Adair, *The History of the American Indians* (London, 1775), 179.
204   Orsamus Turner, *Pioneer History of the Holland Purchase of Western New York* (Buffalo, New York, 1850), 668-69.
205   Orsamus Turner, *Pioneer History of the Holland Purchase of Western New York* (Buffalo, New York, 1850), 668-69.
206   *The Natural and Aboriginal History of Tennessee* (1823) was an attempt to prove that the native tribes of Tennessee were descendants of ancient Hebrews.
207   Haywood later concludes in his text: "since we can trace this art into Egypt prior to the exodus . . . there seems to be incontrovertible evidence that the inscriptions in America were made by people of the old world." John Haywood, *The Natural and Aboriginal History of Tennessee* (Nashville, 1823), pp. 82, 328-30, 372.

End Notes

208 According to scholars such as John W. Hoopes—Acting Director for the Center of Latin American Studies (1996-97) and Director of the Global Indigenous Nations Studies Program (2008-11)—the emergence of metallurgy in pre-Columbian Mesoamerica occurred relatively late in the region's history, with distinctive works of metal apparent in West Mexico by about 700 AD. Jeffrey Quiltes & John W Hoopes (2003). Gold and Power in Ancient Columbia, Panama and Costa Rica. Harvard: Dumhurton Oakes, pp. 220–223. This dates is obviously after 421 AD, which is when the Book of Mormon ends.

209 An extremely thin gold plate with writing on it was extracted from a well in Chichén Itzá, Mexico. However, when Chichén Itzá as a city doesn't appear to fall in the Book of Mormon timeline. Chichen Itza was a major focal point in the Northern Maya Lowlands from the Late Classic (c. 600-900 AD) through the Terminal Classic (c. 800-900 AD) and into the early portion of the Postclassic Period (c. 900-1200 AD). Discovered at Oaxaca, Mexico, in 1932 by Alfonso Caso, were gold items including a gold headdress with plates attached to it, thin gold discs, and other gold items. Monte Albán was founded toward the end of the Middle Formative period at around 500 BC, by the Terminal Formative (ca.100 BC-200 AD). It is unclear when the gold plates were made, but it could have been made during the time of the Book of Mormon.

210 *The War of 1812*, Diana Childress, p. 30.

211 World of the American Indian, by Jules B. Billard, National Geographic Society; First Printing edition (1974), Washington, D.C.

212 Charles C. Willoughby, The Turner Group of Earthworks, Hamilton County, Ohio (Cambridge, Mass., 1922), p. 61; George K. Neumann, "Evidence for the Antiquity of Scalping from Central Illinois," American Antiquity, 5 (1940), pp 287-289; Wilda Anderson Obey, "The Arvilla People," Minnesota Archaeologist, 33:3,4 (1974), 1-33; N. S. Ossenberg, "Skeletal Remains from Hungry Hall Mound 2," MS (1964), Royal Ontario Museum, Toronto, Canada, table 1, 21; Walter Raymond Wood, Nanza, The Ponca Fort (Madison, Wis., 1960), pp. 86-87; Donald E. Wray, "The Kingdom Lake Sequence," MS (n.d.), Illinois State Museum, Springfield; Lucile E. Hoyme and William M. Bass, Human Skeletal Remains from the Tollifero (Ha6) and Clarksville (MC140 Sites, John H. Kerr Reservoir Basin, Virginia, Smithsonian Inst., Bur. of Am. Ethnol., Bull.182 (Washington D. C., 1962), pp. 378-80, oks 102:4.

213 Scott, George Ryley (2003). History of Torture Throughout the Ages. Kessinger Publishing. p. 211.

214 Josephus said, "[T]he entire body of the people of Israel remained in that country; wherefore there are but two tribes in Asia and Europe subject to the Romans, while the ten tribes are beyond Euphrates till now, and are an immense multitude, and not to be estimated by numbers." *Antiquities of the Jews*, Book XI, Chapter 5, Section 2, Flavius Josephus.

215 Joseph F. Smith, *Gospel Doctrine*, p. 409.

216 The alacrity of Joseph Smith and swiftness of the dispatch of Latter-day Saint missionaries to North America are a great contrast to what the prophet did for indigenous Mesoamericans: absolutely nothing. His actions in North America, and apparent lack of urgency to connect with Mesoamerica, corroborate the view that North America is the predominant location of the Book of Mormon.

217 *The Hebrew Pharaohs of Egypt: The Secret Lineage of the Patriarch Joseph*, By Ahmed Osman, p. 15.

218 *The Hebrew Pharaohs of Egypt: The Secret Lineage of the Patriarch Joseph*, By Ahmed Osman, p. 15.

219 Before the discovery of Tutankhamen's lavish treasures, the tomb belonging to both Yuya and Tuyu was one of the most significant burials in the Valley of the Kings. Discovered on February 5, 1905, by Egyptologist James Quibell and excavator Theodore M. Davis, the Egyptian tomb contained one of the most complete and beautifully assembled sets of funerary equipment then known. The many intact items from the tomb indicate the high status of Yuya and his wife, Tuya. Rather than receive a burial in the Valley of the Nobles (Sheikh Abd el-Qurna), where some of the powerful courtiers and other noteworthy figures were buried, he was interred alongside two Pharaohs, Ramses III and Ramses XI, in the Valley of the Kings! His status was high enough to receive a Pharaoh-like burial and yet few people have ever heard of him.

220 Ahmed Osman was born in Cairo in 1934. Osman was a law student at Cairo University before becoming a journalist. He has a Master's Degree in Egyptology.

End Notes

221 Autosomal refers to any of the chromosomes other than the sex-determining chromosomes (i.e., the X and Y) or the genes on these chromosomes.
222 Semitic man (Irano-Afghan Iranid).
223 They both have large aquiline noses, high nasal bridges, long faces, and long-headed (dolichocephalic) heads. Yuya (cephalic index= 70.3), Thutmose II and the TT320-CCG61065 mummy show dolichocephaly, with cephalic indices of 73.4 and 74.3, respectively. Akhenaten = 81.0? (brachycephalic?); Tutankhamun = 83.9 (brachycephalic).
224 *The Funerary Papyrus of Iouiya,* Theodore M. Davis (1915). Henri Naville is quoted.
225 Arthur Weigall quoted in *Tomb of Yuaa and Thuiu,* James Edward Quibell.
226 *Tomb of Yuaa and Thuiu,* James Edward Quibell.
227 *Tomb of Yuaa and Thuiu,* James Edward Quibell.
228 Yuya's mummy was discovered within 3 coffins, like the mummy of Pharaoh Tutankhamun.
229 *The Hebrew Pharaohs of Egypt: The Secret Lineage of the Patriarch Joseph,* Ahmed Osman (September 19, 2003).
230 The photos in between pp. 70-71 in *The Hebrew Pharaohs of Egypt* by Ahmed Osman.
231 British Egyptologist Percy E. Newberry said that "Around the neck and upper part of the chest [of Yuya] is a broad necklace in gold." Percy E. Newberry, *The Tomb of Iouiya and Touiyou.*
232 *The Funerary Papyrus of Iouiya,* Theodore M. Davis (1915), p. 2. Henri Naville is quoted. *The Hebrew Pharaohs of Egypt,* Ahmed Osman.
233 Sūrat Yūsuf—the 12th sura (chapter) of the Koran Muhammad, the prophet of Islam, shared similar temperament traits with Joseph of Egypt.
234 According to a study mentioned in a 2015 PressReader article by Sarah Knapton entitled "Why some people are just more attractive", "Women are unconsciously . . . attracted to a partner with a big jaw, a broad chin, an imposing brow. . . . [and] cheekbone prominence." "Why some people are just more attractive." Apr. 25, 2015 PressReader article by Sarah Knapton.
235 Mummy of Yuya (Joseph of Egypt) found in a coffin in the Valley of the Kings. P2MR type: Catalyst (Out-Curviplex-Circle-Point). Refer to *Recognizing People* by Alexander T Paulos.
236 According to a study mentioned in a 2015 PressReader article by Sarah Knapton entitled "Why some people are just more attractive", "Women are unconsciously . . . attracted to a partner with a big jaw, a broad chin, an imposing brow. . . . [and] cheekbone prominence." "Why some people are just more attractive." Apr. 25, 2015 PressReader article by Sarah Knapton.
237 Mummy of Yuya (Joseph of Egypt) found in a coffin in the Valley of the Kings. P2MR type: Catalyst (Out-Curviplex-Circle-Point). Refer to *Recognizing People* by Alexander T Paulos.
238 Rhys Ifans' P2MR type: Catalyst (Out-Curviplex-Circle-Point). Refer to *Recognizing People* by Alexander T Paulos. A large percentage of Welsh males (83.5%) belong to a paternal lineage known as Haplogroup R1b (Y-DNA).
239 According to a study mentioned in a 2015 PressReader article by Sarah Knapton entitled "Why some people are just more attractive", "Women are unconsciously . . . attracted to a partner with a big jaw, a broad chin, an imposing brow. . . . [and] cheekbone prominence." "Why some people are just more attractive." Apr. 25, 2015 PressReader article by Sarah Knapton.
240 Fernando Torres' P2MR type: Catalyst (Out-Curviplex-Circle-Point). Refer to *Recognizing People* by Alexander T Paulos.
241 A large percentage of Spanish males from Spain (69 %) belong to a paternal lineage known as Haplogroup R1b (Y-DNA). Over 8/10 men tested from Catalonia, Spain (82.5%) belong to Haplogroup R1b. Thirteen percent of Sephardic Jews tested belong to Haplogroup R1b (Y-DNA).
242 The translations of Yuya's titles are contained in Theodore M. Davis' book, *The Funeral Papyrus of Iouiya* [Yuya].
243 *The Funeral Papyrus of Iouiya* [Yuya], Theodore M. Davis.
244 *The Funeral Papyrus of Iouiya* [Yuya], Theodore M. Davis.
245 The translations of Yuya's titles are contained in Theodore M. Davis' book, *The Funeral Papyrus of Iouiya* [Yuya].
246 *The Funeral Papyrus of Iouiya* [Yuya], Theodore M. Davis.
247 Emphasis added to the word "his." *The Funeral Papyrus of Iouiya* [Yuya], Theodore M. Davis.
248 Emphasis added to the word "his." *The Funeral Papyrus of Iouiya* [Yuya], Theodore M. Davis.
249 *The Funeral Papyrus of Iouiya* [Yuya], Theodore M. Davis.

End Notes

250 In the context of a meritocracy where viziers were chosen in this way, the story of Joseph of Egypt becomes more plausible.
251 *The Hebrew Pharaohs of Egypt,* Ahmed Osman, p. 125.
252 According to Theodore Davis, Rekhit seem to have been a privileged caste; what we would call by the modern names of peers. They took part in the coronation. It is to them that the heir was presented; they were the first to pay him homage. According to Egyptologist Ahmed Osman, Yuya is also referred to as "Praised One who came forth from the Body Praised," "Plentiful in Favors in the House of the King," "Plentiful of Favors under his Lord," "Great of Love," "Enduring of Love under his Lord," "Praised of the King," "Favorite of the Good God [Pharaoh]," "Confidant of the Good God [Pharaoh]," "First among the King's Companions," "One made rich by the King of Lower Egypt," and "One made great by the Lord of the King of Lower Egypt." All of these titles of Yuya could easily be applied to Joseph of Egypt. *The Hebrew Pharaohs of Egypt,* Ahmed Osman.
253 *Stranger in the Valley of the Kings,* Ahmed Osman, San Francisco: Harper & Row, (1987).
254 *The Hebrew Pharaohs of Egypt,* Ahmed Osman.
255 Genesis 46:29 reads, "And Joseph made ready his chariot, and went up to meet Israel his father, to Goshen, and presented himself unto him; and he fell on his neck, and wept on his neck a good while." In Genesis 50: 8-9, it states, "And all the house of Joseph, and his brethren, and his father's house: only their little ones, and their flocks, and their herds, they left in the land of Goshen. And there went up with him both chariots and horsemen: and it was a very great company."
256 Emphasis added to the word "his." *The Funeral Papyrus of Iouiya* [Yuya], Theodore M. Davis.
257 *The Hebrew Pharaohs of Egypt,* Ahmed Osman.
258 Grey granite. Karnak. 18th dynasty. Cairo, Egyptian Museum. www.touregypt.net/museum/xviii41.htm
259 *The Hebrew Pharaohs of Egypt,* Ahmed Osman, p. 125.
260 "The Book of Mormon is a record of the forefathers of our [United States] ... tribes of Indians ... By it, we learn that our ... Indians, are descendants from that Joseph [who] was sold into Egypt, and that the land of America is a promised land unto them. . . . The city of Zion spoken of by David, in the one hundred and second Psalm, will be built upon the land of America." Joseph Smith to N. C. Saxton, 4 Jan., 12 Feb. 1833, in Dean C. Jessee, comp. and ed., *The Personal Writings of Joseph Smith* (Salt Lake City: Deseret Book, 1984), pp. 273, 275.
261 Zahi Hawass, an Egyptian Egyptologist and former Secretary General of Egypt's Supreme Council of Antiquities in charge of the study of some Amarna mummies appears to have been one the people behind the possible conspiracy (information suppression). The Amarna mummies, which include Yuya, Tuya, Amenhotep III, Akhenaten, and Tut, all had their DNA tested, but despite public and scientific interest, the names of the haplogroups (i.e. groups of genetic markers) were never published or made public.
262 Watch the episode here: www.discovery.com/tv-shows/other-shows/videos/king-tut-unwrapped-king-tuts-paternal-line.htm
263 The haplogroup of King Tut turned out to be one that is prominent among the Swiss, which could explain their interest in revealing this information. It is likely that this DNA analysis was suppressed because the data contradicts the paradigm in support of an Afrocentric genetic provenance of famous ancient Egyptians such as King Tut.
264 Unless powerful people are able to influence the Egyptians, DNA information about other prominent ancient Egyptians will likely remain hidden. Just as the analysis of the DNA of King Tut is being suppressed by researchers and Egyptian Egyptologists, Yuya's DNA, which has definitely been tested by geneticists, has not been released to the public (or even to the scientific community). There must be significantly more to Yuya than Egyptologists want the world to know. Since modern Egyptian Egyptologists are Afrocentric, it is very possible that they would want to hide evidence that an Israelite was buried in the Valley of the Kings.
265 The leaked analysis of Yuya's DNA (which is uncommon for an Egyptian) matches up with the DNA of indigenous North Americans. Using the 8 marker Short Tandem Repeat (STR) profile of the Egyptian mummy Yuya, DNA Tribes—a genetic ancestry analysis company—analyzed his autosomal DNA. A letter signed by Lucas Martin, a DNA Tribes employee. "DNA Tribes Digest analysis of the Amarna mummies . . . Yuya's 8 marker STR profile was also found to some extent in indigenous North and Central American regions (particularly the North Amerindian and Athabaskan regions)." Yuya's microsatellite marker: D13S317 (11, 13);

End Notes

D7S820 (6, 15); D2S1338 (22, 27); D21S11 (29, 34); D16S539 (6, 10); D18S51 (12, 22); CSF1PO (9, 12); FGA (20, 26). Yuya also has an approximate Match Likelihood Index (MLI) of 35.53 for the African Great Lakes region and 34.48 for the South Africa region.

266   Kihue, also known as Bill Moose Crowfoot, was never at a loss for words. His tales would range from camping along the Scioto and Olentangy Rivers with his family in the 1840's to knowing Buffalo Bill Cody and Annie Oakley. Bill Moose spent nine years working for the Sells Brothers Circus, going to many places in America, and also into Canada and Australia. He never married.

267   Joseph Smith, Jr. (Jan. 6, 1836). DHC 2:357.

268   "Out of Egypt," an educational forum based on new historical and scientific discoveries, Ahmed Osman. *The Hebrew Pharaohs of Egypt: The Secret Lineage of the Patriarch Joseph*, By Ahmed Osman.

269   Some scholars doubt the very existence of Joseph of Egypt. Others note the possible inconsistencies that exist between the stories of Yuya and Joseph of Egypt. Some individuals claim that Joseph of Egypt and Imhotep are one and the same. Is Yuya Joseph of Egypt or is he merely a Semitic man whose story somehow mirrors Joseph in Egypt's? Either way, Yuya is a Semitic-looking man who was buried in the Valley of the Kings in Egypt and looks extremely similar to members of North American Indian tribes and shares a portion of his genome with them. Those facts in and of themselves are significant. However, if for some reason Yuya is not Joseph of Egypt, the main premise of this book still remains intact. If Yuya is in fact Joseph of Egypt, it is merely icing on the cake rather than the cake itself of this book. The cake, in other words the main portion of this book, is that certain North American Indians look like Semites and have Semitic DNA (i.e. the two groups are closely related). This main argument remains unharmed even if the connection made in this book between Yuya and Joseph of Egypt is somehow off. In other words, the truthfulness of the main premise of this book does not hinge upon the fact that Yuya is Joseph of Egypt. It is just an interesting connection that makes the premise of this book a little sweeter if it is in fact true because of how important of a figure Joseph of Egypt is in the Book of Mormon and to Latter-day Saints.

Although the story of Yuya shares uncanny similarities with the Israelite Joseph, Yuya being Joseph of Egypt is yet to be accepted by mainstream Egyptology. Academics remain unconvinced that Yuya is Joseph of Egypt for a few reasons. First of all, some scholars doubt Joseph of Egypt ever even existed. One scholar has claimed that Yuya, no matter if he is Joseph of Egypt or not, was genetically an Egyptian rather than Semitic. According to the opposition, burial locations, time periods, and children of Yuya and Joseph of Egypt also don't match up according to the Bible and historical records.

In his book, *The Struggle of the Nations*, French Egyptologist Gaston Maspero asserts his belief that both Yuya and his wife Tuya were Egyptian, even though the majority of his colleges do not support this idea. If Yuya is Egyptian, with no Semitic DNA, then Maspero's case is valid. However, Yuya possesses Semitic DNA common to certain certain Middle Easterners, European Semites, and North American Indian tribes.

Joseph of Egypt's possible burial location, as discussed in Joshua of the Old Testament, can possibly present a hole in the Yuya-is-Joseph-in-Egypt theory. The Bible reads, And the bones of Joseph, which the children of Israel brought up out of Egypt, buried they in Shechem, in a parcel of ground which Jacob bought of the sons of Hamor the father of Shechem for an hundred pieces of silver: and it became the inheritance of the children of Joseph (Joshua 24:32).

This Bible verse implies that Joseph's mummified body was exhumed and transported by the Israelites to Canaan—details that differ greatly from Yuya's burial. It is said that Joseph was buried at Shechem, but Yuya was buried in the Valley of the Kings. This discrepancy is possibly one of the reasons why biblical scholars and other academics haven't given much credence to the Yuya-Joseph connection. Canadian Egyptologist Donald B. Redford, one of the skeptics of the Yuya-Joseph theory, wrote a scathing review in the *Biblical Archaeology Review* about *Stranger in the Valley of the Kings: Solving the Mystery of an Ancient Egyptian Mummy* that was written by Egyptologist Ahmed Osman—the man who apparently first published the Yuya-Joseph connection in writing. *Review of Stranger in the Valley of the*

End Notes

Kings by Ahmed Osman. BAR 15/2 p.8, review by Donald B. Redford. Writer Deborah Sweeney also expressed great doubt toward Osman's proposed identification of Yuya. Deborah Sweeney stated that the title "God's father of the Lord of the Two Lands" is an extension of the title "God's Father" but is not exclusive to Yuya. Review: Osman, "Stranger in the Valley of the Kings", review by Deborah Sweeney, *The Jewish Quarterly Review*, New Series, Vol. 82, No. 3/4 (Jan. - Apr., 1992), pp. 575-579. Both Redford and Sweeney disagreed with many points of Osman's argument and remain unconvinced that Yuya was Joseph of Egypt.

If Joseph's body was actually transported to Canaan, then Yuya is definitely not Joseph of Egypt. However, if Joseph's family, or others, were unable to find and retrieve Joseph's body from Egypt, then the possibility of Yuya being Joseph increases.

An account in Genesis states that before Joseph's death, Joseph had his brothers (i.e. the sons of Israel) promise to carry his bones out of Egypt to Canaan (See Genesis 50:25). The same chapter states what the Egyptians characteristically did with royal bodies: "Joseph died . . . and they embalmed him, and he was put in a coffin in Egypt." Genesis 50:26. The word "coffin" in Hebrew (bā·'ā·rō·wn) means "mummy-chest." One would assume that Joseph's brothers kept their promise to Joseph, especially after he saved their lives. However, the task of obtaining Joseph's body from powerful Egyptians might have been proven too difficult, and so they may have failed their brother.

In Exodus, Moses is said to have done his best to keep the promise of the sons of Israel by gathering Joseph's bones and taking them with him during the great Exodus from Egypt. The Bible reads, "And Moses took the bones of Joseph with him: for he had straitly sworn the children of Israel, saying, God will surely visit you; and ye shall carry up my bones away hence with you" (Exodus 13:19). According to some believers, Moses used special powers to raise the coffin of Joseph so he could take Joseph's bones with him. Whether or not Moses accomplished the tasks of locating Joseph's tomb and embalmed mummy, and retrieving his actual bones, is uncertain. If Joseph's body was embalmed in Egypt, as the Bible says, no bag of bones would be available to take. Instead, a perfectly preserved body wrapped extremely well in linens would have been the only thing available. Many embalmed Egyptian bodies are preserved incredibly well and so Moses would have had to have taken a well-preserved mummy with him on his Exodus from Egypt rather than mere "bones."

The unlikelihood of Joseph's bones leaving their tomb becomes more apparent when the importance of Egyptian rites and rituals necessary to travel to the afterlife are brought to the forefront. First, Egyptians who wish to move on the afterlife must be properly prepared for the afterlife. Embalming was essential. Their belongings were stored with them in their tomb, and their name was preserved in writing. They were to take their things with them to the life after death. Next, Egyptians who wished to move on the afterlife had to be buried on Egyptian soil. Since these rites and rituals were so important to the Egyptians, it seems implausible for them to allow Moses to destroy one of their former leader's chances of safely remaining in the afterworld by removing Joseph's bones.

The use of the word "bones" in Exodus is suspect. It makes one question Moses' acquisition of Joseph's body because embalmed bodies were not piled up in stacks of "bones." This is why Moses was most likely unable to obtain the "bones," or any part of Joseph's body, from Joseph's tomb. Also, tombs of Pharaohs and viziers, like Joseph's, were difficult to break into. Extraction of entire mummies from sarcophagi in tombs was not an easy feat. Even tomb raiders struggled greatly to extract large pieces from tombs. It is why they almost often broke into ancient tombs and stole small, but highly valuable, items that were easily transportable. Moses would have had as much difficulty as a tomb raider locating Joseph of Egypt's tomb, breaking into it, and extracting Joseph of Egypt's "bones." From biblical texts, it appears that the author of Exodus must have been unaware of the incredible preservation abilities of the Egyptians, or else the usage of the word "bones" is more than one might think it is.

Some Jews and Muslims claim that Moses was successful in retrieving Joseph's body and that Joseph's "bones" were in fact buried in Shechem. However, no solid evidence exists to prove that Joseph's body was removed from his Egyptian burial site. A monument stands to honor

End Notes

Joseph, but it doesn't automatically mean that Joseph is buried in, or near, it. The body of a man, said to be Joseph, is purportedly buried there, but no DNA tests have been performed on the bones. There is no known way of really knowing if the bones belong to Joseph of Egypt anyway. The Itinerarium Burdigalense (333 AD) states that: 'At the foot of the mountain itself, is a place called Sichem. Here is a tomb in which Joseph is laid, in the parcel of ground which Jacob his father gave to him." Schenke 1967, p. 175 n.49: 'Inde ad pede montis ipsius (Gerizim) locus est, cui nomen est Sechim. Ibi positum est monumentum, ubi positus est Joseph in villa, quam dedit ei Jacob pater eius.' Eusebius of Caesarea in the 4th-century records in his Onomasticon: "Suchem, city of Jacob now deserted. The place is pointed out in the suburb of Neapolis. There the tomb of Joseph is pointed out nearby." Wolf, C. Umhau (2006) [1971]. The Onomasticon of Eusebius Pamphili Compared with the Version of Jerome and Annotated. The Tertullian Project. Retrieved 8 September 2011. Schenke 1967, p. 175 n.50:Συχὲμ . .νῦν ἔρημος. δείκνυται δὲ ὁ τόπος ἐν προαστείοις Νέας πόλεως, ἔνθα καὶ ὁ τάφος δείκνυται τοῦ Ἰωσήφ, καὶ παράκειται.'

Both Theodosius I and Theodosius II ordered a search for Joseph's bones, much to the utter dismay of the Samaritan community, Sivan, Hagith (2008). Palestine in late antiquity. Oxford University Press.p. 115 n.24 An imperial commission was dispatched to retrieve the bones of the Patriarchs around 415 CE, and on failing to obtain them at Hebron, sought to at least secure Joseph's bones from Shechem. No gravestone marked the exact site, possibly because the Samaritans had removed one to avoid Christian interference. The officials had to excavate the general area where graves abound and, on finding an intact marble sepulchre beneath an empty coffin, concluded that it must contain Joseph's bones, and sent the sarcophagus to Byzantium, where it was incorporated into Hagia Sophia. Crown, Alan David (1989). "The Byzantine period and Moslem Period". In Alan David, Crown. The Samaritans. Mohr Siebeck. p. 70. Schenke, Hans-Martin (1967). "Jacobsbrunnen-Josephsgrab-Sychar. Topographische Untersuchungen und Erwägungen in der Perspektive von Joh. 4,5.6". Zeitschrift des Deutschen Palästina-Vereins 84 (2): 177.

Another possible problem with the Yuya-Joseph theory is the dates when the two men lived. The time period associated with Yuya does not appear to correspond with the traditional biblical date for Joseph. Yuya is said to have lived during the 18th Dynasty, and the traditional date for Joseph of Egypt places him in the middle of the 12th Dynasty (Redford, Donald B. (1970). A study of the biblical story of Joseph: (Genesis 37–50). Supplements to Vetus Testamentum 20. Leiden: Brill., p. 188.), roughly comparable to the time of the invasion of Egypt by the Hyksos. Modern-day scholarship no longer accepts such a remote dating for Joseph. Sperling, S. David (2003). *The Original Torah: The Political Intent of the Bible's Writers*. NYU Press. p. 98.

Despite the fact that the time periods for Yuya and Joseph of Egypt do not match, it does not mean that Yuya is not Joseph of Egypt. The traditional date for Joseph of Egypt is not set in stone. Scholars are not positive about all dates and locations of many biblical events. What is more, calendars often changed throughout history and chronologies were often inaccurate. Exact dates are difficult to ascertain for biblical figures, including Joseph of Egypt.

Some scholars have even stated that because the time period of the life of Joseph of Egypt is ambiguous, the figure of Joseph itself is often considered to be a "personification of a tribe," rather than an historic person (Schenke 1968, p. 174: "Joseph" ist ein personifizierter Stamm.). For the sake of argument, one must neglect this consideration and focus on the assertion that Joseph of Egypt actually existed.

The Bible mentions a woman named Asenath who became the wife of Joseph of Egypt. It is assumed by many religious scholars that she was Joseph's only wife. However, such a claim cannot be substantiated. With Asenath, the Bible states that Joseph had two children: Ephraim and Manasseh. He may have had more children, but none are mentioned in the Genesis account. Yuya is said to have been married to a woman named "Tuya" rather than Asenath. This detail could discredit the Yuya-Joseph hypothesis. Yuya and Tuya had a daughter named Tiy, who became the queen of Amenhotep II—the seventh Pharaoh of the 18th dynasty of Egypt. According to scholars, DNA tests prove that Tiy is Yuya's biological daughter.

End Notes

These details about Tuya and Tiy do not line up with the Yuya-Joseph hypothesis, but it does not necessarily mean that they disprove it either. Just because Yuya's daughter Tiy was not mentioned in the Bible, it doesn't mean that Joseph of Egypt never had a daughter. It is not unreasonable for Joseph to have had more than one wife and many concubines. There is a possibility that he even more than two kids, especially since polygamy and large families were common among Semites of his time. Joseph of Egypt's father, Israel, is known to have had many wives and concubines. Israel had a favorite wife Rachel, so it is not unreasonable for Joseph to have had his "favorite" wife as well, and other wives. Except for Joseph's brother Benjamin, his other brothers came from three other mothers: Leah, Bilhah, and Zilpah.

Asenath could be Tuya, but merely a different name or spelling given to her. Both women are said to have been Egyptian. From Tuya's mummy, it is clear that she has Egyptian physical traits unlike those of Yuya, which are definitely Semitic.

270   Joseph Fielding Smith, *Improvement Era*, Oct. 1923, p. 1149.
271   The Latter-day Saints' Millennial Star (Saturday, September 18, 1852), vol. 14, no. 30, p. 469.
272   "How Your Looks Betray Your Personality," Feb. 11, 2009, *New Scientist*.
273   Scientific studies have long demonstrated that physical characteristics are almost certainly correlated with personality and temperament traits. According to a 1981 study by Stanford University, psychologists Roger W. Squier and John R. C. Mew: "A variety of findings . . . suggest there may be a relationship between personality dimensions and facial structure. Squier and Mew found that subjects with long, angular faces were more responsive, assertive, and genuine than other types of faces."

In 2007, psychologists Anthony C. Little and David Ian Perrett corroborated these findings by learning for themselves that "Several studies have demonstrated some accuracy in personality attribution using only visual appearance." "Using composite images to assess accuracy in personality attribution to faces," *British Journal of Social Psychology* (BJSP). A number of scientists are discovering the exact same thing. Most research is supporting the idea that an observable link exists between appearance and genetics. According to a 2009 article in the *Association for Psychological Science*: "[N]ew research from a team of Brock University psychologists suggests that we subconsciously ascertain possible aggression in others based on their facial structure." A team consisting of psychologists Justin M. Carré, Cheryl M. McCormick, and Catherine J. Mondloch found in 2009 that "[F]acial structure is a reliable cue of aggressive behavior." Studies from 2015 found a link between conscientiousness and angular faces, especially angular faces with symmetry. "Angry Faces: Facial Structure Linked To Aggressive Tendencies, Study Suggests," November 2, 2009. Association for Psychological Science. Studies from 2015 found a link between conscientiousness and angular faces, especially angular faces with symmetry. "The science of sexiness: why some people are just more attractive." Sarah Knapton, Science Editor, April 8, 2015.

It is most common to find highly conscientious individuals with angular faces. Conversely, highly impulsive individuals who fail to transcend their nature tend to have round faces. According to Angelina R. Sutin of the Department of Behavioral Sciences and Social Medicine at Florida State University: "People with personality traits of high neuroticism and low conscientiousness are likely to go through cycles of gaining and losing weight throughout their lives, according to an examination of 50 years of data in a study published by the American Psychological Association. Impulsivity was the strongest predictor of who would be overweight, the researchers found. Study participants who scored in the top 10 percent on impulsivity weighed an average of 22 lbs. more than those in the bottom 10 percent, according to the study. . . . Among their other findings: Conscientious participants tended to be leaner and weight did not contribute to changes in personality across adulthood." Personality Plays Role in Body Weight, According to Study. Impulsivity strongest predictor of obesity. American Psychological Association. July 18, 2011. "Personality and Obesity Across the Adult Life Span," Angelina R. Sutin, PhD, Luigi Ferrucci, MD, PhD, Alan B. Zonderman, PhD, and Antonio Terracciano, PhD, National Institute on Aging, National Institutes of Health, Department of Health and Human Services, Journal of Personality and Social Psychology, Vol. 101, No. 3.

End Notes

274 "Does the shape of my face show that I have a genetic disorder?" Josh Clark, *Science,*

275 A forensic physical anthropologist can assist in the identification of deceased individuals whose remains are decomposed, burned, mutilated or otherwise unrecognizable. Forensic anthropologists draw on highly heritable morphological features of human remains (e.g. cranial measurements) to aid in the identification of the body, including in terms of race. Sauer, Norman J. (1992). "Forensic Anthropology and the Concept of Race: If Races Don't Exist, Why are Forensic Anthropologists So Good at Identifying them". Social Science and Medicine 34 (2): 107–111. doi:10.1016/0277-9536(92)90086-6. PMID 1738862.

276 In a 1992 article, anthropologist Norman Sauer noted that anthropologists had generally abandoned the concept of race as a valid representation of human biological diversity, except for forensic anthropologists. Sauer, Norman J. (1992). "Forensic Anthropology and the Concept of Race: If Races Don't Exist, Why are Forensic Anthropologists So Good at Identifying them". Social Science and Medicine 34 (2): 107–111. doi:10.1016/0277-9536(92)90086-6. PMID 1738862.

277 Hole-in-the-Sky's P2MR type is Oracle (In-Curviplex-Circle-Point). Refer to *Recognizing People* by Alexander T Paulos.

278 This Ojibwe man's P2MR type is Dreamer (In-Line-Circle-Wave). Refer to *Recognizing People* by Alexander T Paulos.

279 "[S]ome Native American founders could have been of Caucasoid [those having Caucasian physical characteristics] ancestry[.]" Antonio Torroni, Italian geneticist (2000). Antonio Torroni, "Mitochondrial DNA and the Origin of Native Americans," America Past, America Present: Genes and Languages in the Americas and Beyond, ed. Colin Renfrew (Cambridge, UK: McDonald Institute for Archaeological Research, 2000), p. 79.

280 Hole-in-the-Sky's P2MR type is Oracle (In-Curviplex-Circle-Point). Refer to *Recognizing People* by Alexander T Paulos.

281 This Ojibwe man's P2MR type is Dreamer (In-Line-Circle-Wave). Refer to *Recognizing People* by Alexander T Paulos.

282 Brigham Young, *Journal of Discourses* 2:268–69. Even though certain Latter-day Saints are direct descendants of Joseph of Egypt, Brigham Young stated that some of Joseph Smith's progenitors may have come from bloodlines other than that of the tribe of Ephraim. Brigham Young, *Journal of Discourses,* 2:268.

283 Tribe of Ephraim. http://en.wikipedia.org/wiki/Tribe_of_Ephraim

284 P2MR type: Oracle (In-Curviplex-Circle-Point). Refer to *Recognizing People* by Alexander T Paulos.

285 Yuya's P2MR type: Catalyst (Out-Curviplex-Circle-Point). Refer to *Recognizing People* by Alexander T Paulos. The main difference between Hyrum Smith and Yuya is that Yuya was more of an extravert than Hyrum.

286 Patriarchal blessing of Hyrum Smith given by Joseph Smith, Sr. (Dec. 9, 1834).

287 Elder Neal A. Maxwell once said of Joseph Smith and Joseph of Egypt, "The comparisons between the two Josephs . . . reflect varying degrees of exactitude . . . but are quite striking. Some similarities are situational, others are dispositional." *A Choice Seer,* Neal A. Maxwell.

288 P2MR type: Orator (Out-Curviplex-Box-Point). Refer to *Recognizing People* by Alexander T Paulos.

289 Irwin Shaw was born to Russian Jewish immigrants.

290 Joseph Smith belonged to a paternal genetic lineage known as Haplogroup R1b (Y-DNA). This particular paternal lineage is related to Haplogroup R1 (Y-DNA)—the second-most prevalent paternal lineage among American Indians. 'R1b' stems from 'R1'. In other words, a man in the past who belonged to 'R1' had a child who was born with a genetic mutation that gave rise to Haplogroup R1b (Y-DNA).Although the others featured above may not have originated from Haplogroup 'R1,' they all appear to have origins in Middle Eastern lineages. Joseph and Hyrum Smith both belonged to a sub-branch (subclade) within the paternal lineage R1b (called R1b1b2a1a2f2) carrying the M222 defining mutation variant and his mother belongs to maternal lineage H. Joseph Smith DNA Revealed: New Clues from the Prophet's Genes, Ugo A. Perego, Msc, Aug. 7, 2008 FAIR conference. Haplogroup R1b1b2a1a2f2 (Y-DNA) and Haplogroup H (mtDNA).

291 A Discourse by President Brigham Young, Delivered in the Tabernacle, Great Salt Lake City, April 8, 1855. Reported by G. D. Watt.

292 Isaiah 11:12-13; 2 Nephi 21:12-13.

293 P2MR type: Expert (In-Curviplex-Box-Point). Refer to *Recognizing People* by Alexander T

End Notes

Paulos. John Philip Cohane, a University of Pennsylvania archaeologist and friend of Jewish-American Scholar Cyrus H. Gordon (1908-2001), a Semitic man who shared many of his same views, has claimed that many geographical names in America are Semitic in origin. Cyrus Herzl Gordon, Before Columbus; links between the Old World and ancient America, Crown, 1971, p. 138. Cohane published *The Indestructible Irish* in 1968 in which he claimed that the original blood stock in England, Ireland, Scotland, and Wales is Semitic. If we use the original definition of the word Semitic: "one descended from Shem, a son of Noah" then Cohane is correct in his assertions. *Éire-Ireland: a Journal of Irish studies*, Volume 5; Volume 5, Irish American Cultural Institute., 1966, p. 145. Eugene R. Fingerhut, Explorers of pre-Columbian America?: the diffusionist-inventionist controversy, Regina Books, 1994, p. 222. The Critic, Volume 27, Issue 6, Thomas More Association, 1969.

294 P2MR type: Expert (In-Curviplex-Box-Point). Refer to *Recognizing People* by Alexander T Paulos.
295 P2MR type: Expert (In-Curviplex-Box-Point). Refer to *Recognizing People* by Alexander T Paulos.
296 P2MR type: Expert (In-Curviplex-Box-Point). Refer to *Recognizing People* by Alexander T Paulos.
297 P2MR type: Expert (In-Curviplex-Box-Point). Refer to *Recognizing People* by Alexander T Paulos.
298 Gabriel Cousens. P2MR type: Catalyst (Out-Curviplex-Circle-Point). Refer to *Recognizing People* by Alexander T Paulos.
299 Yuya (Joseph of Egypt). P2MR type: Catalyst (Out-Curviplex-Circle-Point). Refer to *Recognizing People* by Alexander T Paulos.
300 Pine Bird, Oglala Sioux (1907). P2MR type: Catalyst (Out-Curviplex-Circle-Point). Refer to *Recognizing People* by Alexander T Paulos.
301 Mico Chlucco the Long Warrior - King of the Seminoles. 1792, Travels through North & South Carolina, Georgia, E & W Florida the Cherokee country, the extensive territories of the Muscogulges or Creek confederacy, and the country of the Choctaws. by William Bartram (1739-1823) American naturalist. This illustration was re-purposed in 1809 as a Sioux warrior in 'The Travels of Capts. Lewis & Clarke'; pub. by H Lester. This portrait resembles portrayals made by Le Moyne of the Timucua, another group of Florida Indians and so it just might be a legitimate depiction of the Creek chief.
302 Chief Red Cloud, Oglala Sioux. P2MR type: Legalist (In-Line-Box-Point). Refer to *Recognizing People* by Alexander T Paulos.
303 Omar Mukhtar (1858-1931), leader of Libyan resistance in Cyrenaica against the Italian colonization.P2MR type: Legalist (In-Line-Box-Point). Refer to *Recognizing People* by Alexander T Paulos.
304 Libyan Y-DNA Haplogroups: E-M81+E-M78+E-M2 (35.88% + 11.07%+8.78%), J-M267+J-M172 (30.53%+ 3.44%), G-M201 (4.20%), R* (3.43%), R-M17 (0.38%). Karima Fadhlaoui-Zid et al. (2013) Genome-Wide and Paternal Diversity Reveal a Recent Origin of Human Populations in North Africa. PLoS One. 2013; 8(11): e80293. See Table S2.
305 P2MR type: Expert (In-Curviplex-Box-Point). Refer to *Recognizing People* by Alexander T Paulos.
306 P2MR type: Expert (In-Curviplex-Box-Point). Refer to *Recognizing People* by Alexander T Paulos.
307 Carmichael, *The Satanizing of the Jews. Origin and Development of Mystical Anti-Semitism* (1992).
308 "Dr. Sami Hanna, an Egyptian, especially *schooled in* the Arabic language, and professor at the University of Utah, was *asked by* the First Presidency of the [LDS] church to translate the Book of Mormon back into its original Semitic cultural format. The following, by personal interview, are his words: 'When I began reading the Book of Mormon, and began making myself familiar with it, I expected to find a very poorly written book, as I had been told by critics of the unschooled nature of the youthful Joseph Smith as he *had purportedly* translated the book. What I found, however, was not a book of poor English; but to the contrary, I found myself reading the most beautiful Semitic book I had ever read!'" Brent G. Yorgason, *Little Known Evidences of the Book of Mormon* (1989).
309 Nephi: "I make a record [the small plates of Nephi] in the language of my father, which consists of the learning of the Jews and the language of the Egyptians" (1 Ne. 1:2). Moroni, the last known Nephite prophet, noted concerning the plates of Mormon that "we have written this

End Notes

record ... in the characters which are called among us the reformed Egyptian, being handed down and altered by us, according to our manner of speech. And if our plates [metal leaves] had been sufficiently large we should have written in Hebrew; but the Hebrew hath been altered by us also.... But the Lord knoweth ... that none other people knoweth our language" (Mormon 9:32-34). In light of these two passages, it is evident that Nephite record keepers knew Hebrew and something of Egyptian.

310   "[T]he Book of Mormon quite clearly reflects a number of Hebrew idioms and contains numerous Hebrew words. This is no doubt due to the fact that the Nephites retained the Hebrew language, albeit in an altered form (See Mormon 9:35). Moreover, it is not impossible that the plates themselves contained Hebrew words, idioms, and syntax written in Egyptian cursive script." John A. Tvedtnes, "Hebraisms in the Book of Mormon: a Preliminary Survey," BYU Studies, vol. 11 (1970-1971), Number 1, Autumn 1970, 50.

311   According to the London-based *Daily Mail,* "Haplogroup R is perhaps the most prominent Y-DNA lineage on Earth today." Daily Mail, London, January 20, 2010. The abundance of paternal lineage 'R' possibly fulfills the prophecy regarding Joseph of Egypt's "fruitful bough" (Genesis 49:22). Because of the many subclades of Haplogroup R (Y-DNA) found among Caucasian Latter-day Saints and select North American Indian tribes, this paternal genetic lineage appears to be somehow connected to the two tribes of Joseph of Egypt: Ephraim and Manasseh.

312   Two figurines of Israelite heads discovered at a site in Israel (600 BC-750 BC) that help one to see what ancient Semitic phenotypes were like. Found by Israel's highway 1 near Jerusalem.

313   Assyrian relief from Sargon's palace at Dur-Sarruke, Levant workmen loading Phoenician boats with cedar logs for transport to Assyria. Relief depicting transportation of cedar wood trunk, from Palace of Sargon, Dur Sharrukin, Iraq / De Agostini Picture Library / G. Dagli Orti / The Bridgeman Art Library. Paris, Musée Du Louvre.

314   King Aretas IV (9 BC - 40 AD), Nabataea (Meshorer, Nabataea 98 var.) struck year 19 AD in Petra Nabataean A 2.

315   The Nabataeans were a Semitic group of traders that made their capital Petra (what is now southern Jordan) and traveled in caravans from Arabia. Petra is a great wonder of the world featured in the Steven Spielberg movie *Indiana Jones and the Last Crusade* (1989).

316   Aretas I (not Aretas IV featured among the coins in the section) is the first known Nabatean king. His name appears on the earliest Nabatean inscription discovered to date, a 168 BC carving found in Halutza. He is also mentioned in 2 Maccabees 5:8. The passage relates that Jason, the high priest who established a Hellenistic polis in Jerusalem, was held prisoner by Aretas I after being forced to leave the city. Rabel I Aretas I's successor, whose reign began c. 140 BC. His name is known from a statue dedicated to him in Petra.

317   Parthia itself has a Hebrew etymology. The consonants of Parthia are P-R-TH, but since we get that name from Greek sources and the Greeks sometimes interchanged their "p's" and "b's" (i.e. they called the Britannic Isles the Pretanic Isles), the name "Parthia" can be rendered as "Barthia" or "Brithia." Whatever form the word is given, the consonants are "B-R-TH," the consonants of the Hebrew word for "Covenant." The ten tribes of Israel received and possessed in perpetuity the "birthright" blessings given to Abraham's seed (Genesis 48:14-20), so they were truly the "Covenant" people. The name of Parthia itself proclaims and preserves this fact.

318   Sabaean Alabaster head of a man (2 views), South Arabian, circa 1st cent. BC -1st century AD (10 ½ inches high).

319   Passport photos of Otto Frank, Anne Frank's father, Jewish genetics (1933). P2MR type of Otto Frank: Orator (Out-Curviplex-Box-Point). Refer to *Recognizing People* by Alexander T Paulos.

320   By the end of 1945, Otto knew he was the sole survivor of his family, and of those who had hidden in the house on the Prinsengracht. *The Hidden Life of Otto Frank*, Carol Ann Lee.

321   Anne Frank is the author of *The Diary of Anne Frank*, a book about how she hid for two years with her family during the Nazi occupation of the Netherlands. After being concealed for two years, Otto's family were betrayed by an anonymous informant in 1944. They were sent to the Dutch transit camp of Westerbork and finally to Auschwitz. It was in Auschwitz that Otto was separated forever from his wife and daughters. He was sent to the men's barracks and found himself in the sick barracks when he was liberated by Soviet troops at the beginning of 1945. He traveled back to the Netherlands over the next six months and set about tracing his arrested family and friends.

322   Abraham Lincoln (1858), by Jackson Calvin, Illinois, ambrotype, Library of Congress.

End Notes

P2MR type: Oracle (In-Curviplex-Circle-Point). Refer to *Recognizing People* by Alexander T Paulos.

323 *Abraham Lincoln and the Jews*, p. 5, Isaac Markers (1909).
324 Abraham Lincoln (1858), by Jackson Calvin, Illinois, ambrotype, Library of Congress. P2MR type: Oracle (In-Curviplex-Circle-Point). Refer to *Recognizing People* by Alexander T Paulos.
325 P2MR type: Oracle (In-Curviplex-Circle-Point). Refer to *Recognizing People* by Alexander T Paulos.
326 Daniel's father was born in Ballintubbert, Queen's County, Ireland, to a Protestant family of Northern Irish and English descent. Daniel's mother, who was born in London, England, was Jewish (of Polish Jewish and Latvian Jewish descent). Jackson, Laura (2005). *Daniel Day-Lewis: the biography*. Blake. p. 3.
327 P2MR type: Oracle (In-Curviplex-Circle-Point). Refer to *Recognizing People* by Alexander T Paulos.
328 P2MR type: Oracle (In-Curviplex-Circle-Point). Refer to *Recognizing People* by Alexander T Paulos.
329 Leonard's paternal grandparents were Lyon Cohen (the son of Lazarus Cohen and Fraidie Garmaise) and Rachel Friedman. Lyon was founding president of the Canadian Jewish Congress, and was born in Budwitcher, Poland. Leonard's mother, Marsha (Masha) Klonitsky, was the daughter of a Talmudic writer, Rabbi Solomon Klonitsky-Kline of Lithuanian Jewish ancestry (the son of Yaakov Klonitsky).
330 P2MR type: Oracle (In-Curviplex-Circle-Point). Refer to *Recognizing People* by Alexander T Paulos.
331 P2MR type: Oracle (In-Curviplex-Circle-Point). Refer to *Recognizing People* by Alexander T Paulos.
332 Of all Italians, Sicilian genetic markers (haplogroups) most closely match Jewish genetic markers (haplogroups). Sicilian Y-DNA haplogroups—R1b (26%), G (8.5%), J2 (23%), and E1b1b (20.5%). Ashkenazi Jewish Y-DNA haplogroups—R1b (15%); J (37%); and E (12%).
333 P2MR type: Expert (In-Curviplex-Box-Point). Refer to *Recognizing People* by Alexander T Paulos.
334 P2MR type: Expert (In-Curviplex-Box-Point). Refer to *Recognizing People* by Alexander T Paulos.
335 P2MR type: Oracle (In-Curviplex-Circle-Point). Refer to *Recognizing People* by Alexander T Paulos.
336 Cyprus is an island country in the Eastern Mediterranean Sea. Greek Cypriot Y-DNA genetic markers (Haplogroups)—I: 8%; Haplogroup R1a (3%); Haplogroup R1b (9%); Haplogroup G (9%); Haplogroup J (37%); Haplogroup E (20%); and Haplogroup T (5%). Ashkenazi Jewish Y-DNA Haplogroups—R1b (15%); J (37%); and E (12%).
www.eupedia.com/europe/european_y-dna_haplogroups.shtml
337 J. Robert Oppenheimer, American with German Jewish parents (Ashkenazi Jews).
338 "Irish DNA originated in Middle East and eastern Europe," The Guardian, Tim Radford, December 28, 2015. Neolithic and Bronze Age migration to Ireland and establishment of the insular Atlantic genome. Lara M. Cassidy et al. (2015).
339 Some Irish maternal lineages are similar to Jews and other Semites, indicating that many Irish women were originally from the Middle East. Maternal lineage (haplogroup) "J," for example, is found at its highest percentages among women in the Near East (12%) and in Ireland (12%). Near Eastern countries: Algeria, Bahrain, Egypt, Iran, Iraq, Israel, Jordan, Kuwait, Lebanon, Libya, Morocco, Oman, Palestinian Territories, Qatar, Saudi Arabia, Syria, Tunisia, United Arab Emirates, Yemen Sephardi Jews have about 15%-29% of haplogroup maternal lineage J (Haplogroup J-M172 (mtDNA)) and Ashkenazi Jews have 15%-23%. Semino, Ornella; Magri, Chiara; Benuzzi, Giorgia; Lin, Alice A.; Al-Zahery, Nadia; Battaglia, Vincenza; MacCioni, Liliana; Triantaphyllidis, Costas et al. (2004). "Origin, Diffusion, and Differentiation of Y-Chromosome Haplogroups E and J: Inferences on the Neolithization of Europe and Later Migratory Events in the Mediterranean Area". The American Journal of Human Genetics 74 (5): 1023–34. doi:10.1086/386295. PMC 1181965. PMID 15069642. Maternal lineage (haplogroup) J was also found in small amounts by D.G. Smith et al. among the American Indians.
340 Museum quality Celtic stone Head Amulet, (300 BC - 1 BC) ca. 4th-1st century BC. The fine grained limestone or marble head of ovoid form with superb classic Celtic features. 1.4 inches.
341 Joseph Smith 1885 reproduction of painting by Charles William Carter.
342 Joseph and Hyrum Smith belonged to paternal lineage R1b (called R1b1b2a1a2f2) carrying

End Notes

the M222 defining mutation variant and his mother belongs to maternal lineage H. Joseph Smith DNA Revealed: New Clues from the Prophet's Genes, Ugo A. Perego, Msc, Aug. 7, 2008 FAIR conference. Haplogroup R1b1b2a1a2f2 (Y-DNA) and Haplogroup H (mtDNA).

343 "Irish DNA originated in Middle East and eastern Europe," The Guardian,Tim Radford, December 28, 2015. Neolithic and Bronze Age migration to Ireland and establishment of the insular Atlantic genome. Lara M. Cassidy et al. (2015).

344 New Study Claims that Irishmen descended from Turkish farmers. Jane Walsh, January 14, 2014. Irish Central.

345 "Sleep Not Longer, O Choctaws and Chickasaws", a speech before a joint council of the Choctaw and Chickasaw nations (1811); also in *The Way: An Anthology Of American Indian Literature* (1972) by Shirley Hill Witt and Stan Steiner. Turner III, Frederick (1978) [1973]. "Poetry and Oratory". *The Portable North American Indian Reader*. Penguin Book. pp. 246–247.

346 *Born To Die: Disease and New World Conquest, 1492-1650 (New Approaches to the Americas)*, pp. 1-11, Noble David Cook.

347 *Adolph Hitler: The Definitive Biography*, p. 202, John Toland.

348 Ever since Europeans claimed the New World for their own, indigenous groups of the Americas have been affected, both in a positive and negative sense. Because of the negative effect European explorers, settlers, and colonizers have had on indigenous groups of the Americas, the location of the Lamanite remnant proves to be a much more difficult task.

It is important to note that this Amerindian depopulation was not genocide. UCLA professor Jared Diamond, author of *Guns, Germs, and Steel: The Fates of Human Societies*: "Throughout the Americas, diseases introduced with Europeans spread from tribe to tribe far in advance of the Europeans themselves, killing an estimated 95 percent of the pre-Columbian Native American population. The most populous and highly organized native societies of North America, the Mississippian chiefdoms, disappeared in that way between 1492 and the late 1600's, even before Europeans themselves made their first settlement on the Mississippi River. ... As for the most advanced native societies of North America, those of the U.S. Southeast and the Mississippi River system, their destruction was accomplished largely by germs alone, introduced by early European explorers and advancing ahead of them" Jared Diamond, author of *Guns, Germs, and Steel: The Fates of Human Societies*, p. 78, 374.

349 *Native American Beliefs and Medical Treatments, During the Smallpox Epidemics: an Evolution*, by Melissa Sue Halverson.

350 "But before the great day of the Lord shall come, . . . [T]he Lamanites shall blossom as the rose, Doctrine and Covenants 49:24 Wilford Woodruff, Journal of Discourses 15:283. The Signs of the Coming of the Son of Man—The Saints' Duties Discourse by Elder Wilford Woodruff, delivered in the 13th Ward Assembly Rooms, Salt Lake City, January 12, 1873. Reported by David W. Evans.

351 *The Origins of Native Americans: Evidence from Anthropological Genetics*, p. 50, Michael H. Crawford.

352 *The Origins of Native Americans: Evidence from Anthropological Genetics*, p. 240, Michael H. Crawford.

353 The Apalachicola moved to Oklahoma in 1836 but were absorbed into the Creek tribe.

354 Chakchiuma remnants incorporated into Chickasaw and Choctaw tribes.

355 Coree. A tribe, possibly Algonquian, formerly occupying the peninsulas of Neuse river, in Carteret and Craven counties, N. C. They had been greatly reduced in a war with another tribe before 1696 AD, and were described by Archdale as having been a bloody and barbarous people.*Handbook of American Indians North of Mexico, Frederick Webb Hodge*, 1906.

356 The Hitchiti were absorbed into the Creek tribe.

357 The remnants of the Koasati joined the Creeks and moved to Oklahoma.

358 Some Oconee absorbed into the Seminole tribe.

359 The Tekesta went to Cuba in 1763.

360 No Tuskegee tribe exists today but remnants may be near Beggs, Oklahoma.

361 *The Origins of Native Americans: Evidence from Anthropological Genetics*, p. 261, Michael H. Crawford.

362 In addition to these three deadly diseases, "a large number of other contagions were brought from Europe and Africa into the New World. These include malaria, yellow fever, chicken pox, whooping cough, scarlet fever, diphtheria, plague, typhoid fever, poliomyelitis, cholera, and

End Notes

trachoma. Some of these diseases exacted higher mortality than did others [in the New World]." *The Origins of Native Americans*, p. 61, Michael H. Crawford.

363   The handful of Mandan smallpox survivors joined the Hidatsa and Arikara. When these three tribes united, all of the remaining Mandan DNA was subsumed into other tribe's DNA groups as they intermarried. Their genetic markers are now mixed. This admixture caused their genetic identities to be somewhat lost.

364   W. H. Jackson, 1877. Annual Report, Part 2, p. 87, United States National Museum.

365   Máh-to-tóh-pa, aka "The Four Bears" (1800-1837), a Mandan (Numakiki) chief.

366   Extended quote of Mandan Chief The Four Bears: "Ever since I can remember, I have loved the whites. I have lived with them . . . since I was a boy, and to the best of my knowledge, I have never wronged a whiteman. On the contrary, I have always protected them from the insults of others. The Four Bears [speaking in third person] never saw a white man hungry, but what he gave him to eat, drink, and a buffalo skin to sleep on, in time of need. I was always ready to die for them, which they cannot deny. I have done everything that a red skin could do for them, and how have they repaid it! With ingratitude! I have never called a white man a dog, but today, I do pronounce them to be a set of black hearted dogs. They have deceived me. Them that I always considered as brothers have turned out to be my worst enemies . . . today I am wounded, and by who? By those same white doge that I have always considered, and treated as brothers. I do not fear death, my friends...but to die with my face rotten, that even the wolves will shrink at horror at seeing me...Listen well what I have to say, as it will be the last time you hear from me. Think of your wives, children, brothers, sisters, friends, and in fact all that you hold dear--are all dead, or dying, with their faces all rotten, caused by those dogs the whites? Think of all that, my friends, and rise all together and not leave one of them alive. The Four Bears will act his part." *Great Speeches by Native Americans*, p. 116, Robert Blaisdell. Annie Heloise Abel. Chardon's Journal at Fort Clark, 1834-1839. Pierre, South Dakota: Department of History, State of South Dakota, 1932. pp. 124-125.

367   Instances of infected wives and husbands jointly committing suicide by jumping off cliffs or stabbing themselves have also been documented.

368   Samuel Smith, *The History of the Colony of Nova-Caesaria, or New Jersey,* Burlington, N. J., 1877, note one p. 100. Clinton A. Weslager, author of *The Delaware Indians: A History,* p. 152.

369   Smallpox spread to eastern and western Indians from Mexico (1520 - 1524). Bubonic plague struck the Southwest Pueblo Indians following a visit from Legendary Spanish explorer Francisco Vázquez de Coronado (1545 - 1548). In 1566 - 1567, the Catawba Indians declined considerably from disease, warfare, and the effects of massive alcohol consumption. In 1585, the Secotan Indians begin to die of diseases of military colonists. Measles struck the Seneca Iroquois Indians from the years 1592 – 1596.

Bubonic plague crept up the northeast coast of North America infecting the Indians who were residing there from the years 1612 - 1619. In 1630, the Huron of Ontario were devastated by disease and warfare. In 1633, the Narragansett Indians of Rhode Island lost 700 to smallpox. In the mid 1600s New Jersey and Pennsylvania and northern tribes lost 90% of their populations to smallpox. An epidemic of smallpox killed thousands of Cherokees in 1738 and 1739. In 1759 AD, smallpox killed about two-thirds of the Catawbas, leaving about 500 of that tribe's population of 1,500.

In 1770, the Sewee Indians of South Carolina were decimated by smallpox. From the years 1780 - 1782 the plains indians were struck and dwindled in numbers by the ravages of smallpox. In 1783, smallpox hit the Cherokee hard. In 1795, the Pamlico tribe in North Carolina were almost completely destroyed by smallpox. In just over a year (1837 - 1838) the Mandan Indians were almost completely decimated by smallpox.

370   The Vancouver expedition encountered likely evidence of the havoc wrought by the epidemic. The expedition's two ships *Discovery* and *Chatham* entered Juan de Fuca Straits and anchored at Port Discovery.

371   *Vancouver,* Vol. 2, p. 229 - 230.

372   Englar, 25.

373   During the 1770s, smallpox eradicated at least 30 percent of the native population on the Northwest coast of North America, including numerous members of Puget Sound tribes. This apparent first smallpox epidemic on the northwest coast coincides with the first direct European

End Notes

contact, and is the most virulent of the deadly European diseases that swept over the region during the next 80 to 100 years. In his seminal work, *The Coming of the Spirit of Pestilence*, historian Robert Boyd estimates that the 1770s smallpox epidemic obliterated more than 11,000 Western Washington Indians, reducing the population from about 37,000 to 26,000.

In 1792, members of the Vancouver Expedition were the first Europeans to witness the effects of the smallpox epidemic along Puget Sound. On May 12, 1792, expedition member Archibald Menzies spoke of "Several Indians pock mark'd – a number of them had lost an eye." *Alaska Travel Journal of Archibald Menzies,* 1793 - 1794, p. 29. Commander George Vancouver (1757 - 1798) stated that two days earlier members of his expedition exploring Hoods Canal spotted "one man, who had suffered very much from the small pox", and went on to say, "This deplorable disease is not only common, but it is greatly to be apprehended is very fatal amongst them, as its indelible marks were seen on many; and several had lost the sight of one eye, which was remarked to be generally the left, owing most likely to the virulent effects of this baneful disorder. *Viewpoints and Visions in 1792: The Vancouver Expedition encounters Indians of Western Washington*, Columbia Magazine, Vol. 2, p. 241-242).

By the 1850s, when the first EuroAmerican settlers arrived at Alki Point and along the Duwamish River, diseases causing death had already taken a devastating toll on various North American Indians and their cultures. During an eighty-year period (from the 1770s to 1850), smallpox, measles, influenza, and other diseases had wiped out an estimated 28,000 North American Indians in Western Washington, leaving about 9,000 survivors. The North American Indian population continued to wane although at a slower rate, till the beginning of the 20th century when it reached its nadir. Since then the American Indian population has been slowly increasing. However, the diversity that once flourished in North America is not the same as it once was.

Smallpox ravaged the Iroquois and the effects of the virus were far-reaching. The greatest impact of the smallpox epidemics was sociocultural change. The loss of massive amounts of individuals within a population hindered subsistence, defense, and cultural roles of North American Indians and the Iroquois were one of the many tribes who suffered these effects. Families, clans, and villages were consolidated, further fragmenting the previous social mores. The population losses also led to the fusion of different residential groups. Eighteen Arikara Indian villages, for example, merged into one consolidated group of three villages in the Middle Missouri River Valley. By 1862, Mandan, Hidatsa, and Arikara Indian tribes were sharing one village. This cultural meshing caused a great diffusion of culture across various populations and new ways of defining personal and populational identity.

374 Passaconaway's native name was "Papoose [or Papisse] Conewa." Later in Passaconaway's life, he became the chief of chiefs (bashaba) of a multi-tribe confederation located in modern-day northern New Hampshire. Because of his legacy, the Daniel Webster Council of the Boy Scouts of America, which serves most of New Hampshire, have shown great deference to him by naming their Order of the Arrow lodge after him.

375 *The Indian heritage of New Hampshire and northern New England,* edited by Tadeusz Piotrowski. Chief Passaconaway farewell speech, 1660 AD.

376 *Letters and Notes on the Manners, Customs, and Condition of the North American Indians* by George Catlin (First published in London in 1844 AD) Letter No. 13. North American Indians: Being Letters and Notes on Their Manners ..., Vol. 1, pp. 121-122, George Catlin.

377 Pontiac, also known as "Obwandiyag" in his native tongue.

378 Amherst's reply to Bouquet, July 16, 1763.

379 William Trent's Journal at Fort Pitt, 1763, ed. by A. T. Volwiler.

380 Sickness, Starvation, and Death in Early Hispaniola, Noble David Cook, Journal of Interdisciplinary History, Vol. 32, No. 3, Winter 2002, pp. 349-386.

381 Early 18th century Charleston, North Carolina Indian.

382 Thomas Jefferson to Benjamin Hawkins, 1786. ME 5:390. Thomas Jefferson, an Oracle type (In-Curviplex-Circle-Point). See *Recognizing People* by Alexander T Paulos.

383 George Catlin, Letters and Notes, vol. II, Campfire Stories with George Catlin.

384 Andrew Jackson in his Sixth Annual Message to Congress, December 1, 1834.

385 *Undaunted Courage: Meriwether Lewis, Thomas Jefferson, and the Opening of the American West,* p. 203, Stephen E. Ambrose.

End Notes

386 Thomas Jefferson to William H. Harrison, 1803. ME 10:368.
387 Thomas Jefferson wrote in a letter to Henry Dearborn in 1807 to express his new stance, We make to [the Indians] this solemn declaration of our unalterable determination, that we wish them to live in peace with all nations as well as with us, and we have no intention ever to strike them or to do them an injury of any sort, unless first attacked or threatened. Thomas Jefferson to Henry Dearborn, 1807. ME 11:344.
388 Jefferson, Thomas (1803). "President Thomas Jefferson to William Henry Harrison, Governor of the Indiana Territory".
389 David Nugent, Joan Vincent (2004). *A companion to the anthropology of politics*. Wiley-Blackwell. p. 407.
390 Since there was non-compliance on the part of the Indians, Jefferson no longer felt that peaceful assimilation was possible. After Jefferson left the US presidency, there were a number of wars between US troops and the Indians.
391 "American Indians," Thomas Jefferson's Monticello.
392 *History of the Church*, 6:206.
393 Joseph Smith, *History of the Church* 4:501. *Joseph Smith Journal*, 25 January 1842.
394 Looking Glass, a Nez Percé chief, on horseback in front of a tepee (1877). P2MR temperament type: Catalyst (Out-Curviplex-Circle-Point). Refer to *Recognizing People* by Alexander T Paulos.
395 Nez Percé ('pierced noses') (Niimíipu) is a term applied by the French to a number of tribes which practiced or were supposed to practice the custom of piercing the nose for the insertion of a piece of dentalium. The term is now used exclusively to designate the main tribe of the Shahaptian family, who have not, however, so far as is known ever been given to the practice. The Nez Percé or Sahaptin of later writers, the Chopuunish (corrupted from Tsútpĕli) of Lewis and Clank, their discoverers, were found in 1805 occupying a large area in what is now western Idaho, northeast Oregon, and south east Washington, on lower Snake river and its tributaries. They roamed between the Blue Mountains in Oregon and the Bitter Root Mountains in Idaho, and according to Lewis and Clark sometimes crossed the range to the headwaters of the Missouri.
396 *America's Military Adversaries: From Colonial Times to the Present,* pp. 301-302, John C. Fredriksen.
397 Chief Joseph, AKA Hin-mah-too-yah-lat-kekt, "Thunder Rolling Down From the Mountain".
398 *America's Military Adversaries,* pp. 301-302, John C. Fredriksen.
399 The Wentworth Letter. Joseph Smith, Official Church Publication *Times and Seasons*, (March 1, 1842) III:707.
400 Samuel George Morton, in *Crania Americana*, summarized his measurement of hundreds of human skulls.
401 "Lamanite" can refer to descendants of Laman and the people who joined his group. It can also refer to an incipient nationality based upon an ideology, with its own genealogical history and religious/political beliefs; (Mosiah 10:12-17. The name "Lamanite"referred to a religious/political faction whose distinguishing feature was its opposition to the church. See Jacob 1:13-14.) or else it can refer to one or more cultures. The Book of Mormon describes several Lamanite cultures and lifestyles, including hunting-gathering (2 Nephi 5:24), commerce (Mosiah 24:7), sedentary herding, a city-state pattern of governance (Alma 17), and nomadism (Alma 22:28). The politicized nature of early Lamanite society is reflected in the way in which dissenters from Nephite society sought refuge among Lamanites, were accepted, and came to identify themselves with them, much as some Lamanites moved in the opposite direction.
402 The Lamanites were descendants of Lehi and Ishmael. Lehi was a descendant of Joseph of Egypt (See 1 Nephi 5:14) and Ishmael might have also descended from that same lineage.
403 *Encyclopedia of Mormonism,* "Lamanite." When the Prophet Joseph Smith spoke of the "Lamanites," he was clearly speaking of the descendants of the people described in the Book of Mormon. The Book of Mormon describes people as being Lamanite either through lineage, or through dissension.
404 "[O]ur seed shall not utterly be destroyed, according to the flesh" (2 Nephi 9:53).
405 "Reframing the Book of Mormon," anonymous author, Sunstone, March 2004, p. 19.
406 Elder Parley P. Pratt remarked: "Thus ended our first Indian mission, in which we had preached the Gospel in its fulness and distributed the record of their forefathers [the Book of Mormon] among the three tribes, viz.; the Catteraugus [sic] Indians [the Seneca of the Iroquois Confederacy] near Buffalo, N. Y.; the Wyandots [Huron], of Ohio, and the Delawares [Lenni

End Notes

Lenape] west of Missouri." *Autobiography of* Parley P. Pratt, pp. 56-61.

407   A memoir of Jacques Cartier, sieur de Limoilou, his voyages to the St. Lawrence, a bibliography and a facsimile of the manuscript of 1534. Baxter, James Phinney, 1831-1921; Roberval, Jean François de La Roque, sieur de, 1501?-1560?; Alfonce, Jean, i.e. Jean Fonteneau, known as, 1483?-1557?

408   Cadwallader Colden, *The History of the Five Indian Nations Depending on the Province of New-York in America* (1727). The Five Nations was a former name for the Iroquois Confederacy. Some capitalized words were changed to lower case in the quote.

409   The "Four Indian Kings" weren't actual Mohawk Indian kings. That name was given to the four Indians sent to England to impress Queen Anne and her subjects. Theyanoguin, who was one of the four Indians sent, was baptized as "Hendrick" by Godfridius Dellius of the Dutch Reformed Church in 1692. The English often referred to him as "Hendrick Peters" or "King Hendrick." Hinderaker. "Searching for Hendrick: Correction of a Historic Conflation", Dean R. Snow, New York History, Vol. 88, No. 3 (Summer 2007), pp. 229-253.

410   *Persuasive Speaking of the Iroquois Indians at Treaty Councils: 1678-1776*, Wynn R. Reynolds.

411   *Forgotten Founders,* Bruce E. Johansen, Benjamin Franklin, the Iroquois and the Rationale for the American Revolution, chapter 3.

412   Bruce Burton, "The Iroquois Had Democracy Before We Did." Indian Roots of Democracy. Northeast Indian Quarterly (1988): 44-48. Print. Grinde, Donald. "It's Time to Take Away the Veil." Indian Roots of Democracy. Northeast Indian Quarterly (1988): 28-33. Print. Johansen, Bruce. "Indian Thought Was Often in Their Minds." Indian Roots of Democracy. Northeast Indian Quarterly (1988): 40-43. Print. Weatherford, Jack. Indian Givers. How the Indians of the Americas Transformed the World. New York: Ballantine Books, 1988. Print.

413   Benjamin Franklin, *Benjamin Franklin Papers* (1782-1783).

414   *American Heritage: Book of Indians*, p. 8, William Brandon, Intro. by John F. Kennedy (1961).

415   The root word "Dar," found in both words "Skanyadariyoh" & "Sikandariyeh" in Hebrew (a close language to Arabic) means "order." "Adar," which is found in the Iroquois title could be of Hebrew origin since "adar" means "to remove what is not needed to create order." This sounds like a fitting title for a leadership position.

Handsome Lake or The Reformed Drunkard Who Talked with Angels, Alfred G. Hilbert. *Indians of Eastern Woodlands,* Sally Sheppard. *Iroquois Culture,* Edith Drumm. *Iroquois Folklore* by Beauchamp. *History of New York Iroquois* by Beauchamp. *Long House of Iroquois* by Spencer L. Adams. *White Woman and Her Valley* by Arch Merrill. *Handbook of North American Indians*, Volume 15, by Sturtevant.

416   Hugh Nibley also said: "Lehi, faced with the prospect of a long journey in the wilderness, sent back for Ishmael, who promptly followed into the desert with a large party; this means that he must have been hardly less adept at moving about than Lehi himself. The interesting thing is that Nephi takes Ishmael (unlike Zoram) completely for granted, never explaining who he is or how he fits into the picture—the act of sending for him seems to be the most natural thing in the world, as does the marriage of his daughters with Lehi's sons. Since it has ever been the custom among the desert people for a man to marry the daughter of his paternal uncle [bint 'ammi], it is hard to avoid the impression that Lehi and Ishmael were related." "Did Father Lehi Have Daughters Who Married the Sons of Ishmael?" Sidney B. Sperry. *Journal of Book of Mormon Studies:* Volume - 4, Issue - 1, Pages: 235-38. Provo, Utah: Maxwell Institute, 1995.

417   Me-Te-a, Pottawatomie chief. Author/Artist, Thomas McKenney and James Hall: The Indian. Indian Tribes of North America with Biographical sketches and anecdotes of the principal chiefs.

418   *History of the Church,* 5:480. Mr. Hitchcock, Letter to John Chambers, in John King to John Chambers, 14 July 1843, Iowa Superintendency , 1838-49, Letters Received by the Office of Indian Affairs, 1824-81, BIA Microfilm #363, 357-60; Helen Mar Whitney, "Life Incidents," Woman's Exponent 9-10 (1880-1882) and "Scenes and Incidents in Nauvoo," Woman's Exponent 11 (1882-83), July 2, 1843. The latter incorrectly cites the Wilford Woodruff journal as its documentary source.

419   Charles Mann, p. 41, *1491: New Revelations of the Americas Before Columbus.* The word culture was capitalized.

420   P2MR type: Catalyst (Out-Curviplex-Circle-Point), the same temperament type as Joseph of Egypt. Refer to *Recognizing People* by Alexander T Paulos.

End Notes

421 Black Beaver or Suck-tum-mah-kway (1806-1880) was a Delaware Indian trapper, scout, and interpreter who became a chief, and later a wealthy rancher in present-day Anadarko, Oklahoma.
422 In 1683, William Penn, an English Quaker, saw signs of the ancient Israelites in the Lenni Lenape (Delaware) Indians. He stated in 1683, Penn believed that the Lenni Lenape were originally from the Jewish race. His reasons include: the Jewish race was set out from Israel to far lands, their appearance was much like the Jewish settlers in Europe, they agreed in Rites, they gave their first fruits and performed a Feast of Tabernacles. They also laid out their altar on twelve stones, their mourning was a year and the customs of the women, (much like the law of Moses). This is an interesting account because two hundred years later Joseph Smith, founder of the LDS church, held the same view that was taken from their Book of Mormon.
423 William Penn on the Leni Lenape (Delaware) Letter to the Committee of the Free Society of Traders, 1683.
424 *A Brief History of the Delaware Indians*, p. 6, Richard C. Adams.
425 *A Brief History of the Delaware Indians*, p. 7, Richard C. Adams.
426 *A Brief History of the Delaware Indians*, p. 6, Richard C. Adams.
427 Kikthawenund was also known by his English name: Chief William Anderson.
428 Parley P. Pratt, *Autobiography of Parley P. Pratt*, pp. 56-61.
429 P2MR type: Catalyst (Out-Curviplex-Circle-Point), the same temperament type as Joseph of Egypt. Refer to *Recognizing People* by Alexander T Paulos.
430 Elder Parley P. Pratt concluded in his journal, "Thus ended our first Indian mission, in which we had preached the Gospel in its fulness and distributed the record of their forefathers among the three tribes, viz.; the Catteraugus Indians near Buffalo, N. Y.; the Wyandots [Huron], of Ohio, and the Delawares west of Missouri." *Autobiography of Parley P. Pratt*, pp. 56-61.
431 In the Huron Tongue . . . Jesous outo etti x'ichie. Outo etti skuaalichi-axe. J chierche axerawensta. D'aotierti xeata-wien. . . . This signifies . . . 'O saving Victim, who art continually sacrificed, and who givest life, thou by whom we enter into Heaven, we are all tempted; do thou strengthen us.' The Jesuit Relations and Allied Documents (1610 - 1791), edited by Reuben Gold Thwaites. Pierre François Xavier de Charlevoix is quoted, pp. 144-145.
432 Pierre François Xavier de Charlevoix, *History of New France* (1744). *A Star in the West*, p. 97, Elias Boudinot (1816).
433 *A Star in the West*, pp. 91-92, Elias Boudinot (c. 1816). Curviplex language abounds in allegory and symbols (See Recognizing People by Alexander T Paulos).
434 *History of the American Nations*, pp. 15-212, James Adair.
435 Ayunini, "Swimmer," a Cherokee man (1888).
436 Indians of the Southern United States, p. 223. Bureau of American Ethnology, Bul. 137. Bartram, 1792, pp. 481-483.
437 William Harlen Gilbert, Jr., pp. 194, 195-196. *The Eastern Cherokees*.
438 St. Louis became George Catlin's base of operations for five trips he took between 1830 and 1836, eventually visiting fifty tribes. George Catlin began his journey in 1830 when he accompanied General William Clark on a diplomatic mission up the Mississippi River into American Indian territory.
439 "[T]he Lamanites had become, the more part of them, a righteous people insomuch that their righteousness did exceed that of the Nephites, because of their firmness and their steadiness in the faith" (Hel. 6:1).
440 At the edge of the United States wild frontier, George Catlin produced some of the most spectacular portraits of his career. Later trips along the Arkansas, Red, and Mississippi rivers, as well as visits to Florida and the Great Lakes, resulted in over 500 paintings and a substantial collection of artifacts.
441 W. H. Jackson, 1877.
442 *Letters and Notes on the Manners, Customs, and Condition of the North American Indians* by George Catlin, pages 93-94. First issued 1841, reprinted by Ross & Haines, Minneapolis, 1965.
443 George Catlin, speaking about the Mandan, said the Mandan were "a very interesting and pleasing people in their personal appearance and manners, differing in many respects, both in looks and customs, from all the other tribes I have seen." *Letters and Notes on the Manners, Customs, and Condition of the North American Indians* by George Catlin, pages 93-94. First issued 1841, reprinted by Ross & Haines, Minneapolis, 1965.
444 George Catlin: "... traders and others who have been amongst them, [refer to them as] 'the polite and friendly Mandan.'... elegance of these people, together with the diversity of

End Notes

complexions, the various colors of their hair and eyes, the singularity of their language, and their peculiar and unaccountable customs, that I am fully convinced that they have sprung from some other origin than that of the North American tribes, or that they are an amalgam of natives with some civilized race." *Letters and Notes on the Manners, Customs, and Condition of the North American Indians* by George Catlin, pages 93-94. First issued 1841, reprinted by Ross & Haines, Minneapolis, 1965.

445   ibid
446   ibid.
447   ibid.
448   W. H. Jackson, 1877. Annual Report, Part 2, p. 87, United States National Museum.
449   *Letters and Notes on the Manners, Customs, and Condition of the North American Indians* by George Catlin (First published in London in 1844) Letter No. 13. North American Indians: Being Letters and Notes on Their Manners ..., Volume 1, p. 206, By George Catlin.
450   George Catlin: "The Mandans are not a warlike people. They seldom, if ever, carry war into their enemies' country, but when invaded, show their valor and courage to be equal to that of any people on earth. Being a small tribe, and unable to contend on the wide prairies with the Sioux and other roaming tribes, who are ten times more numerous, they have very judiciously located themselves in a permanent village, which is strongly fortified. . . . they . . . have supplied their lodges more abundantly with the comforts, and even luxuries, of life than any Indian nation I know of. The consequence of this is, that this tribe have taken many steps ahead of other tribes in manners and refinements (if I may be allowed to apply the word refinement to Indian life); and are therefore familiarly (and correctly) denominated, by the Traders and others, who have been amongst them, 'the polite and friendly Mandans." *Letters and Notes on the Manners, Customs, and Condition of the North American Indians* by George Catlin, pages 93-94. First issued 1841, reprinted by Ross & Haines, Minneapolis, 1965.
451   *Letters and Notes on the Manners, Customs, and Condition of the North American Indians* by George Catlin, pages 93-94. First issued 1841, reprinted by Ross & Haines, Minneapolis, 1965.
452   W. H. Jackson, 1877. Annual Report, Part 2, p. 87, United States National Museum.
453   *Letters and Notes on the Manners, Customs, and Condition of the North American Indians* by George Catlin, pages 93-94. First issued 1841, reprinted by Ross & Haines, Minneapolis, 1965.
454   *Letters and Notes on the Manners, Customs, and Condition of the North American Indians* by George Catlin, pages 93-94. First issued 1841, reprinted by Ross & Haines, Minneapolis, 1965.
455   Red Fly, Oglala Sioux chief. P2MR type: Catalyst (Out-Curviplex-Circle-Point). Refer to *Recognizing People* by Alexander T Paulos.
456   This photo is of Standing Bear (1904). There is a chance it might not be Luther Standing Bear (Ota Kte), but another "Standing Bear." This Indian's P2MR type: Oracle (In-Curviplex-Circle-Point). Refer to *Recognizing People* by Alexander T Paulos.
457   Standing Bear was a wise man. His wisdom manifests itself in his writings: "It is now time for a destructive order to be reversed, and it is well to inform other races that the aboriginal cultures of North America were not devoid of beauty. Furthermore, in denying the Indian his ancestral rights and heritages the white race is but robbing itself. America can be revived, rejuvenated, by recognizing a Native School of thought." Chief Luther Standing Bear (Mochunozhin or Ota Kte), Lakota (Sioux), *Land of the Spotted Eagle* (1933).
458   Spotted Tail, Sioux (1833-1881), photograph from Encyclopædia Britannica. Kids.eb.com. P2MR type: Catalyst (Out-Curviplex-Circle-Point). Refer to *Recognizing People* by Alexander T Paulos.
459   Short Bull, Oglala Sioux, P2MR type: Catalyst (Out-Curviplex-Circle-Point). Refer to *Recognizing People* by Alexander T Paulos.
460   *The Genius of Charles Darwin, God Strikes Back*, part 2/5, Richard Dawkins (2009).
461   Sesardic, Neven (2010). "Race: A Social Destruction of a Biological Concept". Biology (Biology & Philosophy) 25 (143): 143–162. doi:10.1007/s10539-009-9193-7.
462   Joseph F. Smith, *The Improvement Era*, Volume 16, Joseph Smith, Jr., as a Translator, Feb. 1913, p. 380.
463   Elder Neal A. Maxwell, "According to the desire of [our] hearts," October 1996.
464   Emma Smith, Joseph Smith, Jr.'s wife, included this tune in the first LDS hymnal in 1835, but it was subsequently removed in 1927.
465   Raghavan; et al. (21 August 2015). "Genomic evidence for the Pleistocene and recent population history of Native Americans" A. Torroni; T. G. Schurr; C. C. Yang; EJE. Szathmary;

End Notes

R. C. Williams; M. S. Schanfield; G. A. Troup; W. C. Knowler; D. N. Lawrence; K. M. Weiss; D. C. Wallace. "Native American Mitochondrial DNA Analysis Indicates That the Amerind and the Nadene Populations Were Founded by Two Independent Migrations". *Center for Genetics and Molecular Medicine and Departments of Biochemistry and Anthropology, Emory University School of Medicine*, Atlanta, Georgia. Genetics Society of America. Vol 130, 153–162. Wang S, Lewis CM Jr, Jakobsson M, Ramachandran S, Ray N, et al. (2007). "Genetic Variation and Population Structure in Native Americans". PLoS Genet. 3 (11): e185. doi:10.1371/journal.pgen.0030185. PMC 2082466. PMID 18039031. Schurr, Theodore G. "Mitochondrial DNA and the Peopling of the New World" (PDF). American Scientist Online May–June 2000.

466 *The Genius of Charles Darwin, God Strikes Back,* part 2/5, Richard Dawkins (2009).
467 In most cases, Down Syndrome occurs when there is an extra copy of chromosome 21 (Trisomy 21).
468 Sesardic, Neven (2010). "Race: A Social Destruction of a Biological Concept". Biology (Biology & Philosophy) 25 (143): 143–162. doi:10.1007/s10539-009-9193-7.
469 Kihue and Yuya of Egypt both possessed similar facial proportions, especially when viewed straight on. As one can see from these pictures, they both had similar spacing between the eyes, high cheekbones, foreheads, noses, ears, and jawlines. Only slight variations exist between them. This strong connection between select North American Indians like Kihue and Semites such as Yuya highlights the physical correlation between both groups.
470 Albany Md 1 Provile Hopewell skull. Bureau of American ethnology Bulletin 33 plate XIII, Albany Illinois. Hopewell skull from the Middle Woodland Period (1AD - 400 AD).
471 Cranial morphology of early Americans from Lagoa Santa, Brazil: Implications for the settlement of the New World, Neves et. al. Proceedings of the National Academy of Sciences of the United States of America, vol. 102 no. 51, Walter A. Neves, 18309–18314, doi: 10.1073/pnas.0507185102.
472 Webb and Snow 1945. Archaeology: The Adena People. William S. Webb and Charles E. Snow. American Anthropologist (Impact Factor: 1.49). 09/1977; 79(3). DOI: 10.1525/aa.1977.79.3.02a00950.
473 Webb and Snow 1945. Archaeology: The Adena People. William S. Webb and Charles E. Snow. American Anthropologist (Impact Factor: 1.49). 09/1977; 79(3). DOI: 10.1525/aa.1977.79.3.02a00950.
474 Albany Md 1 Provile Hopewell skull. Bureau of American ethnology Bulletin 33 plate XIII, Albany Illinois. Hopewell skull from the Middle Woodland Period (1AD - 400 AD).
475 Albany Md 1 Provile Hopewell skull. Bureau of American ethnology Bulletin 33 plate XIII, Albany Illinois. Hopewell skull from the Middle Woodland Period (1AD - 400 AD).
476 77 Algonkin crania in the Academy of Natural Sciences, Philadelphia, 53 are dolichocephalic, 14 mesocephalic, and 10 brachycephalic. *The American Race: A Linguistic Classification and Description of the Native Tribes of North and South America*, p. 75, Daniel G. Brinton.
477 *The American Race: A Linguistic Classification and Description of the Native Tribes of North and South America*, p. 81, Daniel G. Brinton.
478 *The American Race: A Linguistic Classification and Description of the Native Tribes of North and South America*, pp. 136-136, Daniel G. Brinton.
479 *Joseph Smith Journal,* November 9, 1835; cited in Dean C. Jessee, Mark Ashurst-McGee, Richard L. Jensen, eds., The Joseph Smith Papers: Journals Volume 1:1832-1839, pp. 88-89. The word "indians" was capitalized.
480 dna.ancestry.com/atFAQ
481 Autosomal genetic distances (Fst) based on SNPs in Tian et al. (2009). Ashkenazim compared to Italians (0.0040), Greeks (0.0042), Spanish (0.0056), Germans (0.0072), Druze (0.0088), Palestinians (0.0093), Irish (0.0109), and Russians (0.0137).
482 The dual origin and Siberian affinities of Native American Y chromosomes. Lell JT, Sukernik RI, Starikovskaya YB, Su B, Jin L, Schurr TG, Underhill PA, Wallace DC Am J Hum Genet. 2002 Jan; 70(1):192-206.
483 "The genome of a Late Pleistocene human from a Clovis burial site in western Montana," by Rasmussen et al. (2014). Raghavan, M. et al. 2014. Upper Palaeolithic Siberian genome reveals dual ancestry of Native Americans, Nature, 505, 87–91. n the 2014 study, Rasmussen and his team tested the genome-wide genetic affinity (i.e. performed an autosomal analysis) of an ancient Montana boy known as Anzick-1. Anzick-1 belonged to what is known as the D4h3a haplogroup, or lineage, and his Y-DNA was haplogroup Q-L54*(xM3).

End Notes

484   Non-LDS scholars claim that the Book of Mormon simply cannot be a true account because the majority of American Indians come from ancient Siberian stock. While it is true that a great majority of modern-day American Indians from North, Central, and South America originate from ancient Siberian lineages, new evidence suggests that not all of them do.
485   *Joseph Smith Journal*, Nov. 9, 1835 (emphasis added); cited in Dean C. Jessee, Mark Ashurst-McGee, Richard L. Jensen, eds., *The Joseph Smith Papers: Journals* Volume 1:1832-1839 (Salt Lake City: Church Historian's Press, 2008), 88-89.
486   Moroni, a Nephite Chief Captain (Alma 46:23–24, 27). Readers of the English Bible might suspect that this terminology was taken from Amos 5:15, where the prophet wrote, "It may be that the Lord God of hosts will be gracious unto the remnant of Joseph." However, close analysis suggests that Amos, as well as those whose words are recorded in the Book of Mormon passages cited above, may have been influenced by an earlier passage of scripture.
487   Title page of the Book of Mormon. Emphasis added to the original text.
488   "The Book of Mormon is a record of the forefathers of our [United States] ... tribes of Indians ... By it, we learn that our ... Indians, are descendants from that Joseph [who] was sold into Egypt, and that the land of America is a promised land unto them. . . . The city of Zion spoken of by David, in the one hundred and second Psalm, will be built upon the land of America." Joseph Smith to N. C. Saxton, 4 Jan., 12 Feb. 1833, in Dean C. Jessee, comp. and ed., *The Personal Writings of Joseph Smith* (Salt Lake City: Deseret Book, 1984), pp. 273, 275.
489   DNA Replication and Causes of Mutation, Leslie A. Pray, Ph.D., 2008, Nature Education. Citation: Pray, L. (2008) DNA replication and causes of mutation. Nature Education 1(1):214.
490   Haplogroup J (Y-DNA) and Haplogroup IJ (Y-DNA).
491   Haplogroup I (Y-DNA), Haplogroup J (Y-DNA) and Haplogroup IJ (Y-DNA).
492   Haplogroup J (Y-DNA)—the prototypically "Jewish" haplogroup— paternal lineage, even though haplogroups I and J both originate from the Middle Eastern Haplogroup IJ (Y-DNA). Both haplogroups I and J, which can be found in statistically significant amounts in certain parts of western Eurasia, are both Middle Eastern in origin. This fact is rarely, if ever, discussed by scientists in scientific journals. A number of scientists state that Haplogroup J (Y-DNA) is Semitic, and Haplogroup I (Y-DNA) is European even though it really is Middle Eastern in origin. Even though Haplogroups I and J come from the same paternal lineage, haplogroup IJ (a Y-DNA lineage thought by scholars to originate in the Middle East), it is claimed that haplogroup I (Y-DNA) is a European paternal lineage. What they fail to mention is that it has become a European paternal lineage, when in fact it originated in the Middle East. The mutation may have occurred in Europe, but the mutation's location does not make the haplogroup less Semitic in origin. Haplogroup I (Y-DNA) and haplogroup J (Y-DNA) are equally Semitic.
493   Balanovsky et al 2011. Sample size, (n=143).
494   Behar et al. (2003).
495   Haplogroup J (Y-DNA). El-Sibai et al 2009. Sample size, (n=62).
496   Marjanović, Damir; et al. "The peopling of modern Bosnia-Herzegovina: Y-chromosome haplogroups in the three main ethnic groups." Institute for Genetic Engineering and Biotechnology, University of Sarajevo. November 2005.
497   Marjanović, Damir; et al. "The peopling of modern Bosnia-Herzegovina: Y-chromosome haplogroups in the three main ethnic groups." Institute for Genetic Engineering and Biotechnology, University of Sarajevo. November 2005.
498   Rootsi S, Magri C, Kivisild T, et al. (July 2004). "Phylogeography of Y-chromosome haplogroup I-M170 reveals distinct domains of prehistoric gene flow in Europe". Am. J. Hum. Genet. 75 (1): 128–37. doi:10.1086/422196. PMC 1181996.
499   Rootsi S, Magri C, Kivisild T, et al. (July 2004). "Phylogeography of Y-chromosome haplogroup I-M170 reveals distinct domains of prehistoric gene flow in Europe". Am. J. Hum. Genet. 75 (1): 128–37. doi:10.1086/422196. PMC 1181996.
500   Paternal lineage (haplogroup) R1-M173 Ojibwe (79.3%). Malhi et al. 2008.
501   "The genome of a Late Pleistocene human from a Clovis burial site in western Montana," by Rasmussen et al. (2014). Raghavan, M. et al. 2014. Upper Palaeolithic Siberian genome reveals dual ancestry of Native Americans, Nature, 505, 87–91.
502   Recent analysis of an arm bone belonging to the ancient skeleton of an indigenous North American boy indicates that his DNA was similar to modern Native Americans, and yet, according to ancient DNA expert Eske Willerslev of the University of Copenhagen, the boy "apparently descended not from East Asians, but from people who had lived in Europe or

End Notes

western Asia [a large area that includes the Middle East.]"

"The finding suggests that about a third of the ancestry of today's Native Americans can be traced to 'western Eurasia'." 24,000 year old boy from Lake Baikal is 'scientific sensation' By The Siberian Times reporter. 28 October 2013.

According to Morten Rasmussen's report, the research may help explain why "European ancestry previously detected in modern Native Americans do not come solely from mixing with European colonists, as most scientists had assumed, but have much deeper roots[.]" 24,000 year old boy from Lake Baikal is 'scientific sensation' By The Siberian Times reporter. 28 October 2013.

The study found that "a portion of the boy's genome is shared only by today's Native Americans and no other groups, showing a close relationship. . . . [T]he child's Y chromosome belongs to a genetic group called Y haplogroup R." The child's Y chromosome belongs to a genetic group called Y haplogroup R, and its mitochondrial DNA to a haplogroup U. 24,000 year old boy from Lake Baikal is 'scientific sensation' By The Siberian Times reporter. 28 October 2013. His remains were discovered in 1920 in Lake Baikal, a rift lake in the south of the Russian region of Siberia.

Although the owner of the arm bone—known currently as the "Lake Baikal boy"—tested as paternal lineage Haplogroup R, and he shared a similar genome with North American Indians, his exact haplogroup is uncommon among North American Indians today. Haplogroup R1 is the most common 'R' haplogroup among indigenous Americans. Haplogroup R—the Lake Baikal boy's haplogroup—is older than Haplogroup R1 and Haplogroup R1 is actually a more recent branch of Haplogroup R. This means that, unless the mutation that changed 'R' to 'R1' occurred in ancient North America, which is unlikely, the two haplogroups arrived in the New World at separate times. This appears to open the door for Haplogroup R1 (Y-DNA) to have arrived at the time Lehi and his caravan made their way to the New World.

503   Zegura, Stephen L. et al 2004, High-Resolution SNPs and Microsatellite Haplotypes Point to a Single, Recent Entry of Native American Y Chromosomes into the Americas.
504   Zegura, Stephen L. et al 2004, High-Resolution SNPs and Microsatellite Haplotypes Point to a Single, Recent Entry of Native American Y Chromosomes into the Americas.
505   Malhi, Ripan Singh et al 2008, Distribution of Y Chromosomes Among Native North Americans: A Study of Athapaskan Population History.
506   *Joseph of Egypt,* p. 16, Mark E. Petersen.
507   "An Analysis of Ancient Aztec mtDNA from Tlatelolco: Pre-Columbian relations and the spread of Uto-Aztecan", Kemp et al. (2005). Gonzalez-Oliver et al. (2001).
508   Chief Standing Bear, also known as Luther Standing Bear. His P2MR type: Oracle (In-Curviplex-Circle-Point). Refer to *Recognizing People* by Alexander T Paulos.
509   Bedouin Arab from Dead Sea area (Jordan). His P2MR type: Oracle (In-Curviplex-Circle-Point). Refer to *Recognizing People* by Alexander T Paulos.
510   Haplogroup R1 (Y-DNA) with the M173 defining characteristic. 50% of the Sioux tested are paternal lineage R1. Malhi, Ripan Singh et al 2008, Distribution of Y Chromosomes Among Native North Americans: A Study of Athapaskan Population History. Balaresque et al. 2009.
511   Haplogroup R1 (Y-DNA) with the M173 defining characteristic. This finding is in harmony with what is written in the Book of Mormon. Lehi and his caravan came from this general area (the Dead Sea region) before leaving for the New World. The Book of Mormon reads, "And he [Lehi] came down by the borders near the shore of the Red Sea; and he traveled in the wilderness in the borders which are nearer the Red Sea; and he did travel in the wilderness with his family" (1 Nephi 2:5). The Dead Sea, which is close to Jerusalem (where Lehi and his family came from), is not so far from the Red Sea. 20/45. Flores, C; Maca-Meyer, N; Larruga, JM; Cabrera, VM; Karadsheh, N; Gonzalez, AM (2005). "Isolates in a corridor of migrations: a high-resolution analysis of Y-chromosome variation in Jordan". Journal of Human Genetics 50 (9): 435–41. doi:10.1007/s10038-005-0274-4. PMID 16142507.
512   Hole-in-the-Day [Sky], Jr. (Po Go Nay Ki-Shig) of the Bear Clan of Ojibwe, St. Paul, Minn., c. 1860 AD, Whitney's Gallery, photographer. The word "ki-shig" or "Ge Shick" can mean "day" or "sky". He was assassinated by Ojibwe men from Leech Lake in 1868. Hole-in-the-Sky's P2MR type is Oracle (In-Curviplex-Circle-Point). Refer to *Recognizing People* by

End Notes

Alexander T Paulos.

513 The Ojibwe are Anishinaabe-speaking peoples, a branch of the Algonquian language family. He was often known as "Hole in the Day."

514 Paternal lineage (haplogroup) R1-M173 Ojibwe (79.3%). Malhi et al. 2008. Surprising to researchers, the seven most distant groups genetically from the ancient Montana boy who belonged to Haplogroup Q (Y-DNA) are: Aleutians, East Greenlanders, West Greenlanders, Chipewyan, Algonquin, Cree, and Ojibwe. Chipewyan, Algonquin, Cree, and Ojibwe. They all share high percentages of R1. From the data in this study, it appears that all tribes of North America, and other groups, who have many members among them who belong to Haplogroup R1 are the least like the Siberian groups. If Haplogroup R1 were more Siberian, then more overlap would most likely occur.

515 Matchis Skank, AKA "Someone Traveling," Ojibwe (1901). P2MR type: Catalyst (Out-Curviplex-Circle-Point). Refer to *Recognizing People* by Alexander T Paulos.

516 Paternal lineage (haplogroup) R1-M173 Ojibwe (79.3%). Malhi et al. 2008.

517 Mico Chlucco the Long Warrior - King of the Seminoles. 1792, Travels through North & South Carolina, Georgia, E & W Florida the Cherokee country, the extensive territories of the Muscogulges or Creek confederacy, and the country of the Choctaws. by William Bartram (1739-1823) American naturalist. This illustration was re-purposed in 1809 as a Sioux warrior in 'The Travels of Capts. Lewis & Clarke'; pub. by H Lester. This portrait resembles portrayals made by Le Moyne of the Timucua, another group of Florida Indians and so it just might be a legitimate depiction of the Creek chief.

518 Examples of Muskogean-speaking tribes: Apalachicola, Chiaha, Guale, Hitchiti, Oconee, Okmulgee, Tamathli, Yamasee, Ais, Apalachee, Chikasaw, Calusa, Chatot, Agua Dulce, Mikasuki, Pensacola, Pohoy, Potano, Saturiwa, Cusabo, Utina, Yui, Yustaga, Alabama, Koasati, Mobile, Muklasa, Muskogee (Creek), Osochi, Sawokli, Taposa, Natchez, Ibitoupa, Houma, Griga, Choctaw, Chakchiuma, and Seminole.

519 50% of the Seminoles tested are paternal lineage R1. Malhi, Ripan Singh et al 2008, Distribution of Y Chromosomes Among Native North Americans: A Study of Athapaskan Population History.

520 Haplogroup R1 (Y-DNA) with the defining characteristic M173. Bolnick et al 2006.

521 Malhi, Ripan Singh et al 2008, Distribution of Y Chromosomes Among Native North Americans: A Study of Athapaskan Population History.

522 Narragansett-Mohegan man. P2MR type: Oracle (In-Curviplex-Circle-Point). Refer to *Recognizing People* by Alexander T Paulos.

523 Haplogroup R1 (Y-DNA) with the M173 defining marker. Of the 155 individuals whose DNA was tested among the Algonquin Indians of Northeast North America, 38.1% belong to Haplogroup R1 (Y-DNA) with the M173 defining marker. Bolnick et al 2006.

524 Bolnick et al 2006.

525 Bitter Man man, Cree, Algonquin peoples (1875).

526 Wizi, Yanktonai Sioux, aka Something Yellow or Old Tipi Yellowed with Smoke (1884).

527 The Yanktonai are a Western Dakota people.

528 50% of the Sioux tested are paternal lineage R1. Malhi, Ripan Singh et al 2008, Distribution of Y Chromosomes Among Native North Americans: A Study of Athapaskan Population History. Balaresque et al. 2009.

529 The great majority of American Indians belonging to the paternal lineage known as Haplogroup Q (Y-DNA), When indigenous males from North America, Mesoamerica, and South America are all clumped together, the great majority of them belong to paternal lineage Haplogroup Q1a defined by the genetic marker M3. especially with the M242 and M3 defining markers, appear East Asian and Siberian. By no coincidence, Haplogroup Q (Y-DNA) with the defining marker M242 originated in Asia, and is widely distributed there. Zegura, S. L.; Karafet, TM; Zhivotovsky, LA; Hammer, MF (2003). "High-Resolution SNPs and Microsatellite Haplotypes Point to a Single, Recent Entry of Native American Y Chromosomes into the Americas". Molecular Biology and Evolution 21 (1): 164–75. doi:10.1093/molbev/msh009. PMID 14595095.

530 Q1a is also the main paternal lineage of Native Americans. The testing of the genome of 12,600 year-old boy (known as Anzick-1) from the Clovis Culture in the USA confirmed that haplogroup Q1a2a1 (L54) was already present on the American continent before the end of the Last Glaciation. The vast majority of modern Native Americans belong to the Q1a2a1a1 (M3) subclade. As this subclade is exclusive to the American continent and the Anzick boy was

End Notes

negative for the M3 mutation, it is likely that M3 appeared after Q1a2a1 reached America.

531   In Europe, haplogroup Q1a is believed to have been brought by the Huns, the Mongols and the Turks, who all originated in the Altai region and around modern Mongolia. Haplogroup Q has been identified in Iron Age remains from Hunnic sites in Mongolia by Petkovski et al. (2006) and in Xinjiang by Kang et al. (2013). Modern Mongols belong to various subclades of Q1a, including by order of frequency Q1a2a1c (L330), Q1a1a1 (M120), Q1a1b (M25) and Q1a2a* (L53).

532   Haplogroup Q (Y-DNA). 90% belonged to Q-M3 (Vajda 2012). Tambets, Kristiina et al 2004, The Western and Eastern Roots of the Saami—the Story of Genetic "Outliers" Told by Mitochondrial DNA and Y Chromosomes.

533   The great majority of American Indians belonging to the paternal lineage known as Haplogroup Q (Y-DNA), When indigenous males from North America, Mesoamerica, and South america are all clumped together, the great majority of them belong to paternal lineage Haplogroup Q1a defined by the genetic marker M3. especially with the M242 and M3 defining markers, appear East Asian and Siberian. By no coincidence, Haplogroup Q (Y-DNA) with the defining marker M242 originated in Asia, and is widely distributed there. Zegura, S. L.; Karafet, TM; Zhivotovsky, LA; Hammer, MF (2003). "High-Resolution SNPs and Microsatellite Haplotypes Point to a Single, Recent Entry of Native American Y Chromosomes into the Americas". Molecular Biology and Evolution 21 (1): 164–75. doi:10.1093/molbev/msh009. PMID 14595095.

534   Zegura, Stephen L. et al 2004, High-Resolution SNPs and Microsatellite Haplotypes Point to a Single, Recent Entry of Native American Y Chromosomes into the Americas.

535   The great majority of American Indians belonging to the paternal lineage known as Haplogroup Q (Y-DNA), When indigenous males from North America, Mesoamerica, and South america are all clumped together, the great majority of them belong to paternal lineage Haplogroup Q1a defined by the genetic marker M3. especially with the M242 and M3 defining markers, appear East Asian and Siberian. By no coincidence, Haplogroup Q (Y-DNA) with the defining marker M242 originated in Asia, and is widely distributed there. Zegura, S. L.; Karafet, TM; Zhivotovsky, LA; Hammer, MF (2003). "High-Resolution SNPs and Microsatellite Haplotypes Point to a Single, Recent Entry of Native American Y Chromosomes into the Americas". Molecular Biology and Evolution 21 (1): 164–75. doi:10.1093/molbev/msh009. PMID 14595095.

536   Haplogroup Q (Y-DNA), Haplogroup R1 (Y-DNA) with the M173 defining characteristic.Zegura, Stephen L. et al 2004, High-Resolution SNPs and Microsatellite Haplotypes Point to a Single, Recent Entry of Native American Y Chromosomes into the Americas. 11.7% belonged to Haplogroup R1 (Y-DNA).

537   Zegura, Stephen L. et al 2004, High-Resolution SNPs and Microsatellite Haplotypes Point to a Single, Recent Entry of Native American Y Chromosomes into the Americas. 12.% of those tested belonged to Haplogroup R1 (Y-DNA).

538   Haplogroup Q (Y-DNA). Zegura, Stephen L. et al 2004, High-Resolution SNPs and Microsatellite Haplotypes Point to a Single, Recent Entry of Native American Y Chromosomes into the Americas.

539   66.4% of them belonged to Haplogroup Q (Y-DNA). Tambets, Kristiina et al 2004, The Western and Eastern Roots of the Saami—the Story of Genetic "Outliers" Told by Mitochondrial DNA and Y Chromosomes.

540   Ingmar Bergman, Jewish man from Sweden.

541   Lacandon Mayan leader, Mexico (1989).

542   An Iroquois man who could be Chief Pau Puk Keewis (1800s).

543   Hugh Nibley, Lehi in the Desert; The World of the Jaredites; There Were Jaredites, ch. 3. W. E. Jennings-Bramley, "The Bedouin of the Sinaitic Peninsula," *PEFQ* (1906), 106, and (1907), 281.

544   *New Witness for God,* BH Roberts (1895).

545   The first Lehi in the Book of Mormon. In 589 BC, Nebuchadnezzar II laid siege to Jerusalem, culminating in the destruction of the city and its temple in 587 BC.

546   Laban was a notable citizen of Jerusalem with the power to command many servants. He had access to great wealth.

547   Hugh W. Nibley, "Lehi and the Arabs."

548   Hugh Nibley, Book of Mormon Semester 1, p. 167. There is a remarkable association between the names of Lehi and Ishmael which ties them both to the southern [Arabian] desert, where the

End Notes

legendary birthplace and central shrine of Ishmael was at a place called Be'er Lehai-ro'i. Hugh Nibley, Lehi In The Desert, F.A.R.M.S., p. 40. John Tvedtnes comments on Nibley's claim that the name Ishmael as well as the names Lehi, Lemuel, Alma and Sam are Arabic in origin (*An Approach to the Book of Mormon* 58-60; *Lehi in the Desert* 44-46). Tvedtnes contends that although Ishmael is indeed the name of the son of Abraham who settled that part of Arabia, Ishmael is also the name of a member of the royal family of Judah from the time of Lehi (Jeremiah 40). Thus the name of Ishmael might have been used by more than one people. John Tvedtnes, "Was Lehi a Caravaneer?," F.A.R.M.S., p. 8. According to LDS tradition, Ishmael was an Ephraimite, a tradition based on a discourse delivered by Apostle Erastus Snow, in Logan, Utah, 6 May 1882. Elder Snow said: "The Prophet Joseph Smith informed us that the record of Lehi was contained on the 116 pages that were first translated and subsequently stolen, and of which an abridgment is given us in the First Book of Nephi, which is the record of Nephi individually, he himself being of the lineage of Manasseh; but that Ishmael was of the lineage of Ephraim, and that his sons married into Lehi's family, and Lehi's sons married Ishmael's daughters." JD 23:184.

549   *Observations On Various Passages Of Scripture Placing Them In A New Light*, p. 246, Thomas Harmer.
550   P2MR type: Expert (In-Curviplex-Box-Point). Refer to *Recognizing People* by Alexander T Paulos.
551   P2MR type: Expert (In-Curviplex-Box-Point). Refer to *Recognizing People* by Alexander T Paulos.
552   Palestinian paternal lineage (Y-DNA)—Haplogroup R1b: 8%; Haplogroup R1a: 1%; Haplogroup J: 20%; Haplogroup E: 20%; Haplogroup I: 6%.
553   A Yemenite Sabaean woman with Semitic physical features and Semitic genetics from Southern Arabia. South Arabia as a general term referring to several regions, yet especially refers to the Republic of Yemen.
554   P2MR type: Expert (In-Curviplex-Box-Point). Refer to *Recognizing People* by Alexander T Paulos.
555   Of the 62 Yemen Arabs tested over 8 out of 10 (82.3%) of their paternal lineages were J (which is most common among Semites), 3.2% were E1b1a, 12.9% were E1b1b, 1.6% were G, 0% were I, 0% were L, N, R1a, R1b, and T (Cadenas 2008).
556   Southern Arabia. Sabaeans. Late 3rd century BC. AR unit (16mm, 5.24 gm, 9h). The Sabaeans were a Semitic-speaking people of ancient Saba in Yemen.
557   The Nabataean kingdom, also known as Nabatea, was an Arabic political state of the Nabataeans which existed during classical antiquity and was annexed by the Roman Empire in AD 106. Its capital was the city of Petra in Jordan—one of the great wonders of the world.
558   "Solving the Enigma of Petra and the Nabataeans," Glenn J. Corbett (2015). Bible Archaeology.
559   The "Flatheads," as explorers Lewis and Clark called them, refer to themselves as the "Salish" meaning, "the people." The Salish (Flathead) never practiced the custom of flattening the skull of their children, but North American Indian tribes living around the Columbia River were known to flatten their heads. Lewis and Clark most likely confused the Salish tribe with other tribes who flattened their heads.
560   Red Fly, Oglala Sioux chief. P2MR type: Catalyst (Out-Curviplex-Circle-Point). Refer to *Recognizing People* by Alexander T Paulos.
561   An Arab man from Yemen. P2MR type: Oracle (In-Curviplex-Circle-Point). Refer to *Recognizing People* by Alexander T Paulos.
562   Little Wound, also known as George Little Wound, a Sioux Indian.
563   Detail of Steeh-tcha-kó-me-co, Great King (called Ben Perryman by white settlers), a Chief, Creek, Muskogee, painted by George Catlin in 1834.
564   *Southeastern Indians: Life Portraits; A Catalogue of Pictures 1564-1860*, p. 116, Emma Lila Fundaburk (1996).
565   Of the paintings Catlin made of the Creek Indians, which include Chieftain Hól-te-mál-te-téz-te-néek-ee and Steeh-tcha-kó-me-co (featured in this section), both men are wearing keffiyehs, traditional Arab headdress fashioned from a square—usually cotton—scarf with various patterns. Individuals from other tribes wore North American keffiyeh's and were painted by George Catlin: Ni-có-man, "The Answer," a Second Chief (1830), Delaware Indian; Kút-tee-o-túb-bee, "How Did He Kill?", Choctaw Indian (1834); Tul-lock-chísh-ko, "Drinks the Juice of the Stone," Choctaw Indian (1834); and Lay-láw-she-kaw, "Goes Up the River," an Aged Chief,

End Notes

Shawnee Indian (1830). Many Turks, Arabs, and Muslims are Semitic peoples and Turkish paternal lineages actually closely resemble Ashkenazi Jews. Y-DNA: 24% Haplogroup J2, 14.7% Haplogroup R1b, 10.9% Haplogroup G, 10.7% Haplogroup E1b1b, 9% Haplogroup J1, 6.9% Haplogroup R1a (Cinnioglu et al. 2004).

566   Hole in the Sky [Day], Ojibwe chief. The word "ki-shig" or "Ge Shick" can mean "day" or "sky", and the name is perhaps more correctly translated as "Hole-in-the-Sky. P2MR type: Oracle (In-Curviplex-Circle-Point). Refer to *Recognizing People* by Alexander T Paulos.

567   The Bedouin woman's long braided hairstyle existed among the Bedouin women before the Muslims began to force their women to cover their hair.

568   The Bedouin woman's P2MR type: Oracle (In-Curviplex-Circle-Point). This temperament type is the same as Mosiah, a seer and a prophet of the Book of Mormon.

569   Bedouin paternal lineage (Y-DNA): Haplogroup R1b: 15%; Haplogroup J: 46%; Haplogroup E; 21%; and Haplogroup I: 6%.

570   Hugh W. Nibley, "Lehi and the Arabs."

571   According to *Native Report*, "Hole in the Sky was known for his assertiveness." Ojibwe author Anton Treuer said of Hole-in-the-Sky, "He was especially bright . . . very smart and he seemed to know when to push, just how far he could push without causing an actual war. He used that just like a chess match with the United States government and was able to get a lot of concessions. . . . He was a strong leader and a capable leader." Native Report on PBS, Anton Treuer, speaking of "Hole-in-the-Day [Sky]."

572   *Indian Heroes and Great Chieftains*, p. 227, Charles Alexander Eastman.

573   *Indian Heroes and Great Chieftains*, pp. 227, 233, and 235, Charles Alexander Eastman.

574   *Indian Heroes and Great Chieftains by Minnesota Dakota* (1916), Dr. Charles A. Eastman.

575   *Indian Heroes and Great Chieftains by Minnesota Dakota* (1916), Dr. Charles A. Eastman.

576   2 Nephi 30:3-4.

577   Brigham Young, Journal of Discourses Vol. 10, p. 359 (October 30, 1864).

578   *Smith's History of New Jersey*, p.14. *A Star in the West*, pp. 85-86, Elias Boudinot (1816).

579   *Letters of Benjamin Hawkins 1796-1806 AD*, p. 168.

580   *Letters of Benjamin Hawkins 1796-1806 AD*, p. 168.

581   *Jewes in America, or Probabilities that the Americans are Jewes,* Thomas Thorowgood (1600-1669 AD). London 1660, 2nd ed.

582   Thomas Jefferson Letter to the Marquis de Chastellux, June 7, 1785.

583   Jefferson, *Writings*, p. 226. *The Lost World of Thomas Jefferson*, p. 67, Daniel Boorstin.

584   John Filson, T*he Discovery, Settlement and Present State of Kentucke* [Kentucky](1789), pp. 98-99.

585   Lewis Hanke, Bartolomé de Las Casas: Bookman, Scholar & Propagandist. (Philadelphia: University of Pennsylvania Press, 1952), pp. 49-50.

586   William Penn's Own Account of the Lenni Lenape or Delaware Indians, edited by Albert Cook Myers, 1971. Narratives of Early Pennsylvania, West New Jersey and Delaware ..., Volume 13.

587   "A Jew, by the fact that he belongs to the chosen people and is circumcised, possesses so great a dignity that no one, not even an angel, can share equality with him. In fact, he is considered almost the equal of God." (Pranaitis, I.B., The Talmud Unmasked, Imperial Academy of Sciences, St. Petersburg, Russia, 1892, p. 60).

588   The practice of circumcision seems to be specific to Hebrews. Compared to the available history of circumcision in the Middle East, there is little verifiable evidence for its history among the Aboriginal Australians and Polynesians. What is known comes from their oral histories and accounts of missionaries and explorers. For Aboriginal Australians and Polynesians, circumcision likely started as a blood sacrifice and a test of bravery, and became an initiation rite with attendant instruction in manhood in more recent centuries. The removal of the foreskin was done with seashells, and it is theorized that the bleeding was stopped with eucalyptus smoke. Doyle D (October 2005). "Ritual male circumcision: a brief history". *The Journal of the Royal College of Physicians of Edinburgh* 35 (3): 279–285. There are a few oral accounts of Aboriginal Australians and Polynesians who anciently practiced circumcision, but the origin of these practices are unknown. It is unclear whether or not Australians and Polynesians came up with circumcision on their own. Christian missionaries may or may not have influenced these peoples.

589   Mariano Edward Rivero and John James von Tschudi in *Peruvian Antiquities* (1857). These men were also known by Mariano Eduardo de Rivero y Ustariz and Johann Jakob von Tschudi.

End Notes

590   *America B.C.* p. 17, Barry Fell.
591   2/50 families of Machapunga Indians practiced circumcision. *The History of North Carolina*, John Lawson, London, 1714. Raleigh, 1860. Remnants of the Machapunga of North Carolina, 1916, Frank G. Speck.
592   Hehaka Isnala (also known as James Lone Elk), 1899, Oglala Lakota Sioux. P2MR type: Oracle (In-Curviplex-Circle-Point). Refer to *Recognizing People* by Alexander T Paulos.
593   Kurdish Jew. P2MR type: Oracle (In-Curviplex-Circle-Point). Refer to *Recognizing People* by Alexander T Paulos.
594   The genetic markers found among Kurdish Jews match those of prototypical Semites. Of the Kurdish Jews who were tested, 20% belonged to a branch of a maternal lineage common to Semites: Haplogroup X (mtDNA). Among the male Kurdish Jews whose DNA was tested, 20.1% belonged to a branch of a paternal lineage common to Semites: Haplogroup R1 (Y-DNA). Kurdish Jews Y-DNA: Haplogroup R1b: 20.1%; Haplogroup J: 37%; and Haplogroup E: 21%. Among the Sioux who were tested, 50% belonged to a branch of a paternal lineage common to Semites: Haplogroup R1 (Y-DNA). These statistics imply that the Sioux and Kurdish Jews may share distant relatives.
595   Two Bulls, Dakota Sioux. P2MR type: Oracle (In-Curviplex-Circle-Point). Refer to *Recognizing People* by Alexander T Paulos.
596   Jonathan Miller, Jewish genetics. P2MR type: Catalyst (Out-Curviplex-Circle-Point). Refer to *Recognizing People* by Alexander T Paulos.
597   William Penn's Own Account of the Lenni Lenape or Delaware Indians, edited by Albert Cook Myers, 1971. Narratives of Early Pennsylvania, West New Jersey and Delaware ..., Volume 13.
598   William Penn's Own Account of the Lenni Lenape or Delaware Indians, edited by Albert Cook Myers, 1971. Narratives of Early Pennsylvania, West New Jersey and Delaware ..., Volume 13.
599   Daniel Day-Lewis' mother was of Jewish descent, and his maternal grandparents' families had emigrated to Britain from Latvia and Poland. Jackson, Laura (2005). *Daniel Day-Lewis: the Biography*. Blake. p. 3. P2MR type: Oracle (In-Curviplex-Circle-Point). See *Recognizing People* by Alexander T Paulos.
600   Drawing of a Huron man. 2MR type: Oracle (In-Curviplex-Circle-Point). See *Recognizing People* by Alexander T Paulos.
601   Adrien Brody, Jewish genetics. P2MR type: Oracle (In-Curviplex-Circle-Point). Refer to *Recognizing People* by Alexander T Paulos.
602   Adrien Brody's father is of Polish Jewish descent and his mother – who was raised as a Catholic – was born in Budapest, Hungary, the daughter of a Catholic Hungarian aristocrat father and a Czech Jewish mother. Leslie Camhi (18 March, 2005). "An Autobiography in Pictures". The Jewish Daily Forward (New York City: forward.com).
603   Adrien Brody, Jewish genetics. P2MR type: Oracle (In-Curviplex-Circle-Point). Refer to *Recognizing People* by Alexander T Paulos.
604   Esh-sup-pee-me-shish (also known as Hairy Moccasin), Crow Indian (1854 - 1922). P2MR type: Professional (Out-Line-Box-Point). Refer to *Recognizing People* by Alexander T Paulos.
605   Oded Fehr speaks Hebrew, English and little German. He has Jewish genetics although he is not a practicing Jew. Fehr's P2MR type: Professional (Out-Line-Box-Point). Refer to *Recognizing People* by Alexander T Paulos.
606   David Duchovny, Jewish-American. P2MR type: Oracle (In-Curviplex-Circle-Point). Refer to *Recognizing People* by Alexander T Paulos.
607   Little Wound, also known as George Little Wound, a Sioux Indian.
608   Hecker, Don R. (September 2, 2003). "Amram Ducovny, 75, Late-Blossoming Novelist". The New York Times..
609   Crow's Heart of the Mandan tribe (Prairie Chicken Clan) was a good warrior and the leader of the old wolves during a war party at the age 19. He gained acclaim as a ceremonial leader owner of a number of rights and bundles. He participated in the *Okipa* by hanging over a cliff. At 23, he first went out to trap eagles. Crow's Heart was a member of the Goose Society Singers. At about the age 30 Crow's Heart bought the right to make fish traps from his clan uncle Old Black Bear who taught him how to make the trap and how to use it.
610   Bust of Josephus (circa 37-100 AD), a famous Jewish historian of the Roman Era. His famous work, *Against Apion*, defends the tenets of Judaism against the accusations of the Graeco-Egyptian writer Apion. One of these allegations was that of blood libel, believed to be

End Notes

the first mention of this slander.
611   P2MR type: Legalist (In-Line-Box-Point). Refer to *Recognizing People* by Alexander T Paulos.
612   Lauren Bacall was born Betty Joan Perske to Jewish parents. Bacall, Lauren. *By Myself and Then Some.*
613   Pretty Flower, a Seneca, 1914. P2MR type: Dreamer (In-Line-Circle-Wave). Refer to Recognizing People by Alexander T Paulos. Also known as Goldie Jamison-Conklin, daughter of Jacob J. Jamsison and Eliza D. Jamison, wife of Charles Conklin.
614   *An Ancient American Setting for the Book of Mormon*, pp. 81-82, John L. Sorenson. Carleton S. Coon, *The Living Races of Men* (1965), pp. 79-80; C.C. Seltzer, Contributions to the Racial Anthropology of the Near East, HUPM 16, no. 2 (1940), pp. 5-9, 11, 60, plates 1, 3.
615   Cutout of a South Arabian limestone pillar stele, c. 3rd century BC-1st century AD.
616   Human face effigy Kentucky, Gallatin County, Warsaw (200 BC - 1 AD). 25.4 cm. National Museum of the American Indian, DC.
617   Lenni Lenape (Delaware) Indians stone effigy, Monroe County Historical Association's permanent display at Stroud Mansion, Monroe County, Pennsylvania..
618   *Autobiography of* Parley P. Pratt, pp. 56-61.
619   Human Head (with typical Hopewell style features), Middle Woodland. Ohio Historical Society, Columbus, #283/140. Hopewell Mounds, Ross Co., Ohio. Ca. 3" tall.
620   Death mask retrieved from a two-thousand-year-old burial mound near the banks of Wisconsin's Red Cedar River (1 AD - 400 AD).
621   Human face effigy, Woodland Period (200 BC - 400 AD), Ross County, Ohio.
622   Adena human effigy pipe Ohio Ross county Adena mound (100 BC -100 AD) pipestone. Photo taken in 1901.
623   Hopewell burial effigy (200 BC - 400 AD) goes back to the earliest days of the Moundbuilders. Illinois, Calhoun County, Knight Mound group, Mound 8 (200 - 400 AD). Painted earthenware, Milwaukee Public Museum.
624   Ohio Intrusive Mound Culture (abt. 700 -900 AD) Henry Ward of Pickaway Co., Ohio, 1930.
625   Keller figurine from the Cahokia Mounds, late Woodland Period (500 - 1000 AD).
626   Mississippian Indian, Copper Repoussé Human Profile Cutout with forked eye decoration, Spiro, Oklahoma.
627   The people of the Spiro Mounds are believed to have been Caddoan speakers, like the modern Wichita, Kichai, Caddo, Pawnee, and Arikara of North America. According to scholars, the site where the Copper Repoussé Human Profile was found remained unoccupied from 1600 until 1832. While Choctaw and Choctaw Freedmen cleared the mound site for farming late in the AD 1800s, they did not allow any major disturbance of the site until the Great Depression.
628   Spiro has been the site of human activity for at least 8000 years, but was a major settlement from 800 to 1450 AD, which corresponds with the Mississippian Culture period of North America. "Spiro Mounds." Oklahoma Historical Society. (retrieved 30 May 2011).
629   The type of metalwork used to create this artifact is known as repoussé, an uncommon style of metalwork in the world in general, but a common style found in the Old World. Repoussé is an ancient metalworking technique in which a malleable metal, such as copper, is ornamented (or shaped) by hammering from the reverse side to make a design in what is known as "low relief." This technique was first introduced to the world in the Middle East around 1900 - 1200 BC (the Late Bronze Age) and spread from there. Ancient Egyptians, Greeks, and Romans heavily used the technique. Repoussé was so popular amongst the wealthy that even King Tut's mummy mask, for example, was made using the metalwork style. It appears that the technique was most likely brought from the Old World to the New World since it was commonly used in the Hopewell and Mississippian eras of the North American Southeast and Midwest. Repoussé copper were fashioned as ritual regalia and eventually used in prestige burials by the Mississippians and the Hopewell (which were in America much earlier than the Mississippians). Power, Susan (2004). Early Art of the Southeastern Indians-Feathered Serpents and Winged Beings. University of Georgia Press.
630   Timucua, detailed view of an archer from plate XXXI, by Jacques Le Moyne, 1564 AD.
631   In 1564, French illustrator and cartographer Jacques Le Moyne de Morgues (1533-1588) accompanied Rene de Laudonniere's ill-fated experiment to colonize Florida. Le Moyne's artwork no longer exists from that time since it was destroyed. However, after returning to France, Le Moyne produced replicas by memory of the artwork he produced of the Florida Indians.

End Notes

632 Timucua, detailed view Plate XI, Jacques Le Moyne de Morgues, 1564.
633 The artist of this reproduction is unknown and it is unclear whether or not this depiction of a Roanoke Indian is correct. It is unclear if the North American Indian was Europeanized by the artist. If the appearance of the Roanoke Indian was not drastically altered, the facial profile, body type, neck, and stance of this American Indian do not indicate a North or East Asian origin. He has a long face, a higher nasal bridge than many Mongoloids and his body type does not appear to be East Asian or Siberian. Histoire de la Virginie ...' by Robert Beverley, 1705 AD reproduction of John White's work of the Roanoke Indians (1585).
634 John White (1540 - 1593). A 1585 - 1586 venture left a valuable record, however, thanks to a scholar, Thomas Hariot, and an artist, John White, depictions of North American Indians were etched and preserved. Hariot published A brief and true report of the new found land of Virginia in 1588 and included depictions of North American Indians.
635 Detail of John Smith's 1612 map engraved by William Hole (or Holle) containing a Susquehannock (sometimes called Conestoga and/or Andastes Indians) Indian chief. This tribe had agriculturally-centered villages.
636 Names of tribes associated with the Susquehannock: Akhrakuaeronon (Atrakwaeronnon), Akwinoshioni, Atquanachuke, Attaock, Carantouan, Cepowig, Junita (Ihonado), Kaiquariegehaga, Ohongeoguena (Ohongeeoquena), Oscalui, Quadroque, Sasquesahanough, Sconondihago (Seconondihago or Skonedidehaga), Serosquacke, Takoulguehronnon, Tehaque, Tesinigh, Unquehiett, Usququhaga, Utchowig, Wyoming, and Wysox.
637 It is ambiguous whether or not William Hole, a skilled English engraver, Europeanized the visage of this Indian. If he did not embellish the etching, and it is a correct representation, then the Indian does not show any predominant North or East Asian physical features.
638 *The Voyages of Captain John Smith* (of Jamestown, Va.) during the Years 1607-1609. The Jesuit Relations and Allied Documents, Edna Kenton and Reuben Gold Thwaites.
639 "[S]ixtie of those gyant-like people came downe."Many of the words in the quote are misspelled. *The Voyages of Captain John Smith* (of Jamestown, Va.) during the Years 1607-9 AD. When John Smith was on Virginia's lower eastern shore June 8-10, 1608, he, or perhaps a scribe, delineated in a journal a description of the natives they met: [T]he Easterne [Eastern] shore . . . The first people we saw were two grim and stout Salvages [savages] upon Cape Charles, with long poles like Javclings [javelins], headed with bone, they boldly demanded what we were, and what we would; but after many circumstances they seemed very kinde [kind], and directed us to Accomack, the habitation of their Werowance [tribal chief], where we were kindly intreated [entreated]. This King was the comliest [comeliest], proper, civill Salvage [civil savage] we incountred [encountered] . . . they spake the language of Powhatan, wherein they made such descriptions of the Bay, Isles, and rivers, that often did us exceeding pleasure. *The Accidents that hap'ned in the Discovery of the Bay of Chisapeack*, John Smith, the first voyage, journal 1608.
640 The constant warfare between Iroquoian-speaking tribes gave the Susquehannock a military advantage over their more peaceful Algonquin neighbors to the east and south. Using canoes for transport, Susquehannock war parties routinely attacked the Lenni Lenape tribes along the Delaware River and traveled down the Susquehanna where they terrorized the Nanticoke, Conoy, and Powhatan living on Chesapeake Bay.
641 *The Voyages of Captain John Smith* (of Jamestown, Va.) during the Years 1607- 1609.
642 *The Voyages of Captain John Smith* (of Jamestown, Va.) during the Years 1607 -1609.
643 "Sasquesahanough" "people at the falls" "people of the muddy river " (Andastes to the French from a Huron word "Andastoerrhonon," Minquas "treacherous" by the Dutch and Swedes from an Algonquian word) people were Iroquoian-speaking Native Americans who lived in areas adjacent to the Susquehanna River and its tributaries from the southern part of what is now New York, through Pennsylvania, to the mouth of the Susquehanna in Maryland at the north end of the Chesapeake Bay. Evidence of their habitation has also been found in West Virginia.
644 "Figure des savages Armouchiquois" Samuel de Champlain, Carte Geographique de la Nouelle Franse (1613) Buchenau.
645 The Armouchiquois were actually several tribes, not a single one, that disappeared around 1631 due to disease and attacks from neighboring tribes, such as the Souriquoi. In Joseph Williason's 1832 *History of Maine,* he said the Armouchiquois were the same as the Malecite tribe living on the St. John's River. However, Champlain had earlier said that their language differed from the Micmac and the Etchimin bands which were also of the Malecite tribe. Some Frenchmen used the term to describe several tribes that the English included under the term

End Notes

"Massachusetts." In Francis Parkman's book, Jesuits in North America, published in 1867, the term included the Algonquin tribes of New England, including the Mohegan, Pequot, Massachusett, Marraganset and others who were in in a chronic state of war with the tribes of New Brunswick and Nova Scotia.

646   See Alma 49; 50:1-24 for examples of fortifications.

647   Artwork circa 1653, Samoset (died c. 1653). Native American of Abnaki people of the Algonquin nation. In spring 1621, as the Pilgrims were still building the Plymouth settlement, Samoset entered calling out 'Welcome' in English. The next day brought Squanto, fluent in English.

648   The Pilgrims wintered aboard the Mayflower after their arrival in the New World in 1620. English Separatist leader of settlers in Massachusetts, William Bradford, wrote in *Of Plymouth Plantation*, "Besides, what could they see but a hideous and desolate wilderness, full of wild beasts and wild men—and what multitudes there might be of them they knew not." *Of Plymouth Plantation: Sixteen Twenty to Sixteen Forty-Seven*, p. 62, William Bradford.

649   *Chronicles of the Pilgrim Fathers,* Alexander Young.

650   Great Sun, a Natchez chief in winter clothing. Engraving. Naturels en Hyver, *Histoire de la Louisiane*, Le Page du Pratz, 1758. LC Prints and Photographs Division.

651   The Natchez are North American Indians originally from the Natchez bluffs region, near the present-day city of Natchez, Mississippi. The Natchez Indians are noted for being the only Mississippian Culture with complex chiefdom characteristics to have survived long after Europeans entered the New World. They retained common customs with the ancient mound-building cultures believing their chief was descended from the sun god.

652   The Natchez, The Library of Congress.

653   The Natchez Massacre was an attack by Natchez Indians against French colonists near present-day Natchez, Mississippi, on November 29, 1729 AD.

654   Logan's Indian name continues to be the subject of some dispute. He has been identified over the past two centuries as Tah-ga-jute, Tachnechdorus, Soyechtowa, Tocaniodoragon and Talgayeeta.

655   Tachnechdorus was known to the white man as Chief James Logan.

656   Portrait of Mary Bernard, a Mi'kmaq woman from Whykokamagh [Whycocomagh], Nova Scotia, painted by Ellen Nutting circa 1840 - 1846.

657   Mainstream epigraphers have labeled Fell's claims "baseless," and yet his claim appears to be true. Goddard & Fitzhugh, Schmidt & Marshall.

658   Haplogroup R1 (Y-DNA) and Haplogroup X (mtDNA).

659   50% (3/6) tested among the Mi'kmaq tribe belonged to Haplogroup X2a (mtDNA). The peopling of the Americas: Genetic ancestry influences health". Scientific American. "Learn about Y-DNA Haplogroup Q" (Verbal tutorial possible). Wendy Tymchuk – Senior Technical Editor. Genebase Systems. 2008. Retrieved 2012-11-21. Of the 28 Northern Ojibwe who were tested, 25% belonged to maternal lineage (haplogroup) X and of the 35 Southern Ojibwe who were tested, 25.7% were maternal lineage (haplogroup) X. Maternal lineages X2a, X2g and X2a1a are found among Ojibwa/Chippewa and W. Chippewa. Of the 28 Northern Ojibwa who were tested 25.0% were X (Torroni et al. 1993a). And of the 35 Southern Ojibwa who were tested 25.7% (Scozzari et al. 1997).

660   Also, Semitic-looking individuals may have already been in the Americas before Book of Mormon peoples arrived in the New World.

661   Janne M. Sjödahl, "Suggested Key To Book of Mormon Geography," *Improvement Era* 30 no. 11 (September 1927).

662   La Venta Monument 13, for example, features an Olmec man with has a beard, sash, and possible Middle Eastern physical features. Also, other anomalies in the prototypical East Asian and Siberian appearance among ancient Mesoamericans appear such as an ancient Mayan incense burner from Guatemala with the neck beard, long nose, and high nasal bridge (400 BC - 800 AD). This artifact that appears Semitic may or may not fit within the Book of Mormon timeline. Incense burner from ancient Guatemala Maya, Guatemala. Late Preclassic/early Classic Period. Or the no longer extant large ancient Monte Alto head of Guatemala (1800 BC) that can only be found in one photograph taken in the 1940s. And the Olmec clay head from Trez Zapotes, Vera Cruz, Mexico, that most likely depicts a Phoenician sailor that made his way to Mesoamerica. The Olmec clay head is very atypical for an Olmec, who almost all appear prototypically East Asian and Siberian, with his pointy goatee and Smurf hat (Phrygian cap) dates back to around 1500 BC - 400 BC turn up.

End Notes

663   Lamanai, mask from the temple in Belize. Lamanai means "submerged crocodile" in Yucatec Maya (400 BC to 100 AD).

664   Offering 4. La Venta. Middle Formative. Miller, Mary Ellen. Art of Mesoamerica. Revised edition. New York- Thames and Hudson, 1996 (1200 BC through 400 BC).

665   A major Olmec cultural and political center from about 1000 BC to 400 BC, La Venta, Mexico, has provided scholars and students with the largest number of sculptural works and beautifully carved stone objects. This goes to show that it is common for prominent Latter-day Saint scholars to believe that the Olmec of Mesoamerica and the Jaredites of the Book of Mormon are one and the same. In a 1992 article, Latter-day Saint scholar John L. Sorenson stated that La Venta, Tabasco, Mexico, "was one of the great centers of Olmec civilization, whose distribution and dates remind us of Jaredite society." Sorenson, John, 1992 article. According to historians, the Olmec were the first major civilization in Mexico. They lived in the tropical lowlands of south-central Mexico, in the modern-day states of Veracruz and Tabasco. According to LDS scholar John Sorenson, "The hill Ramah of the Jaredites, which is the same as the hill Cumorah of the Nephites, was where the final extermination of both peoples took place; that hill corresponds in all relevant parameters to Cerro El Vigía in the Tuxtla mountains of south-central Veracruz state, Mexico. . . . A decline of the society in which the Jaredites lived took place over a period of several centuries before their extinction around 600 BC. The Olmec cultural tradition declined and disappeared from the culture history of Mexico at the same time." John L. Sorenson, *An Ancient American Setting for the Book of Mormon* (Salt Lake City, Utah : Deseret Book Co.; Provo, Utah: Foundation for Ancient Research and Mormon Studies, 1996 [1985]). Kaminaljuyu is a Pre-Columbian site of the Maya civilization that was primarily occupied from 1500 BC to AD 1200, which means that Sorenson was implying that Nephites were Mayans.

666   La Venta Complex #3 also has a very interesting shrine in it. It is called # 4 and is composed of 16 miniature statues and six inscribed celts.

667   Detail from the Tablet 2 of the 96 Glyphs, in the tower of the "Palace" at Palenque, Mexico. The Palenque ruins date back to 226 BC to around 799 AD.

668   Ancient Maya artifact of Palenque, Chiapas, Mexico. The Palenque ruins date back to 226 BC to around 799 AD.

669   "Extract from Stephens' 'Incidents of Travel in Central America'," *Times and Seasons* 3 no. 22 (15 September 1842), 911 - 915.

670   Ancient Maya artifact of Palenque, Chiapas, Mexico. The Palenque ruins date back to 226 BC to around 799 AD.

671   Ancient Maya artifact of Palenque, Chiapas, Mexico. The Palenque ruins date back to 226 BC to around 799 AD.

672   This is a depiction of Itzamná, a Mesoamerican bearded ancestral deity of the Itza-Maya. In Yucatec Maya mythology, Itzamna was the name of an upper god and creator deity thought to be residing in the sky. Detail of the north face of Stela E, depicting K'ak' Tiliw Chan Yopaat holding a God K sceptre. This stela was dedicated on 24 January 771 by K'ak' Tiliw Chan.

673   "We are not going to declare positively that the ruins of Quirigua are those of Zarahemla, but when the land and the stones, and the books tell the story so plain, we are of opinion, that it would require more proof than the Jews could bring to prove the disciples stole the body of Jesus from the tomb, to prove that the ruins of the city in question, are not one of those referred to in the Book of Mormon ". *Times and Seasons*, October 1, 1842.

674   Hypothetically, if Quiriguá and Zarahemla are one and the same, then the hypothetical heavy Semitic influence vanished quickly. All Semitic-looking effigies from the time of the Book of Mormon, if they existed, have disappeared and the only effigies that remain are Far Eastern in appearance. The seed of Zedekiah, also known as the Mulekites, were a Semitic group who had independently made their way from the Near East to the Americas and set up the city of Zarahemla. The name "Mulek " of "Mulekite" is believed by some to be a discrete version of "MalkiYahu son of the King Zedekiah" found in Hebrew Bible: See for instance Coon, W. Vincent, Choice Above All Other Lands, pp. 125–126. Coon cites Jeremiah 39:6 from Hebrew scripture.

675   Stela H statue at Mayan ruins in Copán, Honduras Classic Period (751 AD).

676   Late Classic Maya Vase K1185, (600 AD-900 AD) This cylindrical vase in roll out form is from the Nakbe Region in Guatemala.

677   A detail of a painting from the Eastern Han Dynasty (25 AD-220 AD), Museum of Fine Arts, Boston

End Notes

678   Notwithstanding the Mayan man's apparent East Asian ancestry, LDS tour guide and academic John L. Lund, and other LDS scholars like him, continue to use Eas Asian-looking Mayans for the covers of their books about Book of Mormon geography. Lund's books *Mesoamerica and the Book of Mormon*, and *Joseph Smith and the Geography of the Book of Mormon* both feature on their covers East Asian-looking Mayans that do not even date back to Book of Mormon times.
679   Detail of an ancient Chinese painting, Five Dynasties (907 AD - 960 AD).
680   Chinese Terracotta Warrior armor (300 - 250 BC), "Guardians of China's First Emperor" exhibit on display at the National Geographic Museum.
681   Robert Heine Geldern, *The Civilizations of Ancient America: The Selected Papers of the XXIXth International Congress of Americanists*.
682   Maize God with headdress in the form of a stylized ear of corn and hair in the form of the silk of the corn (600 AD-800 AD). Copán Ruinas, Honduras.
683   Ancient Chinese Buddha, Buddha Mudra # 1- ABHAYA - No Fear.
684   Copán is an archaeological site of the Maya civilization located in the Copán Department of western Honduras, not far from the border with Guatemala. It was the capital city of a major Classic Period kingdom from about 401 - 899 AD, a time frame that somewhat overlaps the Book of Mormon.
685   Human face effigy Kentucky, Gallatin County, Warsaw (200 BC-01 AD). 25.4 cm. National Museum of the American Indian, DC.
686   Pre-Columbian Art, Olmec Stone Mask - PF.5534 (900 BC-500 BC) Mexico Olmec stone mask.
687   Olson, J.S. (1992). *Historical Dictionary of the Spanish Empire*, 1402-1975. New York: Greenwood Press.
688   The earliest known Maya mural painting (100 BC - 1 BC). Late Preclassic Maya. San Bartolo mural painting, north wall. Ancient Mesoamerica, Petén, Guatemala.
689   Olmec colossal head from San Lorenzo Tenochtitlán (c. 1200 BC-900 BC).
690   Nayarit, Mexico 2, Proto-Classic, 100 BC - 250 AD seated figure, seated figure, terracotta, height 40.7cm.
691   Totonac, Pantepec, State of Puebla, Mexico.
692   The Totonac people are one of the possible builders of the Pre-Columbian city of El Tajín, and further maintained quarters in Teotihuacán—a city which they claim to have built.
693   Otomi man from Huixquilukan, Mexico. Otomies are of little stature and mesaticephalic (moderate headed).
694   Otomis are indigenous peoples of Mexico.
695   Olmec colossal head (two views) number 6, from San Lorenzo Tenochtitlán, Mexico. Taken at the museum of anthropology at Xalapa, Vera Cruz, Mexico (1200 - 400 BC).
696   "The city of Nephi was probably the archaeological site of Kaminaljuyu, which is now incorporated within suburban Guatemala City;" John L. Sorenson, *An Ancient American Setting for the Book of Mormon* (Salt Lake City, Utah : Deseret Book Co.; Provo, Utah: Foundation for Ancient Research and Mormon Studies, 1996 [1985]). Kaminaljuyu is a Pre-Columbian site of the Maya civilization that was primarily occupied from 1500 BC to 1200 AD, which means that Sorenson was implying that Nephites were Mayans.
697   *The Lost Book of Mormon: A Journey Through the Mythic Lands of Nephi, Zarahemla, and Kansas City, Missouri*, Avi Steinberg. Granted, history states that King Solomon's temple was large and impressive, but the size and ornateness were uncommon for Israelite temples. Furthermore, King Solomon's temple was quite diminutive in comparison to many of the many ancient stone structures found today in Mesoamerica.
698   Hugh Nibley, *An Approach to the Book of Mormon* (Melchizedek Priesthood manual, 1957), appendix section titled "Looking for the Wrong Things", pp. 440-441.
699   Hugh Nibley, *An Approach to the Book of Mormon* (Melchizedek Priesthood manual, 1957), appendix section titled "Looking for the Wrong Things", pp. 440-441.
700   Hugh Nibley, *An Approach to the Book of Mormon* (Melchizedek Priesthood manual, 1957), appendix section titled "Looking for the Wrong Things", pp. 440-441.
701   Hugh Nibley, *An Approach to the Book of Mormon*, p. 431.
702   ibid.
703   ibid.
704   A reminiscent account by Latter-day Saint William B. Pace. The Prophet Joseph Smith's discourse was delivered on June 24, 1844. During this discourse, Joseph Smith laid his hands

End Notes

upon the head of Latter-day Saint Levi W. Hancock and gave an injunction to the Elders of Israel. Joseph Smith, Jr., as quoted by William Bryan Pace, 1832-1907, Autobiography (1832-1847), Typescript, HBLL, chapter One: Birth and Boyhood. *Journal of Mormon History,* Vol. 19, Issue 1, 1993. Pace was apparently quoting Alfred Bell of Lehi, Utah who reportedly made a transcript of Smith's address. For other versions of the speech, which claimed to be copies of William Clayton's report, see Wilford Woodruff, Affidavit, 18 November 1878, and John S. Fullmer, Statement, 28 April, 1881, John S. Fulmer Book; both in LDS Church Archives. "A Prophecy of Joseph the Seer", found in *The Fate of the Persecutors of the Prophet Joseph Smith*, p. 154, 156.). Journal of Mormon History, Vol. 19, Issue 1, 1993.

705 Orson Pratt, *Journal of Discourses*, Volume 9, Discourse 33; Orson Pratt lived from 1811 to 1881, Spoken July 15th, 1855.

706 Redemption of Zion—Persecution—Baptism of Indians—Second Coming of Christ—Every Jot and Every Tittle of Divine Revelation Will Be Fulfilled. Discourse by Elder Orson Pratt, delivered in the Twentieth Ward Meetinghouse, on the Evening of Sunday, February 7, 1875. Reported by David W. Evans.

707 Redemption of Zion—Persecution—Baptism of Indians—Second Coming of Christ—Every Jot and Every Tittle of Divine Revelation Will Be Fulfilled. Discourse by Elder Orson Pratt, delivered in the Twentieth Ward Meetinghouse, on the Evening of Sunday, February 7, 1875. Reported by David W. Evans.

708 Elder Orson Pratt also said: "After they [the Lamanite remnant] are all gathered, . . . Zion will be redeemed, and all among the Gentiles who believe will assist this remnant of Jacob in building the New Jerusalem. . . . Not asking God to redeem Zion before he has redeemed a portion of the remnants of Joseph . . . Zion must needs be redeemed by power, with an outstretched arm, the angel of the Lord going before the camp of this people, and they will return, and a remnant of the Lamanites with them to build up the city of Zion in Jackson County." Redemption of Zion—Persecution—Baptism of Indians—Second Coming of Christ—Every Jot and Every Tittle of Divine Revelation Will Be Fulfilled. Discourse by Elder Orson Pratt, delivered in the Twentieth Ward Meetinghouse, on the Evening of Sunday, February 7, 1875. Reported by David W. Evans.

709 A reminiscent account by Latter-day Saint William B. Pace. The Prophet Joseph Smith's discourse was delivered on June 24, 1844. During this discourse, Joseph Smith laid his hands upon the head of Latter-day Saint Levi W. Hancock and gave an injunction to the Elders of Israel. Joseph Smith, Jr., as quoted by William Bryan Pace, 1832-1907, Autobiography (1832-1847), Typescript, HBLL, chapter One: Birth and Boyhood. *Journal of Mormon History,* Vol. 19, Issue 1, 1993. Pace was apparently quoting Alfred Bell of Lehi, Utah who reportedly made a transcript of Smith's address. For other versions of the speech, which claimed to be copies of William Clayton's report, see Wilford Woodruff, Affidavit, 18 November 1878, and John S. Fullmer, Statement, 28 April, 1881, John S. Fulmer Book; both in LDS Church Archives. "A Prophecy of Joseph the Seer ", found in *The Fate of the Persecutors of the Prophet Joseph Smith*, p. 154, 156.). Journal of Mormon History, Vol. 19, Issue 1, 1993.

710 Joseph Smith, Jr., quoted by William Bryan Pace, 1832-1907, Autobiography (1832-1847), Typescript, HBLL, Chapter One: Birth and Boyhood.

711 Wilford Woodruff, *Journal of Discourses* vol. 9, pp. 227-228. "Preaching the Gospel to, and Helping the Lamanites—Obedience to Counsel." Remarks by Elder Wilford Woodruff, made in the Bowery at Provo, Utah, July 15, 1855. Reported by J. V. Long.

712 Wilford Woodruff, 1855.

713 "But before the great day of the Lord shall come, . . . [T]he Lamanites shall blossom as the rose, Doctrine and Covenants 49:24 Wilford Woodruff, Journal of Discourses 15:283. The Signs of the Coming of the Son of Man—The Saints' Duties Discourse by Elder Wilford Woodruff, delivered in the 13th Ward Assembly Rooms, Salt Lake City, January 12, 1873. Reported by David W. Evans.

714 John Taylor, *Millennial Star* 44:33, October 18, 1882.

715 Spencer W. Kimball, Conference Reports, Oct. 1947.

716 "In this I have great faith." (Spencer W. Kimball "Our Paths Have Met Again," Ensign, Dec. 1975, pp. 5, 7 .)

717 Wilford Woodruff, St. George Conference, June 12th and 13th, 1892.

718 A reminiscent account by Latter-day Saint William B. Pace. The Prophet Joseph Smith's discourse was delivered on June 24, 1844. During this discourse, Joseph Smith laid his hands upon the head of Latter-day Saint Levi W. Hancock and gave an injunction to the Elders of

End Notes

Israel. Joseph Smith, Jr., as quoted by William Bryan Pace, 1832-1907, Autobiography (1832-1847), Typescript, HBLL, chapter One: Birth and Boyhood. *Journal of Mormon History,* Vol. 19, Issue 1, 1993. Pace was apparently quoting Alfred Bell of Lehi, Utah who reportedly made a transcript of Smith's address. For other versions of the speech, which claimed to be copies of William Clayton's report, see Wilford Woodruff, Affidavit, 18 November 1878, and John S. Fullmer, Statement, 28 April, 1881, John S. Fulmer Book; both in LDS Church Archives. "A Prophecy of Joseph the Seer", found in *The Fate of the Persecutors of the Prophet Joseph Smith*, p. 154, 156.). Journal of Mormon History, Vol. 19, Issue 1, 1993.

719   Kurdish Muslims Y-DNA—Haplogroup R1b: 17%; Haplogroup R1a: 12%; Haplogroup J: 40%; and Haplogroup E: 7%.

720   Kurdish Jews Y-DNA—Haplogroup R1b: 20%; Haplogroup J: 37%; and Haplogroup E: 21%.

721   Sephardic Cohanim Y-DNA—Haplogroup J: 75%; CMH (Cohanim Modal Haplotype) 56%.

722   Ashkenazi Jews Y-DNA—Haplogroup R1b: 15%; Haplogroup R1a: 4%; Haplogroup J: 37%; and Haplogroup E: 12%. Ashkenazi Cohanim (Haplogroup J) 87%, CMH 45%.

723   Bedouin Y-DNA—Haplogroup R1b: 15%; Haplogroup J: 46%; Haplogroup E: 21%; and Haplogroup I: 6%.

724   Palestinian Y-DNA—Haplogroup R1b: 8%; Haplogroup R1a: 1%; Haplogroup J: 20%; Haplogroup E: 20%; and Haplogroup I: 6%.

725   Yemeni Muslims Y-DNA—Haplogroup R1b: 0%; Haplogroup J: 82%; and Haplogroup E: 16%.

726   Albanian Y-DNA—Haplogroup R1b: 16%; Haplogroup R1a: 9%; Haplogroup J2: 19.5%; Haplogroup I: 15.5%; HaplogroupE: 27.5%; Haplogroup N: 0%; Haplogroup G: 1.5%. www.eupedia.com/europe/european_y-dna_haplogroups.shtml

727   Macedonian Y-DNA—Haplogroup R1b: 10.5%; Haplogroup R1a: 14.5%; Haplogroup J2: 12.5%; Haplogroup G: 4%; Haplogroup I: 31%; HaplogroupE: 20.5%. www.eupedia.com/europe/european_y-dna_haplogroups.shtml

728   Italians (Sicily) Y-DNA—Haplogroup R1a: 4.5%; Haplogroup R1b: 30%, (J) 26.5%, (I) 5%. www.eupedia.com/europe/european_y-dna_haplogroups.shtml

729   Cyprus (an island country in the Eastern Mediterranean Sea) Y-DNA—Haplogroup I: 8%; Haplogroup R1a: 3%; haplogroup R1b: 9%; Haplogroup G: 9%; Haplogroup J2: 37%; Haplogroup E1b1b: 20%; and Haplogroup T: 5%. www.eupedia.com/europe/european_y-dna_haplogroups.shtml

730   Bulgarian paternal DNA (Y-DNA)—Haplogroup I: 25.5%; Haplogroup R1a: 17%; Haplogroup R1b: 10.5%; Haplogroup G: 5%; Haplogroup J: 14%; Haplogroup E1b1b: 24%; Haplogroup T: 1.5%. www.eupedia.com/europe/european_y-dna_haplogroups.shtml

731   French paternal DNA (Y-DNA)—Haplogroup J: 8%; haplogroup I: 15.5%; Haplogroup R1a: 2.5%; Haplogroup R1b: 61%; Haplogroup G: 5%; and Haplogroup E1b1b: 7%. www.eupedia.com/europe/european_y-dna_haplogroups.shtml

732   German paternal DNA (Y-DNA)— Haplogroup J: 4.5%; Haplogroup I: 22%; Haplogroup R1a: 16%; Haplogroup R1b: 44.5%; Haplogroup G: 5%; and Haplogroup E1b1b: 5.5%. www.eupedia.com/europe/european_y-dna_haplogroups.shtml

733   Eighty percent of the Kubachi and Dargins from Dagestan in the Northeast Caucasus who were tested belonged to Haplogroup J1 (Y-DNA). In Arabic countries, J1 hits its peak among the Marsh Arabs of South Iraq (81%), the Sudanese Arabs (73%), the Yemeni (72%), the Bedouins (63%), the Qatari (58%), the Saudi (40%), the Omani (38%) and the Palestinian Arabs (38%). Reasonably high percentages are also observed in the United Arab Emirates (35%), coastal Algeria (35%), Jordan (31%), Syria (30%), Tunisia (30%), Egypt (21%) and Lebanon (20%). Most of the Arabic J1 belongs to the J1c3 variety. The world's highest frequency of J2 is found among the Ingush (88% of the male lineages) and Chechen (56%) people in the Northeast Caucasus. Other high incidence of Haplogroup J2 are found in many other Caucasian populations, including the Azeri (30%), the Georgians (27%), the Kumyks (25%), and the Armenians (22%). The Rothschild family belongs to Haplogroup J2a1-L210 (a subclade of M67). According to a study conducted by L.A. Feryodoun Barjesteh van Waalwijk van Doorn and Sahar Khosrovani published in Qajar Studies, Journal of the International Qajar Studies Association, volume VII (2007), Qajar dynasty, the Iranian royal family who ruled over Persia from 1785 to 1925, belonged to Haplogroup J1.

734   Approximately 20% of Ashkenazi Jews belong to Haplogroup E1b1b. Outside Europe, E1b1b is found at high frequencies in Morocco (over 80%), Somalia (80%), Ethiopia (40% to 80%), Tunisia (70%), Algeria (60%), Egypt (40%), Jordan (25%), Palestine (20%), and Lebanon

End Notes

(17.5%). On the European continent it has the highest concentration in Kosovo (over 45%), Albania and Montenegro (both 27%), Bulgaria (23%), Macedonia and Greece (both 21%), Cyprus (20%), Sicily (20%), South Italy (18.5%), Serbia (18%) and Romania (15%). Caravaggio—Haplogroup E1b1b1; William Harvey—Haplogroup E; Napoleon Bonaparte—Haplogroup E; The Wright Brothers—Haplogroup E-V13; Albert Einstein—Haplogroup E-Z830; Adolf Hitler—Haplogroup E1b1b; Lyndon B. Johnson—Haplogroup E1b1b1; and Sir David Attenborough— Haplogroup E1b1b1.

735 Haplogroup T is found among groups who possess Semitic physical features in the Middle East, Europe, and parts of East Africa. The current highest amounts of Haplogroup T are found among the Kazakhs (38.8%), Greeks from Crete and southern Aegaeans (33.3%), German Stilfser/Tyrolese (23.5%), Venetians of Italy (22.2%), Balearics of Spain (16.7%), Northeastern Portuguese Jews (15.7%). Haplogroup T is also found among the Fulani people of Cameroon (18% of the population). Thomas Jefferson belonged to Haplogroup T-M184.

736 Currently, G2a is found mostly in mountainous regions of Europe, for example, in the Apennine mountains (15 to 25%) and Sardinia (12%) in Italy, Cantabria (10%) and Asturias (8%) in northern Spain, Austria (8%), Auvergne (8%) and Provence (7%) in south-east France, Switzerland (7.5%), the mountainous parts of Bohemia (5 to 10%), Romania (6.5%) and Greece (6.5%). Among 77 samples taken in Georgia (2003), 31% were Haplogroup G (Y-DNA). Nasidze I, Sarkisian T, Kerimov A, Stoneking M (March 2003). "Testing hypotheses of language replacement in the Caucasus: evidence from the Y-chromosome". Human Genetics 112 (3): 255–61. doi:10.1007/s00439-002-0874-4. PMID 12596050. A famous member of Haplogroup G (Y-DNA) was Joseph Stalin (G2a1), who was of Georgian origin. Joseph Stalin looks like he's from the Middle East. Most of the Plantagenets monarchs belonged to Haplogroup G2 or R1b-U152. www.eupedia.com/europe/origins_haplogroups_europe.shtml

737 B. Arredi, E. S. Poloni and C. Tyler-Smith (2007). "The peopling of Europe". In Crawford, Michael H. Anthropological genetics: theory, methods and applications. Cambridge, UK: Cambridge University Press. p. 394.

738 'Lost tribe of Israel' found in southern India, Canadian Jewish News, 7 October 2010.

739 Isaac Galland, was best known for selling large tracts of land around Commerce, Illinois to Joseph Smith in 1839.

740 History of the Church, 4:8–9; punctuation modernized; from a letter from Joseph Smith to Isaac Galland, Sept. 11, 1839, Commerce, Illinois.

741 President Spencer W. Kimball wrote: "The Lamanite is a chosen child of God, but he is not the only chosen one. There are many other good people including the Anglos, the French, the German, and the English, who are also of Ephraim and Manasseh. They, with the Lamanites, are also chosen people, and they are a remnant of Jacob. The Lamanite is not wholly and exclusively the remnant of Jacob which the Book of Mormon talks about. We are all of Israel! We are of Abraham and Isaac and Jacob and Joseph through Ephraim and Manasseh. We are all of us remnants of Jacob." The Teachings of Spencer W. Kimball, Salt Lake City: Bookcraft, 1982, pp. 600–601.

742 Brigham Young, *Millennial Star,* vol. 26, p. 7.

743 England (68.8%) R1b. www.eupedia.com/europe/european_y-dna_haplogroups.shtml

744 http://isogg.org/ffdna.htm

745 Wales n=65 (R1b: 92.3%). Balaresque et al. (2009). Balaresque, Patricia; Bowden, Georgina R.; Adams, Susan M.; Leung, Ho-Yee; King, Turi E. et al. (2010). Penny, David, ed. "A Predominantly Neolithic Origin for European Paternal Lineages". PLOS Biology (Public Library of Science) 8 (1): e1000285. doi:10.1371/journal.pbio.1000285. PMC 2799514. PMID 20087410. Retrieved August 19, 2014.

746 Basque of Spain n=116, (R1b: 87.1%). Balaresque et al. (2009). Balaresque, Patricia; Bowden, Georgina R.; Adams, Susan M.; Leung, Ho-Yee; King, Turi E. et al. (2010). Penny, David, ed. "A Predominantly Neolithic Origin for European Paternal Lineages". PLOS Biology (Public Library of Science) 8 (1): e1000285. doi:10.1371/journal.pbio.1000285. PMC 2799514. PMID 20087410. Retrieved August 19, 2014.

747 Basque of Spain n=116, (R1b: 87.1%). Balaresque et al. (2009). Balaresque, Patricia; Bowden, Georgina R.; Adams, Susan M.; Leung, Ho-Yee; King, Turi E. et al. (2010). Penny, David, ed. "A Predominantly Neolithic Origin for European Paternal Lineages". PLOS Biology (Public Library of Science) 8 (1): e1000285. doi:10.1371/journal.pbio.1000285. PMC 2799514. PMID 20087410. Retrieved August 19, 2014.

748 Ireland, n=796 (R1b: 85.4%) Moore et al. (2006). Moore et al.; McEvoy, B; Cape, E; Simms,

End Notes

K; Bradley, DG (2006). "A Y-Chromosome Signature of Hegemony in Gaelic Ireland". American Journal of Human Genetics 78 (2): 334–8. doi:10.1086/500055. PMC 1380239. PMID 16358217.

749   Ireland, n=796 (R1b: 85.4%) Moore et al. (2006). Moore et al.; McEvoy, B; Cape, E; Simms, K; Bradley, DG (2006). "A Y-Chromosome Signature of Hegemony in Gaelic Ireland". American Journal of Human Genetics 78 (2): 334–8. doi:10.1086/500055. PMC 1380239. PMID 16358217.

750   Cruciani et al. 2010, Human Y chromosome haplogroup R-V88: a paternal genetic record of early mid Holocene trans-Saharan connections and the spread of Chadic languages.

751   Grugni et al. (2012). Grugni, Viola et al 2012, Ancient Migratory Events in the Middle East: New Clues from the Y-Chromosome Variation of Modern Iranians PLoS ONE 7(7): e41252. doi:10.1371/journal.pone.0041252.

752   Siwa Berber males of Egypt (28% Haplogroup R1b). Dugoujon et al. (2009).

753   Kurdish Jews of Iran (20.2% Haplogroup R1b). Nebel, A; Filon, D; Brinkmann, B; Majumder, P; Faerman, M; Oppenheim, A (2001). "The Y Chromosome Pool of Jews as Part of the Genetic Landscape of the Middle East". The American Journal of Human Genetics 69: 1095–112. doi:10.1086/324070. PMC 1274378. PMID 11573163.

754   Of the 146 Arabs tested from Jordan, 17.8% of belonged to Haplogroup R1b (Y-DNA). Abu Amero et al. (2009). Abu-Amero, Khaled K; Hellani, Ali; González, Ana M; Larruga, Jose M; Cabrera, Vicente M; Underhill, Peter A (2009). "Saudi Arabian Y-Chromosome diversity and its relationship with nearby regions". BMC Genetics 10: 59. doi:10.1186/1471-2156-10-59. PMC 2759955. PMID 19772609.

755   Of the Ashkenazi Jews whose DNA was tested, 15% belonged to Haplogroup R1b. Nebel, A; Filon, D; Brinkmann, B; Majumder, P; Faerman, M; Oppenheim, A (2001). "The Y Chromosome Pool of Jews as Part of the Genetic Landscape of the Middle East". The American Journal of Human Genetics 69: 1095–112. doi:10.1086/324070. PMC 1274378. PMID 11573163.

756   Female dignitary. Maya Culture, Guatemala, Late Classical Period (300-800 AD). White stucco with brick-red and turquoise paint. H. 12.8 in – W. 11.8 in – D. 8.7.

757   Haplogroup Q (Y-DNA).

758   Almost 7/ 10 (68.2%) belonged to Haplogroup A (mtDNA). Fagundes, Nelson J.R.; Ricardo Kanitz, Roberta Eckert, Ana C.S. Valls, Mauricio M. Bogo, Francisco M. Salzano, David Glenn Smith, Wilson A. Silva, Marco A. Zago, Andrea K. Ribeiro-dos-Santos, Sidney E.B. Santos, Maria Luiza Petzl-Erler, and Sandro L.Bonatto (2008). "Mitochondrial Population Genomics Supports a Single Pre-Clovis Origin with a Coastal Route for the Peopling of the Americas" (pdf). American Journal of Human Genetics 82 (3): 583–592. doi:10.1016/j.ajhg.2007.11.013. PMC 2427228. PMID18313026. Retrieved 2009-11-19.

759   Starikovskaya et al. (2005).

760   "An Analysis of Ancient Aztec mtDNA from Tlatelolco: Pre-Columbian relations and the spread of Uto-Aztecan", Kemp et al. (2005). Shurr et al (1990); Torroni et al. (1992).

761   Nearly 9/10 (87.5%). An Analysis of Ancient Aztec mtDNA from Tlatelolco: Pre-Columbian relations and the spread of Uto-Aztecan, Kemp et al. (2005). Gonzalez-Oliver et al. (2001).

762   "An Analysis of Ancient Aztec mtDNA from Tlatelolco: Pre-Columbian relations and the spread of Uto-Aztecan", Kemp et al. (2005). Carlyle et al. (2000).

763   Starikovskaya et al. (2005).

764   Wen et al. (2005).

765   "An Analysis of Ancient Aztec mtDNA from Tlatelolco: Pre-Columbian relations and the spread of Uto-Aztecan", Kemp et al. (2005). Shurr et al (1990); Torroni et al. (1992). Other indigenous Mesoamerican groups belonged to Haplogroup B (mtDNA). Thirteen percent the 43 Aztec women of Tlatelolco (Mexico) whose mitochondrial DNA was tested were Haplogroup B (mtDNA).

766   Tabbada et al. (2010).

767   Haplogroup C (mtDNA) is common in Mongolia (15%) and most of the populations of central Asia (7–18%), but occurs as rarely as 1–5% in China, Korea, Japan, Thailand, Island southeastern Asia, and India. Haplogroup C (mtDNA) is virtually absent in Africa and western Europe and is detected at a very low frequency in many populations located in eastern and central Europe. "Origin and Post-Glacial Dispersal of Mitochondrial DNA Haplogroups C and D in Northern Asia." Derenko et al. 2010.

768   "An Analysis of Ancient Aztec mtDNA from Tlatelolco: Pre-Columbian relations and the

End Notes

spread of Uto-Aztecan", Kemp et al. (2005). Merriwether et al. (1997).
769 "An Analysis of Ancient Aztec mtDNA from Tlatelolco: Pre-Columbian relations and the spread of Uto-Aztecan", Kemp et al. (2005). Shurr et al (1990); Torroni et al. (1992).
770 "An Analysis of Ancient Aztec mtDNA from Tlatelolco: Pre-Columbian relations and the spread of Uto-Aztecan", Kemp et al. (2005). Shurr et al (1990); Torroni et al. (1992).
771 Ville N Pimenoff, David Comas, Jukka U Palo et al., "Northwest Siberian Khanty and Mansi in the junction of West and East Eurasian gene pools as revealed by uniparental markers", European Journal of Human Genetics (2008) 16, 1254–1264; doi:10.1038/ejhg.2008.101
772 Volodko et al. 2008.
773 "An Analysis of Ancient Aztec mtDNA from Tlatelolco: Pre-Columbian relations and the spread of Uto-Aztecan", Kemp et al. (2005). Carlyle et al. (2000).
774 Many Ashkenazi Jews are haplogroup K (K2a, K4a1, K1a9) but not necessarily K1. Approximately one-third (32%) of Ashkenazi Jews tested belong to Haplogroup K (mtDNA).Doron M. Behar et al. "MtDNA evidence for a genetic bottleneck in the early history of the Ashkenazi Jewish population." Of the Druze from Syria, Lebanon, Israel, and Jordan who were tested, 16% belonged to Haplogroup K. Israeli Druze are located mostly in Galilee (81%), around Haifa (19%), and in the Golan Heights."Press Release: The Druze Population of Israel" (DOC). Israel Central Bureau of Statistics. 2009-04-23. (Hebrew). Haplogroup K (mtDNA) was also found in significant amounts among Palestinian Arabs. 12% of the Irish tested were maternal lineage K. 14% of Belgium, 10% Netherlands, 10% Iceland, 9% Denmark, 8.5% France. Haplogroup K most likely originated in the Middle East or Northern Italy. Haplogroup K (mtDNA): Steven Pinker, a Canadian-Jew, belongs to Haplogroup K (subclade K2a2a). So do Meryl Streep, Stephen Colbert, and Mike Nichols.
www.colbertnation.com/the-colbert-report-videos/183232/august-14-2007/dr--spencer-wells
775 Of the 300 Irish tested, 133 (44%) were H1 (mtDNA). Of the 129 Libyan Tuareg of the Fezzan region who were tested, 61% were H1. "Mitochondrial Haplogroup H1 in North Africa: An Early Holocene Arrival from Iberia." Ottani et al 2010. The origin of the maternal lineage H is the Caucasus or Southern France. H is common in Europe, North Africa, and the Middle East. 54% of people from Spain were H. 51.5% of Albanians tested were H maternal lineage, 48% of the Swedes tested were H, 47% of Hungarians tested were H, 46% of the Polish tested were H, 42.5% of Scottish were H. Queen Victoria, who was Queen of the United Kingdom of Great Britain and Ireland from June 20, 1837 until her death, belonged to Haplogroup H.
776 Haplogroup I (mtDNA): Dargins (6.5%), Chechens (6%), Kumyks (5.5%), and Mordovia (6%).
777 Mitochondrial DNA haplogroup N1b is rare in most European populations, was found to comprise nearly 10% of the Ashkenazi mitochondrial DNA pool.
778 Of the Udmurts of Russia, 24% belonged to Haplogroup T2 (mtDNA). Many Udmurts have red hair. Of the Chechen-Ingush of Daghestan (12.5%), Netherlands (12%), Sardinia, Italy (10%), Iceland (10%), and Switzerland (9.5%) all belonged to Haplogroup T2 (mtDNA).
779 Ten percent of Udmurts tested belong to Haplogroup U2. Haplogroup U2: Mordvins (7%), as well as the Karachay-Balkars (4.5%), Nogays (3.8%), North Ossetians (3.6%), Adyghe-Kabardin (3.6%) and Dargins (3.6%) in the North Caucasus, and the Latvians (3.5%) in the East Baltic. The highest percentages of U3 are observed in Jordan (15%), Syria (5%), Lebanon (5%), Cyprus (5%), Iraq (5%), Armenia (5%), Georgia (4.5%), Azerbaijan (3.5%), Turkey (3.5%), Greece (3.5%) and Egypt (3%), as well as among the various ethnic groups of the North Caucasus, including the Karachay-Balkars (7.5%), Adyghe-Kabardin (7%), Chechen-Ingush (6%), Kumyks (6%), Nogays (4.5%), and North Ossetians (3.5%). Haplogroup H4: Its highest frequency is observed among the Chuvash (16.5%), Bashkirs (15%) and Tatars (7%) of the Volga-Ural region of Russia, followed by Latvia (8.5%), Georgia (8.5%), Serbia (7%), and southern Daghestan (6.5%). Haplogroup U5: Nearly half of all Sami of Norway and one fifth of Finnish maternal lineages belong to U5. Other high frequencies are observed among the Mordovians (16%), the Chuvash (14.5%) and the Tatars (10.5%) in the Volga-Ural region of Russia, the Estonians (13%), the Lithuanians (11.5%) and the Latvians in the Baltic, the Dargins (13.5%), Avars (13%), the Chechens (10%) in the Northeast Caucasus, the Basques (12%), the Cantabrians (11%) and the Catalans (10%) in northern Spain, the Bretons (10.5%) in France, the Sardinians (10%) in Italy, the Slovaks (11%), the Croatians (10.5%), the Poles (10%), the Czechs (10%), the Ukrainians (10%), and the Slavic Russians (10%). Haplogroup U6: It is most common in North-West Africa, especially among the Mozabites, who are Ibadi Muslims, (28%) and Kabyles (18%) of Algeria, as well as Mauritanians (14%) and Canary Islanders (13.5%).

End Notes

780 Haplogroup V (mtDNA): Saami (42%) of northern Scandinavia and Finland, and the Cantabrians (19%), isolated in mountains in northern Spain. Bono of U2 and Benjamin Franklin: Haplogroup V.

781 Of those from Finland tested, 9.5% belonged to Haplogroup W, 8% of northern Pakistanis and 5% of Hungarians belonged to Haplogroup W (mtDNA).

782 King Richard III of England belonged to Haplogroup J (mtDNA). Edward IV of England and his brother Richard III of England, both sons of Cecily Neville, Duchess of York, would have shared the same mtDNA haplogroup J1c2c. Rachel, Ehrenberg (6 February 2013). "A king's final hours, told by his mortal remains". Science News. Society for Science & the Public. Retrieved 8 February 2013.

783 Druze woman from northern Israel. P2MR type: Possibilitarian (In-Curviplex-Circle-Wave). Refer to *Recognizing People* by Alexander T Paulos.

784 Falling Star, Abenaki, 1900. P2MR type: Possibilitarian (In-Curviplex-Circle-Wave). Refer to *Recognizing People* by Alexander T Paulos.

785 Of the Druze whose DNA was tested, 39 of 41 were Haplogroup X (mtDNA). The Druze: A Population Genetic Refugium of the Near East, Shlush et al (2008). www.plosone.org/article/info:doi/10.1371/journal.pone.0002105

786 3/6 tested among the Mi'kmaq tribe belonged to Haplogroup X2a (mtDNA). The peopling of the Americas: Genetic ancestry influences health". Scientific American. "Learn about Y-DNA Haplogroup Q" (Verbal tutorial possible). Wendy Tymchuk – Senior Technical Editor. Genebase Systems. 2008. Retrieved 2012-11-21. Of the 28 Northern Ojibwe who were tested, 25% belonged to maternal lineage (haplogroup) X and of the 35 Southern Ojibwe who were tested, 25.7% were maternal lineage (haplogroup) X. Maternal lineages X2a, X2g and X2a1a are found among Ojibwa/Chippewa and W. Chippewa. Of the 28 Northern Ojibwa who were tested 25.0% were X (Torroni et al. 1993a). And of the 35 Southern Ojibwa who were tested 25.7% (Scozzari et al. 1997).

787 14.6% of the Sioux who were tested belonged to Haplogroup X2a (mtDNA), Bianchi and Bailliet (1997). Among the Sioux X2a1a was found (6113G marker). Among North American Indians, the Chippewa (Ojibwe), Navajo, and Mi'kmaq were also X2a. Fagundes, Nelson J.R.; Ricardo Kanitz, Roberta Eckert, Ana C.S. Valls, Mauricio R. Bogo, Francisco M. Salzano, David Glenn Smith, Wilson A. Silva, Marco A. Zago, Andrea K. Ribeiro-dos-Santos, Sidney E.B. Santos, Maria Luiza Petzl-Erler, and Sandro L.Bonatto (2008). "Mitochondrial Population Genomics Supports a Single Pre-Clovis Origin with a Coastal Route for the Peopling of the Americas". American Journal of Human Genetics 82 (3): 583-592.doi:10.1016/j.ajhg.2007.11.013. PMC 2427228. PMID 18313026. The peopling of the Americas: Genetic ancestry influences health". Scientific American. "Learn about Y-DNA Haplogroup Q" (Verbal tutorial possible). Wendy Tymchuk – Senior Technical Editor. Genebase Systems. 2008. Retrieved 2012-11-21.

788 Nuu-Chah-Nulth (n=15) 13.3% Haplogroup X (mtDNA), Torroni et al. (1993a). Navajo (n=92) 6.5% Torroni et al. (1993a). Other studies have the Navajo at 11-13% Haplogroup X (mtDNA). Yakima (n=42) 4.8% Haplogroup X (mtDNA), Shields et al. (1993). Fagundes, Nelson J.R.; Ricardo Kanitz, Roberta Eckert, Ana C.S. Valls, Mauricio R. Bogo, Francisco M. Salzano, David Glenn Smith, Wilson A. Silva, Marco A. Zago, Andrea K. Ribeiro-dos-Santos, Sidney E.B. Santos, Maria Luiza Petzl-Erler, and Sandro L.Bonatto (2008). "Mitochondrial Population Genomics Supports a Single Pre-Clovis Origin with a Coastal Route for the Peopling of the Americas". American Journal of Human Genetics 82 (3): 583–592. doi:10.1016/j.ajhg.2007.11.013. PMC 2427228. PMID 18313026.

789 Haplogroup X (mtDNA). *Indians in the Americas: The Untold Story,* p. 16, William Marder, Paul Tice.

790 The strong presence of X2 around the Caucasus, progressively fading towards the Near East and Mediterranean, hints that it could be related to the spread of Y-DNA haplogroup G2a. R1b1b and G2a both having origins around the Caucasus it is unsurprising to find X2 alongside these two Y-DNA haplogroups. www.eupedia.com/europe/origins_haplogroups_europe.shtml

791 www.crystalinks.com/romapeople.html Origins and divergence of the Roma (Gypsies) Gresham et al 2001.

792 William Penn's Own Account of the Lenni Lenape or Delaware Indians, edited by Albert Cook Myers, 1971. Narratives of Early Pennsylvania, West New Jersey and Delaware ..., Vol. 13.

793 Latter-day Saint scholar Hugh Nibley made some connections between the Gypsies and Cain

End Notes

(who most likely was not a Sub-Saharan black man because of his genetic lineage): "Cain, 'Qayin,' is the wandering smith in Arabic, Hebrew, and Aramaic. A qayin is a blacksmith. He blackens his face professionally because he works at the forge. This is a mark of his profession, the blackened face. It advertises his profession, and he wanders. You find these, and they are great metal workers, as we will see Cain's descendants are. Their rites are secret, and they intermarry. You think of the Gypsies, of course. The Gypsies belong to that particular class of people. They are always wandering. . . . They tell fortunes and have all sorts of insights. They can really tell them too. I've had some beautiful fortunes told, and they hit it 'right on the button.'" Hugh Nibley, *Ancient Documents and the Pearl of Great Price*, p.3.

The word "Gypsy" was originally "gipcyan," which is short for Egyptian. Gypsies were thought to have originally come from Egypt. Gypsies speak "Romany," a language that is related to Hindi. The language is currently believed to have originated in South Asia. However, it is probable that the Gypsies and their language originally came from the Middle East or the surrounding area because of their Middle Eastern appearance and haplogroups. About half of the Gypsy population belong to Haplogroup M, and more specifically M5 (reflected by Y-Haplogroup H1a), which is otherwise exclusive to South Asia. The other mtDNA haplogroups found among the Gypsy community are mostly of Eastern European, Caucasian or Middle Eastern origin. MtDNA haplogroups of the Gypsies: H (H1, H2, H5, H9, H11, H20, among others), J (J1b, J1d, J2b), T, U3, U5b, I, W et X (X1b1, X2a1, X2f) (sources). The same diversity exist on the Y-DNA side (45% of Haplogroup H1a, followed by I1, I2a, J2a4b, E1b1b, R1b1b, R1a1a).

794 U. Perego et al., Distinctive Paleo-Indian Migration Routes from Beringia Marked by Two Rare mtDNA Haplogroups, Current Biology, Volume 19, Issue 1, (13 January 2009), Pages 1-8.
795 Of the 43 Aztecs of Tlatelolco (Mexico) whose mitochondrial DNA was tested, 0% possessed Haplogroup X An Analysis of Ancient Aztec mtDNA from Tlatelolco: Pre-Columbian relations and the spread of Uto-Aztecan", Kemp et al. (2005). Carlyle et al. (2000).
796 Of the 24 ancient Maya of Xcaret in Mexico (Yucatan, Quintana Roo) who were tested, 0% possessed X. "An Analysis of Ancient Aztec mtDNA from Tlatelolco: Pre-Columbian relations and the spread of Uto-Aztecan", Kemp et al. (2005). Gonzalez-Oliver et al. (2001).
797 Of the 9 ancient Maya of Copán, Honduras whose mitochondrial DNA was tested, 0% belonged to X. "An Analysis of Ancient Aztec mtDNA from Tlatelolco: Pre-Columbian relations and the spread of Uto-Aztecan", Kemp et al. (2005). Merriwether et al. (1997).
798 Of the 27 modern-day Mexican Maya of the Yucatan who were tested, 0% possessed Haplogroup X. "An Analysis of Ancient Aztec mtDNA from Tlatelolco: Pre-Columbian relations and the spread of Uto-Aztecan", Kemp et al. (2005). Shurr et al (1990); Torroni et al. (1992).
799 Among the Israeli Druze can be found Haplogroup X2. X2 is also found in Egypt and Tunisia. Since the Druze have the highest percentages of X2, X2 is most likely Semitic and spread among the Egyptians and Tunisians. Among the Israeli Druze, Haplogroup X2e (mtDNA) is found. Among Siberians (i.e. Altaian-Kickhis, Buryats, Teleuts) X2e is found. Since certain Siberians are Semitic-looking and Israeli Druze don't look Siberian it is easy to see how Siberia did not possess X2e first. Among the Israeli Druze, Haplogroup X2b (mtDNA) can be found. It can also be found among Sardinians of Italy and Moroccans. Since Sardinians and Moroccans often have Semitic appearances and Semitic DNA, odds favor that X2b came from the Middle East.
800 According to current research, subgroups X2a and X2g are found in North America, and nowhere else. Miraslava et al have stated, "To extend the survey of Asian mtDNAs for the presence of haplogroup X, we screened the mtDNAs of a total of 790 individuals for the RFLP markers that define this lineage. Haplogroup X mtDNAs were detected only in Altaians, at a frequency of 3.5%. It should also be noted that none of the Altaian X mtDNAs harbored the 225A variant, which is a marker for a major part of haplogroup X. However, the X mtDNAs that we detected in the Altaian sample do not bear the 16213A and 200G variants that are characteristic of most American Indian haplogroup X mtDNAs." Miroslava V. Derenko, Tomasz Grzybowski, The Presence of Mitochondrial Haplogroup X in Altaians from South Siberia, American Journal of Human Genetics, 69:237-241, 2001.
801 U. Perego et al., Distinctive Paleo-Indian Migration Routes from Beringia Marked by Two Rare mtDNA Haplogroups, Current Biology, Volume 19, Issue 1, (13 January 2009), Pages 1-8.
802 forums.skadi.net/showthread.php?p=831496

End Notes

803 An mtDNA Analysis In Ancient Basque Populations: Implications for Haplogroup V as a Marker or a Major Paleolithic Expansion from Southwestern Europe by N. Izagirre and C. de la Riia. American Journal of Human Genetics (1999), 65.199-207.

804 Of the 47 ancient Illinois Hopewell Indians (200 BC - 500 AD) sampled, one belonged to Haplogroup X (2.1%). The Hopewell Culture is an ancient American Indian civilization that arose in Ohio and other parts of eastern North America during the Middle Woodland Period, perhaps as early as 100 BC. The Hopewell Culture, like the culture of the Nephites, ended by 400 AD. Haplogroup X was found at mound 11 of the Pete Klucnk mound group, associated with the Hopewell Indians. Bolnick & Smith (2007), Migration and social structure among the Hopewell (Indians), p. 635.

https://umdrive.memphis.edu/amicklsn/ESCI_7310/Articles/Bolnick_and_Smith_2007.pdf
Thirty-four ancient mtDNA sequences were retrieved in North America: 41% haplogroup A, 9% haplogroup B, 29% haplogroup C, 21% haplogroup D and 0% haplogroup X (mtDNA). Lisa Ann Mills, Philosophy, Ohio State University, Anthropology, 2003. The results were provided on pages 97 – 102. www.ohiolink.edu/etd/view.cgi?osu1054605467

805 One possible problem with Haplogroup X (mtDNA) being linked to Book of Mormon peoples is that it was present in North America before they arrived. Among the Windover Bog Mummies—7000 year old mummies found in a Florida cemetery is unique DNA. One of the mummies has 2 haplogroup X markers (16223 T and 16278 T), Smith et al 1999. A skull of a 9,000 year old Caucasoid skull (Caucasian skull type) discovered in North America dubbed, Kennewick Man, found in the Kennewick River in Washington State, USA has Haplogroup X (mtDNA). NOVA Online | Mystery of the First Americans | Dr. Robson Bonnichsen.

806 The Latter-day Saints' Millennial Star (Saturday, September 18, 1852), vol. 14, no. 30, p. 469.

807 "Heritability of the big five personality dimensions and their facets: a twin study," Jang KL et al. (1996).

808 *Doctrine and Covenants* 46:11.

809 "Those people who get high scores on agreeableness are typically altruistic, helpful, friendly, tenderminded, credulous and empathetic; those people who get low scores are, in turn, typically selfish, distrustful, competitive and antagonistic." "The Impact of Personality Factors on the Experience of Spatial Presence", Ana Sacau et al. (2005).

810 "Atheists are Disagreeable and Unconscientious." March 4, 2010. Saroglou, V. (2009). Religiousness as a Cultural Adaptation of Basic Traits: A Five-Factor Model Perspective Personality and Social Psychology Review, 14 (1), 108-125 DOI: 10.1177/1088868309352322

811 Many Latter-day Saints in this day and age are most likely unaware of why topics like "believing blood" and "literal lineages of Israelites" matter. Blogger and Latter-day Saint Kevin Barney, who founded the blog "By Common Consent," said in 2006, "I consider myself to be a universalist, and I don't really understand what the role of Israelite descent is supposed to be beyond Paul's spiritual take that the true Christian is of Israel." By Common Consent, "The Gathering of Israel," November 24, 2006, Kevin Barney. http://bycommonconsent.com/2006/11/24/the-gathering-of-israel/ Reid L. Neilson, managing director of the LDS Church History, also gave his opinion about the subject of "believing blood" in 2010: "[T]he racialist idea of the potency of the 'believing blood' of Israel is thankfully fading in the general Mormon consciousness[.]" Reid L. Neilson. *Early Mormon Missionary Activities in Japan*, 1901-1924. Salt Lake City: University of Utah Press, 2010, 214 pp. Photographs, endnotes, bibliography, and index. Paperback: $29.95. ISBN: 978–0–87480–989–3. Reviewed by Andrew R. Hall. While Latter-day Saints are correctly taught that God loves everyone regardless of genetics, these statements demonstrate a lack of understanding regarding the importance of certain temperament traits—particularly those related to "believing blood"—existing prominently in a covenant-keeping population.

812 Elder Bruce R. McConkie, *A New Witness for the Articles of Faith*, pp. 38-39.

813 Charles W. Penrose (First Presidency) Conference Report, Oct. 1922, pp. 21–22.

814 "Who Is My Neighbor?" Elder Spencer W. Kimball Of the Council of the Twelve Apostles, Spencer W. Kimball, Conference Report, April 1949, pp. 103-113.

815 "Who Is My Neighbor?" Elder Spencer W. Kimball Of the Council of the Twelve Apostles, Spencer W. Kimball, Conference Report, April 1949, pp. 103-113.

816 Faust, James E. "Heirs to the Kingdom of God." Ensign (May 1995).

817 'Rock solid in faith', Conference messages are broadcast to nine states By Gerry Avant. Church News.

End Notes

818 'Rock solid in faith', Conference messages are broadcast to nine states By Gerry Avant. Church News.
819 "The Tapestry of God's Hand." Divine Hand Evident in Life of Joseph Smith, Elder Ballard Says. Elder M. Russell Ballard at the Joseph Smith Memorial Fireside at the Logan Utah Institute of Religion on Sunday, February 13. The institute is located adjacent to Utah State University.
820 "The Tapestry of God's Hand." Divine Hand Evident in Life of Joseph Smith, Elder Ballard Says. Elder M. Russell Ballard at the Joseph Smith Memorial Fireside at the Logan Utah Institute of Religion on Sunday, February 13. The institute is located adjacent to Utah State University.
821 Tame individuals tend to be people who are born highly conscientious.
822 Conscientiousness leads to orderliness rather than sloppiness, industriousness rather than laziness, reliability rather than unreliability, decisiveness rather than indecisiveness, impulse control and cautiousness rather than carelessness and reckless abandon, formalness as opposed to casualness, and less non-conventionality. A second study examined the factor structure of 36 different scales assessing aspects of conscientiousness, drawn from seven major personality inventories, which included the most widely used questionnaires in existence (e.g., the NEO-PI-R, California Psychological Inventory, Multidimensional Personality Questionnaire, and the like; Roberts, Chernyshenko, et al., 2005).
823 *The Emotional Brain,* p. 135, Joseph LeDoux.
824 "Predicting School Success: Comparing Conscientiousness, Grit, and Emotion Regulation Ability," Journal of Research Personality (2014), Ivcevic, Z, Brackett, M. A. Statistics from Adjective Check List (ACL) Gough & Heilbrun, 1983. "THIS Is The Most Valuable Personality Trait To Have, Says Science," Christine Schoenwald.
825 Watson, David (2001). "Procrastination and the Five-factor model: a facet level analysis" (PDF). Personality and Individual Differences. 30: 149–158. doi:10.1016/s0191-8869(00)00019-2. Dyscontrol EP Chik & Kellerman, 1974; impulsive, imaginative=ENP Omnibus Personality inventory (OPI) (Heist, Yonge, Connelly, and Webster, 1968) (Males and females).
826 Preaching and Testimony—Gathering Israel—The Blood of Israel and the Gentiles—The Science of Life A Discourse by President Brigham Young, Delivered in the Tabernacle, Great Salt Lake City, April 8, 1855. Reported by G. D. Watt. p. 268-269.
827 John A. Tvedtnes, "My First-Born in the Wilderness," in *Journal of Book of Mormon Studies,* Spring 1994, F.A.R.M.S., p. 208. 1 Nephi 18: 9-19; 2 Nephi 2:1. *Links between the Book of Mormon and the Hebrew Bible*, David Bokovoy, John A. Tvedtnes, p. 55.
828 "[T]he Lamanites had become, the more part of them, a righteous people insomuch that their righteousness did exceed that of the Nephites, because of their firmness and their steadiness in the faith" (Helaman 6:1).
829 Hogan, Robert; Johnson, John; Briggs, Stephen (1997). Handbook of Personality Psychology. Academic Press. p. 856.
830 The existence of psychopathy in the world is a good illustration of how some people are born with less of a chance to bear tame fruit. Many psychopaths are known to be cruel and manipulative because of their inability to feel empathy for others. It is extremely difficult for psychopaths to not take advantage of others.
831 Merriam-Webster dictionary.
832 Studies indicate that tame people who are conscientious tend to be more religious, more successful, and honest. Hogan, Robert; Johnson, John; Briggs, Stephen (1997). Handbook of Personality Psychology. Academic Press. p. 856. Administrative skills and covenant-keeping often go hand in hand with high conscientiousness. These types of individuals often have the gift of the Spirit known as the gift of administration (D&C 46:15; see also 1 Corinthians 12:5).
833 Some people who aren't born with high conscientiousness can be highly conscientious, but it is unnatural for them. Individuals who feign high conscientiousness display pseudo-conscientiousness, which is a common occurrence in work settings due to work demands. Studies indicate that people regularly exhibit pseudo-conscientiousness at work, given that it is both socially desirable and a major driver of success in the workplace. "The Cost of Faking Your Personality at Work," Melissa Dahl, nymag.com.
834 "Here's the personality trait that predicts success for employees — and entrepreneurs," Drake Baer (2015).
835 Roberts, Chernyshenko, Stark, & Goldberg, 2005 [Abstract].

End Notes

836 Raynor & Levine, 2009.
837 Radosevich, Levine, & Kong, 2009.
838 Gramzow, Sedikides, Panter, Sathy, Harris, & Insko, 2004.
839 "Heritability of the big five personality dimensions and their facets: a twin study," Jang KL et al. (1996).
840 Although some studies cite many third-world countries as boasting higher numbers than the US, UK, Scandinavia, and Israel in high conscientiousness (Tanzania 53.27, Congo 55.71, Ethiopia 54.36, Botswana 50.27, Zimbabwe 51.75), this data has to result from self-assessment error. Zimbabwe is the third poorest country in the world because of the poverty ratio of 80%. The Democratic Republic of the Congo is the seventh poorest country on earth with $484.20 Gross National Product per capita (PPP). Tanzania is the forty-seventh poorest country on earth. Ethiopia is the seventeenth poorest country on earth with $484.20 Gross National Product per capita (PPP). High concentrations of highly conscientious individuals often leads to the creation of first world countries. It is the countries with the most people with low conscientiousness that are often the poorest.
841 Before Elders Pratt and Peterson visited the Seneca, Joseph Smith called Latter-day Saint Elder Peter Whitmer, Jr. to accompany his brother-in-law Oliver Cowdery on a mission to the Lamanites with instructions to "open thy mouth to declare my Gospel" and to "give heed unto the words and advice of thy Brother" who had been given power "to build my Church among thy Brethren the Lamanites". See D&C 30:5-6. Revelation, September 1830-D, JSP (*Joseph Smith Papers*). Even though Elders Whitmer and Cowdery were called first, Elders Pratt and Peterson appear to have visited as missionaries the first Lamanites on record. The Smiths—Joseph Smith, Sr. and Don Carlos Smith—preached in villages north of the St. Lawrence River in Upper Canada (September 1830).
842 Elders Heber C. Kimball and Orson Hyde first preached in Preston, England in 1837.
843 Orson Hyde spent a week in Rotterdam and Amsterdam preaching to Jewish rabbis.
844 Orson Hyde preached in Constantinople.
845 Orson Hyde preached in Jerusalem and dedicated Palestine for the return of the Jews.
846 Ma'at wearing a feather of truth, prominent Egyptian goddess (approx. 2375 BC - 1081 BC).
847 Princess Red feather, Wampanoag, Chappaquiddick band (1945).
848 Also featured in this section is Princess Red Feather of the Wampanoag tribe. Her headband has a single feather.
849 *The Gods of the Egyptians*, E. A. Wallis. Budge www.touregypt.net/featurestories/headdress.htm#ixzz2gQvP6ocP
850 Headdress. www.touregypt.net/featurestories/headdress.htm#ixzz2gQvP6ocP
851 Detail of an Assyrian relief depicting the defeat of an Egyptian fort (the captive, most likely a Nubian, has a feather in a headband).
852 Rain-in-the-Face (Ité Omáǧažu) Lakota Sioux.
853 A foreign-looking crown of feathers standing upright in a close ring ... That such is the signification of Anqet is indicated by the crown of feathers, by the meaning of her name "to surround," and by the determinative hieroglyphic of her name, a serpent, signifying "knowledge". *The Correspondences of Egypt*, C. TH. Odhner.
854 Remler, P., 2010, *Egyptian Mythology, A to Z*, p. 18.
855 Headdress. www.touregypt.net/featurestories/headdress.htm#ixzz2gQvP6ocP
856 *The Creation Myths*, Akhet. www.touregypt.net/featurestories/headdress.htm#ixzz2gQvP6ocP
857 *Swastika the Earliest known Symbol and its Migrations*, p. 912, Thomas Wilson (1896). "Several of the skeletons in these mounds bore unmistakable marks of the ravages of syphilis." Jones, Joseph, 1876, Explorations of the Aboriginal Remains of Tennessee.
858 *Swastika the Earliest known Symbol and its Migrations*, p. 912, Thomas Wilson (1896). "Several of the skeletons in these mounds bore unmistakable marks of the ravages of syphilis." Jones, Joseph, 1876, Explorations of the Aboriginal Remains of Tennessee.
859 Cyrus H. Gordon (1908 - 2001), a Jewish-American scholar.
860 Cyrus H. Gordon believed in the authenticity of the controversial Bat Creek Stone, which was found in a Cherokee burial mound in Tennessee (1883). The stone contains, what appears to be, an ancient Semitic language, like Paleo-Hebrew on it. Pace, Eric (2001-04-09). "Cyrus Gordon, Scholar of Ancient Languages, Dies at 92". *The New York Times*.
861 *Lehi in the Desert; The World of the Jaredites; There Were Jaredites*, Hugh Nibley, pp. 255, 263.
862 *Lehi in the Desert; The World of the Jaredites; There Were Jaredites*, Hugh Nibley, pp. 255,

End Notes

263.

863 Hamsa eye-in-hand gorget, neck pendant (upside down), Spiro site, Oklahoma (about 800 AD). Prehistoric Designs, Spiro, plate 28. Fragments of designs engraved on Spiro shells; various cult symbols (All these are after Hamilton, The Spiro Mound, 1952 AD, and reproduced, courtesy, The Missouri Archaeological Society.) Fig. 84, C.V. Stone collection.

864 It is possible that the Oklahoma Spiro site hand-and-eye (hamesh) represents the ability to reach and see into the future and might have been worn by a religious figure.

865 Ceremonial object, sandstone "Rattlesnake Disk" with hamsa, Moundville, Alabama (about 1277 AD). F found at the prehistoric site of Moundville, Alabama. Moundville, Moore, 1905, fig. 7. Stone Pallete, plate 93, 12.5" diameter. The rattlesnake disc is probably the most famous item found at Mound State Monument. It is thought that the rattlesnakes bound at two points mean war.

866 Hamsa eye-in-hand engraved shell cup, long nails, Phillips and Brown 1978 infer shamanistic practice of defleshing bones of elite dead. "Nonconforming eye" vriety desc. by Phillips and brown right hand.

867 Sheet-copper pendant symbol from Moundville, Alabama including a right hand hamsa (pendant of sheet-copper, with Burial no. 164). This artifact appears to have been used as jewelery since it has two holes at the top of it.

868 Robert L. Rands, "Comparative Notes on the Hand-Eye and Related Motifs," American Antiquity 22 (1957 AD):247-57. p. 126.

869 Double-hand shell gorget 1250 - 1450 AD Middle mississippian period, Craig mound, Spiro site, LeFlore county Oklahoma, marine shell, 3.31 inches. A.J. Waring, Jr. and Preston Holder, "A Prehistoric Ceremonial Complex in the Southeastern United States," American Anthropologist, vol. 47, no. 1, January-March 1945, pp 1-34.

870 Hopewell sheet-mica hand, Ross County, Ohio (Mound 25), 100 BC - 500 AD.

871 Etruscan hand, sheet-bronze artifact, Italy (650 BC).

872 *Cahokia*, p. 21, Timothy R. Panketat.

873 Spiro has been the site of human activity for at least 8000 years, but was a major settlement from 800 to 1450 AD. "Spiro Mounds." Oklahoma Historical Society. (retrieved 30 May 2011).

874 Cernunnos, horned god of Celtic polytheism, drawing of the antlered figure depicted on plate A of the Gundestrup Cauldron found in Gundestrup, Himmerland, Denmark. 200 BC - 300 AD.

875 The *Gundesdtrup Cauldron* represents a collection of thirteen plates found deposited in a dry section of the Raevemose peat bog, Gundestrup, Himmerland, Denmark. This remarkable collection of 97%-pure silver plates (some of which are partly gilded) was discovered in 1891 and is kept in the National Museum of Denmark, Copenhagen. The plates themselves are thought to date from the late Celtic La Tène period (late second to first century BC).

876 Hopewell skull with copper horns, hopewell mound, Ross County, Ohio, Moorehead, "primitive man in Ohio" (300 BC - 500 AD).

877 Pre-Columbian Art / Hopewell Green Slate Ceremonial Swastika. Origin The Mississippi Valley, USA Circa 200 BC to 200 AD. The swastika is made up of two crossed serpents. Image from The Barakat Gallery.

878 The 6th Century BC King of Persia, Darius the Great, in an inscription in Naqsh-e-Rostam wrote: "I am Darius the great King, ... A Persian, son of a Persian, an Aryan, having Aryan lineage." Many Persians are from Iran of the Middle East and have the same kinds of Semitic genetic profiles as Ashkenazi Jews.

879 The term "Aryan" derives from the ancient East Indian (Vedic Sanskrit) and Persian (Avestan) term *arya* for "noble." The Aryan tribes in India called their land *Aaryaa varta* or *Aryan expanse*. The Aryans in Iran similarly named their land *Airyanem Vaejah*, or *Iranian Expanse*, today known as *Iran*, which is a variant of "Aryan." 1896 book, *The Swastika: The Earliest Known Symbol and its Migrations*, Thomas Wilson, former curator of the Department of Prehistoric Anthropology in the U.S. National Museum.

880 Pre-Nazi Germany archaeologist Heinrich Schliemann (1822-1890) was convinced the swastika was originally a central Aryan religious symbol that spread out to the world. Concerning the Aryan race, Walter Goffar who wrote *Barbarian Tides: The Migration Age and the Later Roman Empire* states, "These people, the Aryans, would bestow their social institutions on the Indian subcontinent and spawn a unique spiritual culture, Hinduism." Before the Germans transformed this beautiful ideogram, the swastika, into a Nazi symbol of death and destruction, it was regarded by some as a symbol of peace. The swastika was one of the first signs discovered in the Euphrates-Tigris valley (including Mesopotamia) and in some areas of

End Notes

the Indus valley, some over 3,000 years old. It was not until approximately 1000 BC that the swastika became a commonly used sign, most likely first commonly used in ancient Troy in the north west of today's Turkey. The Swastika did not originate in India. Instead, Semitic peoples brought the sign to India. Before settling in Europe, some scholars have asserted that nomadic "Aryans" (Caucasians) traveled to India from the north, bringing the framework for Hinduism with them and introducing Hinduism into the subcontinent, where it has flourished ever since.

881   *The Swastika: The Earliest Known Symbol & Its Migrations; with Observations,* p. 981, Thomas Wilson.

882   Copper swastika (cataloged as # 58205 by archeologists), depicting the four Cardinal or Semi-cardinal Directions of the cosmos, plus its spin, but without its Center (100 BC-500 AD). From the Hopewell earthwork, Mound 25, Copper Deposit of symbols. *The Scioto Hopewell and Their Neighbors:Bioarchaeological Documentation and Cultural Understanding* (Hardcover) by D. Troy Case and Christopher Carr and Cheryl A. Johnston and Beau Goldstein.

883   Hohokam village ruins (approx. 400 AD) near Phoenix, Arizona; Swastika on Arizona Hohokam vessel found in the ruins.

884   According to the United States National Park Service website, Hohokam is a Pima (O'odham) word used by archaeologists to identify a group of indigenous Americans who lived in the Sonoran Desert of North America.

885   Jews worldwide share genetic ties, Alla Katsnelson, Scientific American. Comment made.

886   Meander swastika pattern in the ancient Byzantine mosaic floor, at the Byzantine church ruins in Shavei- Zion, Israel. Of the Byzantine Palestine era (circa 360s AD to 636 AD).

887   Meander swastika pattern in the ancient Byzantine mosaic floor, at the Byzantine church ruins in Shavei- Zion, Israel. Of the Byzantine Palestine era (circa 360s AD to 636 AD).

888   Meander swastika found in the Katzrin Synagogue (built in the 6th century AD atop a more modest 4th-5th century synagogue). Synagogue located in the Golan Heights (Israel's mountainous northern region on the outskirts of the Israeli settlement Katzrin. It features the reconstructed remains of a Talmudic-era village.

889   Second Ann. Rep. Bureau of Ethnology, 1880-81, p. 276, pl. 56, figs. 1, 2.

890   Swastika the Earliest known symbol and its migrations, p. 912, Thomas Wilson (1896). "Several of the skeletons in these mounds bore unmistakable marks of the ravages of syphilis." Jones, Joseph, 1876, Explorations of the Aboriginal Remains of Tennessee.

891   Celtic bronze disc with yin yang whorl in a triskele arrangement spawning three stylized bird heads; find spot Longban Island, Derry; pre-Christian period; on display at Ulster Museum in Belfast.

892   Engraved Shell Disk, Tennessee, Three-armed Volute (triskelion) in ancient North American mound.

893   Hesse or the Rhineland, Stater (Rainbow Cup), 2nd century BC (200 BC-100 BC), obverse (left), reverse (right).

894   Stumpf, Gerd (Munich). "Rainbow cup." Brill's New Pauly. Brill Online, 2013 AA. Reference. 23 June 2013. The term "cup" in numismatic verbiage means a "bowl-shaped convex coin" and is a common style among Celtic coins.

895   A Mississippian Culture Nashville I style shell gorget found by William Myer at the Castalian Springs Mound Site in Sumner County, Tennessee and now part of the collection of the National Museum of the American Indian. Castalian Springs Mound Site (also known as Bledsoe's Lick Mound and Cheskiki Mound). It was occupied from 1100 AD to 1450 AD.

896   Iron Age (1200 BC-550 BC) Castro Culture triskelion, reused in a barn. Airavella, Allariz, Galicia (Spain).

897   Park, Edwards; "Where Did Chief Joseph Get a Cuneiform Tablet?" Smithsonian Magazine, 9:36, February 1979 (1 in. x 1 in.).

898   The Nez Percé—a tribe that originated in Bear's Paw, Montana, and ended up being relocated to the northern parts of Idaho and Oregon.

899   Mary Gindling of Helium's *History Mysteries.* "Chief Joseph Carried the Star of Ashur," By Benjamin Daniali, *AssyriaTimes.*

Made in the USA
Charleston, SC
17 December 2016